*Thermobacteriology in
Food Processing*

SECOND EDITION

Thermobacteriology in Food Processing

C. R. STUMBO

Department of Food Science and Technology
University of Massachusetts
Amherst, Massachusetts

SECOND EDITION

Academic Press
San Diego New York Boston
London Sydney Tokyo Toronto

ACADEMIC PRESS, INC.
A Division of Harcourt Brace & Company
525 B Street, Suite 1900, San Diego, California 92101-4495

United Kingdom Edition published by
ACADEMIC PRESS, INC. (LONDON) LTD.
24/28 Oval Road, London NW1 7DX

ISBN 0-12-675352-0

LIBRARY OF CONGRESS CATALOG CARD NUMBER: 72-7692

Transferred to digital printing 2006
96 QW 9 8 7

This book is dedicated to our great food processing industry

Contents

CHAPTER 1

Introduction

THERMOBACTERIOLOGY

CHAPTER 2

Organisms of Greatest Importance in the Spoilage of Canned Foods

CHAPTER 3

Bacteriological Examination of Spoiled Canned Foods

CHAPTER 4

Organisms of Greatest Importance in Food Pasteurization

CHAPTER 5

Contamination and Its Control

CHAPTER 6

Producing, Harvesting, and Cleaning Spores for Thermal Resistance Determinations

CHAPTER 7

Death of Bacteria Subjected to Moist Heat

CHAPTER 8

Thermal Resistance of Bacteria

THERMAL PROCESS EVALUATION

CHAPTER 9

Important Terms and Equations

CHAPTER 10

Basic Considerations

CHAPTER 11

The General Method

CHAPTER 12

Mathematical Methods

CHAPTER 13

Conversion of Heat Penetration Data

CHAPTER 14

Typical Process Determination Problems — Canned Foods

CHAPTER 15

Evaluation and Equivalency of Pasteurization Processes

CHAPTER 16

Quality Factor Degradation

APPENDIX

Foreword

This fine text and reference volume will surely find its way onto book shelves of all professionals who must deal with the problems of producing safe, acceptable, and nutritious canned foods. It sets forth both the principles and practices required to preserve foods in hermetically sealed containers by heat sterilization. To my knowledge in no other one volume are there all of the facts and figures which permit a person to determine the heat processing necessary to assure a safe canned food product with minimal damage to organoleptic quality and nutritive value. I am convinced that it will prove to be an essential reference book for all those whose responsibility it is to establish heat processing requirements. In addition it will serve as a valuable text for university courses which deal with this topic.

This work is a "must" for scientists and technologists who deal with canned foods. It is short and concise, yet contains all of the essential information needed for the job at hand.

G. F. STEWART

Preface to the Second Edition

The first edition of this work dealt primarily with the basic principles involved in establishing adequate sterilization processes for canned foods. This edition has been expanded to include similar consideration of processes for the pasteurization of foods. It also includes procedures for determining the extent of degradation during thermal processing of quality factors which, as bacteria, are destroyed exponentially with time of heating at a constant temperature.

To implement this expansion, tables of $f_h/U:g$ relationships have been greatly expanded with the aid of computer techniques. This computerization resulted in some refinements, although these have been found rather minor. However, all examples and sample problems appearing in the first edition have been revised in accordance with these refinements.

Finally, use of the book as a classroom text during the past seven years indicated some minor interstitial changes in the interest of clarity. These have been made.

It should be noted that with respect to principle and procedure coverage in the first edition has been retained. As indicated above, coverage has been expanded to make the text more versatile and inclusive.

C. R. STUMBO

Preface to the First Edition

The inspiration for writing this text stemmed from a deep realization of its need during many years of teaching the subject matter it covers and from still more years of responsibility for industrial procedures and problems it treats. Perhaps the many scientific disciplines involved, and the intricate interrelationships of these disciplines, explains why a book has not heretofore appeared which could serve both classroom and industrial interests. Comprehension of the subject has required major self-education in the application of various scientific principles in the solution of problems relating to food preservation by heat.

Though the text relates specifically to food preservation, the principles discussed apply equally to moist heat sterilization of other materials, e.g., bacteriological culture media, media for larger scale antibiotic production, pharmaceutical preparations, surgical supplies, hospital equipment, etc.

The subject matter has been arranged in order of logic and intensity. This should be of special benefit to individuals who have little or no opportunity for classroom training. Also, it has been arranged so that an individual may delve to whatever depth in the subject he wishes. The early chapters deal with microorganisms of greatest concern in sterilization and with procedures for their detection and control. These are followed by discussions relating to the destruction of bacteria subjected to moist heat. Then it is shown how parameters characterizing bacterial resistance to heat are integrated with parameters describing heat transfer and heat intensity in the graphical and mathemathical calculation and evaluation of sterilizing processes.

Examples are given showing how each mathematical equation is employed. Finally, typical sterilization problems are posed and solved.

An attempt has been made to make the book self sufficient with regard to elements essential in the solution of sterilization problems. It was designed not only as a textbook, but also as a work book.

November 1965   C. R. STUMBO

Acknowledgment

I gratefully acknowledge the valuable suggestions made by Drs. W. L. Hunting, D. A. Evans, and G. H. Snoeyenbos during preparation of this edition. Deeply appreciated is the untiring assistance of Dr. Kailash S. Purohit in the computer programming which led to expansion and refinement of the tables of essential process evaluation parameters. Special thanks are due my wife, Vi, without whose patience, love, and devotion this work would have been impossible.

1

Introduction

This text is designed to acquaint the student with the basic principles of thermobacteriology and thermal process determination. For the sake of brevity and continuity, references to published works contributing to our present-day knowledge of the subject are purposely kept to the minimum. It is not the intention to avoid giving credit where credit is due. The objective is to present the subject as clearly and concisely as possible. Detailed review, or even mention of many important contributing works, is beyond the scope of this purpose. Also beyond the scope of this objective is a detailed historical review of events leading to our present-day state of knowledge—only a "birds-eye" view of these can be given.

Nicholas Appert, in 1810, won a 12,000-franc prize for successfully preserving, for the first time, a variety of perishable food products by heat-processing them in glass jars and bottles. Progress since that eventful year has turned the discovery into one of the most important advances in the history of mankind.

For many decades heat has been the most widely used agent in food preservation. Foods preserved in this manner, at times and locations of their abundance, for distribution and later consumption throughout the world, are essential to the existence of our present-day world society. They have become so commonplace in the human diet that the health of the world population now depends in great measure on the quality of these foods.

Since the time of Appert, the increase in man's knowledge relating to the preservation of foods by heat, if plotted against time, may well be likened to the classical bacterial growth curve. During the first hundred years (the lag phase), progress was extremely slow. Gradually, dormancy was broken by the application of knowledge gained from important advances in science and technology, to turn the curve upward (the accelerated growth phase). Though difficult of definition, this phase lasted perhaps some twenty-five or thirty years. Continued nurturing by science and technology kept the growth curve bending upward, and by 1940 it was obvious that the logarithmic phase of growth had been reached. Since then, the acceleration of growth has been an exponential function of time.

What advances were instrumental in breaking the long dormancy, in nurturing progress through the accelerated growth phase, and in supporting it during the logarithmic phase of growth? They have been many and varied. Mention will be made of the more notable only—or, rather, what the author considers the more notable.

The Accelerated Growth Phase

Two things stand out as most responsible for breaking the long dormancy. First is the pioneering work of Pasteur, Prescott, Underwood, and others, during the latter part of the nineteenth century and the early part of this one. Their research on the microbiology of heat processed foods, and the application of principles thus established, marked the introduction of science into the preservation of foods by heat processing. Progress was slow at first, because the territory was virgin and unexplored, but through untiring perseverance it was made definite—the growth curve finally began to bend upward.

The second most important thing contributing to a break in dormancy was the commercial introduction, about 1900, of the now well-known "sanitary" can. Prior to this, all tin containers used commercially were either of the hand-soldered open-top style or the "hole and cap" type. Introduction of the "sanitary" can paved the way not only for high-speed economical can production but also for high-speed hermetic packaging of foods for heat processing.

Knowledge of the microbiology of heat processed foods continued to increase. Technological advances in can manufacturer and methods of applying heat were notable. However, it was soon realized that these were only two of many growth factors. A third, most potent one was soon supplied by the pioneering work of Bigelow, Ball, and others, about 1920. The thermocouple was applied to measure temperatures during heating and cooling of foods. The information so gained, combined with that relating to heat resistance of food-spoilage bacteria, led to the development of graphical and mathematical procedures for estimating the amount of heating required to produce "commercially sterile" and adequately pasteurized food products. The application of these methods resulted in important refinements in the thermal processing of foods.

The era 1910 to 1940 was one of vitamin discovery and description. During the latter part of this period, many studies were devoted to ascertaining the heat stability of vitamins essential to human health. Though some were found to be quite stable, others showed varying degrees of sensitivity, thiamine being one of the more sensitive. Now it was obvious that heat applied for the purpose of preserving foods was deleterious not only to certain organoleptic properties but to nutritive properties as well. This finding sharply emphasized the principle that no more heat should be applied for preservation than the minimum necessary to free foods of microorganisms that might spoil the foods or endanger the health of consumers.

This, in turn, emphasized the need for greater precision in calculating and evaluating heat processes for food preservation. To accomplish this, it was necessary to improve methods of temperature measurement, methods of measuring heat resistance of food-spoilage bacteria, and methods for evaluating and calculating thermal processes. Thus a chain reaction was started which today is still very much in the logarithmic phase of acceleration.

The recognition of the effect of heat on organoleptic and nutritive properties also emphasized the need for processing methods that would minimize overprocessing. Temperature-distribution studies demonstrated that, for canned foods that heated by conduction, the greater portion of the container's contents was being severely overprocessed in order to sterilize a small volume about the geometrical center. Many of the foods that heated by conduction were found to heat by convection if agitated during processing, thereby almost eliminating the overprocessing that resulted from nonuniform temperature distribution. The real value of "agitating" cooking was thus established, and consequently two types of this method have since grown continually in popularity. These are generally referred to as "axial" and "end over end."

In studies relating to the heat stability of nutritive and organoleptic qualities of foods, one of the most important discoveries in the history of the food industry was made—that is, that higher temperatures applied for short times, equivalent in sterilizing value to lower temperatures applied for longer times, were significantly less destructive of these qualities. It was soon established that, under conditions of virtually uniform heating, the degree of destruction of desirable organoleptic and nutritive properties was directly related to the temperature employed. High-temperature short-time (HTST) processing was the obvious key to improved quality.

The use of higher temperatures in methods of "agitating" cooking was definitely indicated. This could be achieved only through marked advances in engineering of cookers and high-speed canning lines, improvement in the "sanitary" can to withstand increased stress, development of methods for measuring food temperatures in containers during agitation, development of methods for measuring the heat resistance of bacterial spores to the higher temperatures, and development of procedures for calculating processes for foods under agitation equivalent in sterilizing value to processes already established as adequate for the same products in "still" cooking.

In spite of foreseeable advances in engineering and technology, the maximum temperatures that could be contemplated for use in "agitating" cooking were still below those required for maximum quality retention in many food products. There remained the need for a method or methods whereby foods could be sterilized by rapidly heating them to still higher temperatures for very short times, followed by rapid cooling. In answer, the method of aseptic canning was conceived—the food would be heated and cooled in specially designed heat exchangers, then transferred aseptically into presterilized containers. Implementation of the concept called not only for design and engineering of special heat exchangers but for development of methods for high-speed sterilization of containers and closures and methods for maintaining asepsis during filling and closing of the containers. And it was soon discovered that certain deteriorative enzymes, like many desirable food qualities, were relatively much more resistant to HTST processes than to lower-temperature processes of equivalent sterilizing capacity. It became evident that, for many products, enzyme destruction would dictate minimum processes. As a result, studies relating to kinetics of enzyme destruction were mandatory if HTST processing were to be successful.

The Logarithmic Phase

By the late 1930's, the increase in man's knowledge relating to the preservation of foods by heat had definitely reached the logarithmic phase of growth. All the basic discoveries and concepts discussed above were a matter of record. Knowledge for further development, refinement, and application of them was accumulating rapidly, though not as rapidly as might have been expected in view of the great stimuli which had been provided. And industry was moving toward application of the accumulating knowledge, though there seemed to be some apparently inexplicable inertia. There was obvious danger that our curve would bend toward the stationary phase at a time when growth had scarcely reached the logarithmic phase. What further was needed?

A small group of men of great vision diagnosed the problem as the lack of an organization to promote food science and technology and to disseminate technical information throughout the food and allied industries. They acted. In 1939 they founded the Institute of Food Technologists, an organization that has been in the logarithmic phase of growth since its founding. Since its inception, the Institute has truly been the circulatory system of the industry. In heart it has conscientiously promoted food science and technology, and its two great journals have been arteries of knowledge nurturing all disciplines. And herein lies a great secret of its success—its membership is not confined to scientists and technologists but includes individuals from virtually every discipline of food and allied industries, from management to plant operator. The Institute has proved to be the most important growth factor responsible for keeping the increase in man's knowledge of food processing in the logarithmic phase of growth.

Other potent growth factors have been research and teaching relating to principles of thermal processing and process calculation. They have been supplied by universities and many industrial organizations associated with food and allied industries. They have functioned not only in gathering and disseminating knowledge but in stimulating interest in the field.

In these years of logarithmic growth, knowledge relating to all the discoveries and concepts of the accelerated growth phase has increased exponentially. Continuous "agitating" cookers and associated equipment have been developed to operate at speeds of more than 1000 cans per minute. Containers have been developed that permit use of temperatures as high as 270°F. Ingenious methods have been devised for measuring the temperatures of products in containers during agitation and for measuring thermal resistance of bacterial spores to temperatures somewhat above 270°F. Mathematical procedures have been developed for calculating equivalent processes for "still" and "agitated" cooks. Improvements in instrumentation, in automation, and in continuous processing have been phenomenal. Heat exchangers have been developed for rapidly heating various types of foods to high temperatures and rapidly cooling them. Methods have been devised for high-speed sterilization of metal containers, as well as systems of asepsis for filling and closing the containers. The aseptic canning technique is now a reality. Methods have been developed for filling and packaging foods at high temperatures in sterile pressurized atmospheres—container sterilization being by heat of the products prior to rapid cooling.

It is the purpose of this text to augment the dissemination of knowledge relating to food preservation by the application of heat. An attempt has been made to reduce the subject matter to its simplest form, and yet to incorporate most elements essential to an understanding of the more complex aspects of the subject.

THERMOBACTERIOLOGY

2

Organisms of Greatest Importance in the Spoilage of Canned Foods

Microbial spoilage of heat-processed canned foods is caused by microorganisms that survive the heat processes or that gain entrance, through container leakage, subsequent to the heat processing. It is virtually impossible to predict what types of microorganisms may gain entrance through container leakage. However, there are important measures for minimizing this type of spoilage. These will be discussed in some detail later. From the standpoint of establishing sterilization processes, spore-bearing bacteria, except in highly acid products, are the organisms of chief concern. Because the type of microbial spoilage, as well as the heat resistance of bacteria, is closely related to the acidity of foods, for convenience of discussion the different foods are often grouped in classes with respect to pH.

Classification of Food with Respect to Acidity

Various classifications have been proposed. Cameron and Esty (1940) suggested the following:

1. Low-acid foods: pH 5.0 and higher.
2. Medium- (or semi-) acid foods: pH 5.0 to 4.5.
3. Acid foods: pH 4.5 to 3.7.
4. High-acid foods: pH 3.7 and lower.

Except for the dividing line between acid and high-acid foods, this classification serves as well perhaps as any. In practice it has been found that pH 4.00 is a more realistic dividing line between acid and high-acid foods. Seldom will spore-bearing bacteria grow in heat-processed foods with pH values of 4.0 or lower. Some of the butyric anaerobes and *Bacillus coagulans* (*B. thermoacidurans*) will grow in laboratory culture or even in foods at pH values as low as 3.7; however, this generally occurs only with very heavy inocula.

Some workers recognize only three classes of foods, namely:

1. Low-acid foods: pH above 4.5.
2. Acid foods: pH 4.0 to 4.5.
3. High-acid foods: pH below 4.0.

The dividing line between low-acid and acid foods is taken as 4.5 because some strains of *Clostridium botulinum* will grow and produce toxin at pH values as low as about 4.6. Some of the highly heat-resistant saccharolytic anaerobes—for example, *Clostridium thermosaccharolyticum*—grow in and cause spoilage of foods in this semi-acid range. Therefore, until more is known of bacterial heat resistance and bacterial growth in the semi-acid foods, they perhaps should remain in the low-acid grouping.

Classification of Spore-Bearing Bacteria with Reference to Oxygen Requirements

In low-acid and acid foods the spore-bearing bacteria are of greatest concern from the standpoint of sterilization. With respect to oxygen requirements they may be classified as follows: (1) obligate aerobes—*Bacillus* spp.; (2) facultative anaerobes—*Bacillus* spp.; (3) obligate anaerobes—*Clostridium* spp.

Obligate Aerobes.   This group includes those species that require molecular oxygen for growth. From the standpoint of food sterilization, it is the least important of the three groups. Under present-day methods of canning, most foods contain very low levels of molecular oxygen, insufficient to support appreciable growth. Beyond this, spores of most obligate aerobes are low in heat resistance compared with spores of a number of organisms in the other two groups. Heat processes designed to free foods of the more-resistant organisms in groups 2 and 3 are generally more than adequate to free foods of the aerobic bacilli. However, as with most things, there are exceptions. In canned cured meat products containing nitrate, the *Bacillus subtilis–Bacillus mycoides* group of organisms may at times be of greater economic importance than some other groups of bacteria (Jensen, 1945).

Some canned cured meat products are given relatively mild heat processes, inhibitory action of the curing agents, and in some cases refrigeration, being depended on to prevent spoilage by organisms in groups 2 and 3. It is not unusual for spores of aerobic bacilli to survive in some of these products. A number of strains of these bacilli are capable of reducing nitrate to obtain oxygen for metabolic activity. If not destroyed by the heat process, and if subjected to favorable temperatures, many of them will grow in the cured meat products and cause gaseous spoilage. It should be noted that the dry spices used in seasoning some of these products are sometimes heavily contaminated with spores of the aerobic bacilli.

Facultative Anaerobes. Spore-bearing bacteria of the facultative anaerobic group are among the more important from the standpoint of food sterilization. Of particular importance in low-acid and acid foods are thermophilic spore-bearing bacilli. Some of these produce spores that are more resistant to heat than are spores produced by most obligate anaerobes. They generally cause what is known as "flat-sour" spoilage; that is, they produce acid but little or no gas. Though the most troublesome ones are classified as thermophiles, many of them have the faculty to grow, though more slowly, at or near ambient food-handling temperatures.

Of greatest importance in low-acid foods are *Bacillus stearothermophilus* and related species. These flat-sour-producing organisms are often not identified with respect to species by the food bacteriologist but simply labeled as belonging to the *B. stearothermophilus* group. A number of these strains and species produce spores characterized by D_{250} values in excess of 4.00 minutes, D_{250} being the time, in minutes, at 250°F. required to destroy 90% of the spores in a given population. This is in contrast to a small fraction of a minute at 250°F. required to destroy 90% of the spores of most obligate aerobic bacilli. The optimum growth temperature for these flat-sour thermophiles varies considerably among strains and species, from about 49° to 55°C. (120° to 131°F.). Most of them seldom grow at temperatures below about 38°C. (100°F.).

Of greatest importance in acid foods are three species of facultative anaerobes—namely, *Bacillus coagulans* (*B. thermoacidurans*), *Bacillus macerans*, and *Bacillus polymyxa*. Of these, *B. coagulans* is the more important, particularly in spoilage of tomatoes and tomato products. *Bacillus macerans* and *B. polymyxa* have been isolated as

causative organisms in spoilage of some fruits and fruit products
(Vaughn *et al.*, 1952). There is some doubt whether spores of these
latter organisms resisted the heat processes given or whether the
organisms gained entrance through container leakage subsequent to
the heat processing.

Bacillus coagulans has been the cause of economically important
spoilage of acid foods, particularly tomatoes and tomato products.
It is quite acid-tolerant, growing at pH values of 4.0 or slightly lower.
It grows well in tomato products and some other semi-acid foods in
the pH range 4.0 to 4.6. It is of little concern in low-acid foods, however, because heat processes for these foods are generally more than
adequate to destroy its spores. The D_{250} values of *B. coagulans* spores
are of the order of 0.1 and less. *Bacillus coagulans* is generally classified as a thermophile; however, it has the faculty to grow quite well
at ambient food-handling temperatures.

Obligate Anaerobes.   Some species of spore-bearing anaerobes produce spores that are relatively quite heat-resistant. With reference
to canned food spoilage these may be classified in two groups, mesophilic and thermophilic. Of the thermophiles, the most important are
the saccharolytic organisms which do not produce hydrogen sulfide.
Clostridium thermosaccharolyticum is generally considered the type
species of the group. These organisms are very saccharolytic, producing large quantities of gas, chiefly carbon dioxide and hydrogen, from
a wide variety of carbohydrates. Consequently, they cause spoilage of
the "swell" or gaseous type. They often produce a butyric or "cheesy"
odor in foods. They generally are of greater importance in the spoilage of semi-acid products (pH 4.5 to 5.0) than are flat-sour types.
Their optimum growth temperature is about 55°C. (131°F.). They
seldom grow at temperatures below about 32°C. (90°F.). They can
be a source of real trouble at temperatures of 35°C. (95°F.) and
above. Some strains produce spores of extremely high heat resistance,
making severe processing necessary if their spores are present in
appreciable numbers.

Thermophilic spore-bearing anaerobes that produce hydrogen sulfide
are responsible for the so-called "sulfur stinker" spoilage of canned
foods. *Clostridium nigrificans* is considered the type species of this
group. These organisms are proteolytic, and hydrogen sulfide is the
only gas they produce in great quantity. They are only weakly, if at
all, saccharolytic. Because the hydrogen sulfide is soluble in the
product, spoiled cans usually remain flat. Many products spoiled by

these organisms become black, owing to the interaction of the hydrogen sulfide and iron. Spoilage by these organisms is comparatively rare for two reasons: Incidence of their spores in most products is generally low, and the spores of most strains are relatively low in heat resistance compared with spores of the saccharolytic thermophilic anaerobes and the flat-sour thermophilic facultative anaerobes. Neither of the above groups is of importance in foods having pH values below about 4.5.

Next in importance, in low-acid foods, to the groups discussed above are spore-bearing mesophilic anaerobes. Because of its public health significance, the toxin-producing organism, *Clostridium botulinum*, should perhaps be considered of greatest importance among organisms of this group. Of the different botulinum types, A, B, and E are of greatest significance. Spores of Types A and B are considerably more heat-resistant than spores of Type E and therefore of greatest concern in the sterilization of canned foods. Even the spores of A and B are not nearly as heat-resistant as the spores of a closely related nontoxic organism identified only as P.A.3679. Whereas the most-resistant spores of Types A and B botulinum are characterized by D_{250} values of the order of 0.10 to 0.20, the most-resistant spores of P.A.3679 are characterized by D_{250} values of the order of 0.50 to 1.50.

Fortunately the incidence of the more-resistant strains of P.A.3679 is very low in most foods. For this reason, canning processes designed to assure a high degree of safety with regard to *C. botulinum* are often adequate to prevent economically important spoilage by even the most resistant of these nontoxic putrefactive anaerobes. Perhaps the greatest importance of P.A.3679 is as a test organism to check the adequacy of canning processes. It is almost ideal for this purpose, not only because of its heat resistance, but because it is nontoxic, it is easy to cultivate, and its spoilage of foods is accompanied by voluminous production of gas.

Other proteolytic or putrefactive organisms most frequently causing spoilage of low-acid and semi-acid foods are *C. putrificum, C. histolyticum, C. bifermentans, C. sporogenes*, and related species. Again, heat processes designed to assure a high degree of safety with regard to *C. botulinum* are often adequate to prevent economically important spoilage by these organisms; however, not infrequently, more severe processes than considered adequate to control *C. botulinum* are necessary to prevent spoilage of a number of food items by the more heat-resistant members of this group, including P.A.3679.

Important in the spoilage of some acid foods are saccharolytic spore-bearing anaerobes of lower heat resistance than the more-resistant putrefactive anaerobes. Of greatest importance in this group are *C. pasteurianum*, *C. butyricum*, and related butyric anaerobes. Some strains of these grow well in foods with pH values in the range of 4.0 to 4.5. However, their spores are usually less resistant to heat than are the spores of the flat-sour organism *B. coagulans*. Therefore, heat processes designed to free acid foods of *B. coagulans* are generally adequate to free them of the butyric anaerobes.

In summary, of the mesophilic obligate anaerobes, in spoilage of low-acid and semi-acid foods (pH 4.5 and above) the putrefactive anaerobes, *C. sporogenes* and related species, are of greatest importance. In spoilage of acid foods (pH 4.0 to 4.5) the butyric anaerobes, *C. pasteurianum* and related species, are of greatest importance. The optimum growth temperature range of these organisms is 25° to 35°C. (77° to 95°F.).

Non-Spore-Bearing Bacteria, Yeasts, and Molds

Except when they gain entrance through container leakage, these organisms are of greatest importance in spoilage of high-acid canned foods (pH below 4.0) given relatively mild heat processes—pickles, grapefruit, citric juices, rhubarb, cranberries, etc. They are of some importance also in spoilage of concentrated and sweetened products given mild heat processes, the high soluble solids content being depended on to inhibit growth of spore-bearing organisms—for example, fruit and vegetable concentrates, jellies, preserves, sweetened condensed milk.

Of the non-spore-bearing bacteria, *Lactobacillus* and *Leuconostoc* spp. are most important. A wide variety of yeasts have been found as spoilage agents of a number of high-acid foods given mild heat processes. The more resistant of these yeasts and bacteria are characterized by D_{150} values of the order of 1.00. Pasteurization processes based on this resistance value usually prove adequate. Molds are generally considered insignificant as spoilage agents in canned foods. A notable exception to this is the species *Byssochlamys fulva*. This organism, described by Oliver and Rendle (1934), is important in the spoilage of canned fruits. It may cause complete disintegration of the fruit due to breakdown of pectinous material. It may produce sufficient carbon dioxide to cause swelling. It is relatively quite heat-resistant, surviving processes of the order of 10 minutes at 87.8°C. (190°F.) (Gillespy and Thorpe, 1962). A related mold, *Byssochlamys nivea,* was reported by Gillespy and Thorpe (1962). The ascospores of this organism were considerably less heat-resistant than those of the species *fulva,* failing to survive 10 minutes at 82.2°C. (180°F.) Williams *et al.* (1941) isolated a heat-resistant penicillium species. They reported that, for sclerotia of this organism, 300 minutes at 85°C. (185°F.) was required for destruction. Its ascospores, however, were destroyed in 15 minutes at 81°C. (177.8°F.) These resistant molds are exceptions and are not highly important in spoilage of foods in this country.

The optimum growth temperature range for most of these non-spore-bearing bacteria, yeasts, and molds is from 20° to 35°C. (68° to 95°F.). A summary of the organisms important in the different classes of foods and their comparative approximate average maximum heat resistance is given in Table 5, Chapter 8. The resistance values given for the spore-bearing bacteria are for their spores.

Organisms Entering via Container Leakage

If a container "breathes" even momentarily, bacteria may be sucked in with contaminated air or water. They are far more likely to enter with contaminated water than with air. In most cases the openings causing leakage are small enough to filter out bacteria-laden dust from the air. This is not the case for bacteria suspended in water. Because of this and because of stresses on can seams and jar closures during the cooling operation, bacteria are most likely to gain entrance during cooling or shortly thereafter, while the containers are still wet. With cans, rough handling on contaminated conveyors, elevators, and lowerators while the cans are wet is a major contributing factor to can leakage and bacterial contamination. The best defenses against spoilage because of container leakage are:

1. Good can seams and jar closures.
2. Careful handling to avoid container damage, such as dents close to can seams.
3. Chlorinated cooling water (3 to 5 p.p.m. active chlorine).
4. Rapid drying of containers after cooling.
5. Dry and sanitary post-cooling can-handling equipment.

Because of varied types of bacteria in contaminated cooling water and on can-handling equipment, many different species of bacteria have been associated with leaker spoilage. Usually, though not always, leaker spoilage of a container is caused by a mixed flora. Commonly found are micrococci, gram-negative rods of the *Pseudomonas–Achro-*

mobacter group, and yeasts. It should be recognized that very minute amounts of heavily contaminated water can cause spoilage. For example, if the water is contaminated to the extent of 1 million bacterial cells per milliliter, only 1 millionth milliliter would have to enter the container to cause spoilage.

SELECTED REFERENCES

Allen, M. B. (1953). The thermophilic aerobic sporeforming bacteria. *Bacteriol. Rev.*, **17**, 125.

Berry, R. N. (1933). Some new heat resistant, acid tolerant organisms causing spoilage in tomato juice. *J. Bacteriol.*, **25**, 72.

Breed, R. S., Murray, E. G. D., and Hitchens, H. P. (1948). "Bergey's Manual of Determinative Bacteriology," 6th ed., 1529 pp. Williams & Wilkins, Baltimore.

Cameron, E. J., and Esty, J. R. (1940). Comments on the microbiology of spoilage in canned foods. *Food Res.*, **5**, 549.

Clark, F. M., and Dehr, Arlin (1947). A study of butyric acid producing anaerobes isolated from spoiled canned tomatoes. *Food Res.*, **12**, 122.

Faville, L. W., and Hill, E. C. (1951). Incidence and significance of microorganisms in citrus juices. *Food Technol.*, **5**, 423.

Faville, L. W., and Hill, E. C. (1952). Acid-tolerant bacteria in citrus juices. *Food Res.*, **17**, 281.

Frazier, W. C. (1958). "Food Microbiology," 472 pp. McGraw-Hill, New York, Toronto, London.

Gillespy, T. G. (1936–1937). The mold *Byssochlamys fulva*. Progress Report. Annual Report, p. 68. University of Bristol, Fruit Preservation Research Station, Campden.

Gillespy, T. G., and Thorpe, R. H. (1962). Survey of *Byssochlamys fulva* ascospore infection on strawberries for canning. Tech. Mem. No. 44. Fruit and Veg. Canning and Quick Freezing Res. Assoc., Campden.

Gordon, R. E., and Smith, N. R. (1949). Aerobic sporeforming bacteria capable of growth at high temperatures. *J. Bacteriol.*, **58**, 327.

Hays, G. L. (1951). The isolation, cultivation and identification of organisms which have caused spoilage in frozen concentrated orange juice. *Proc. Florida State Horticultural Soc.*, **64**, 135.

Hersom, A. C., and Hulland, E. D. (1964). "Canned Foods—An Introduction to Their Microbiology" (Baumgartner), 5th ed., 291 pp. Chem. Publ. Co., New York.

Jensen, L. B. (1945). "Microbiology of Meats," 2nd ed., 389 pp. Garrard Press, Champaign, Illinois.

McClung, L. S. (1935). Studies on anaerobic bacteria. IV. Taxonomy of cultures of a thermophilic species causing swells in canned foods. *J. Bacteriol.*, **29**, 189.

Oliver, M., and Rendle, T. (1934). A new problem in fruit preservation. Studies on *Byssochlamys fulva* and its effect on tissues of processed fruit. *J. Soc. Chem. Ind. (Brit.)*, **53**, 166T.

Smith, N. R., Gordon, R. E., and Clark, F. E. (1946). Aerobic mesophilic sporeforming bacteria. *U. S. Dept. Agr. Misc. Publ.*, **559**, 112 pp.

Speigelberg, C. H. (1940). *Clostridium pasteurianum* associated with spoilage of acid canned fruit. *Food Res.*, 5, 115.

Townsend, C. T. (1939). Spore-forming anaerobes causing spoilage in acid canned foods. *Food Res.*, 4, 231.

Townsend, C. T., Somers, I. I., Lamb, F. C., and Olson, N. A. (1956). "A Laboratory Manual for the Canning Industry," 2nd ed., Chapters 6 and 7. National Canners Association Research Laboratories, Washington, D. C.

Vaughn, R. H., Kreulevitch, I. H., and Mercer, W. A. (1952). Spoilage of canned foods caused by the Bacillus *macerans-polymyxa* group of bacteria. *Food Res.*, 17, 560.

Weinziral, J. (1919). The bacteriology of canned foods. *J. Med. Res.*, 39, 349.

Werkman, C. H. (1929). Bacteriological studies on sulfide spoilage of canned vegetables. *Iowa State Coll. Agr. Expt. Sta. Res. Bull.*, 117.

Werkman, C. H., and Weaver, H. J. (1927). Studies in the bacteriology of sulfur stinker spoilage of canned sweet corn. *Iowa State Coll. Jr. Sci.*, 2, 57.

White, L. S. (1951). Spoilage bacteria in tomato products. *Food Res.*, 16, 422.

Williams, C. C., Cameron, E. J., and Williams, O. B. (1941). A facultatively anaerobic mold of unusual heat resistance. *Food Res.*, 6, 69.

CHAPTER **3**

Bacteriological Examination of Spoiled Canned Foods

The incidence of spoilage of commercially canned foods is very low. However, when it occurs, identification of the type or types of causative organisms greatly aids in devising steps leading to its elimination. Techniques for examining spoiled cans vary somewhat among different laboratories. The ones suggested here have been employed with reasonably good success; however, there are alternative methods that are no doubt just as good. The number of containers examined will depend on circumstances. If possible, at least five containers of a spoiled product should be examined.

Pertinent Information

First, all that is possible should be learned of the heat process which the product has been given. If it is a formulated product, consideration of the different ingredients is often helpful in judging what to suspect. Information regarding the speed of cooling and the temperature to which the product was cooled subsequent to heat processing is sometimes helpful, as is information regarding storage temperatures to which the product may have been subjected. Knowledge of the age of the product at the time spoilage occurred is often important. All this information should be recorded, along with the following:

1. Name and nature of product.
2. Container size.
3. Container code.

4. Condition of cans (evidence of leakage).
   a. Pinholes or rusting.
   b. Dents, particularly near the double seams.
   c. Signs of buckling and paneling.
5. Condition of cans (evidence of type of spoilage) as described by one of the following:
   a. *Flat*—No evidence of swelling.
   b. *Flipper*—Appears flat but when it is brought down sharply on a flat surface, one end will flip out; when lightly pressed, this end will flip back in.
   c. *Springer*—One end flat and one end permanently bulged; when pressure is applied to the bulged end it will flip in and the other end will flip out.
   d. *Soft swells*—Both ends are bulged but not tightly; they yield to thumb pressure.
   e. *Hard swell*—Both ends are bulged tightly; they will not yield readily to thumb pressure.
6. Condition of glass containers.
   a. Record and describe condition of closure, particularly with respect to evidence of leakage.
   b. Record and describe appearance of product.
7. After bacteriological examination of contents, seams of cans should be stripped and any evidence of faulty seaming recorded. Similarly for glass containers, the closure should be carefully examined, particularly the smoothness and condition of the finish.

Opening the Container

Reference here is to one end of a can or to the closure end of a glass container. First, the container end, including an inch or so of the container wall, should be thoroughly cleaned. Soap and water is best for this purpose, though alcohol often will suffice. Next, the outside surface of the container end should be sterilized. For flat, flipper, and springer cans this may be done by direct flaming. For soft-swollen and hard-swollen cans there is danger of explosion from flaming due to expansion of gases. With glass containers, flaming may injure the finish, making accurate appraisal of the condition of the container difficult. Therefore, ends of soft-swollen and hard-swollen cans and glass containers should be sterilized by a strong disinfectant such as 5% phenol solution or a chlorine solution (200 p.p.m.). A satisfactory procedure is to place a clean cloth over the upright container, and then pour disinfectant liberally on the towel covering the top of the container. The disinfectant should be allowed to act for at least 15 minutes before the container is opened.

Caution should be exercised in opening hard-swollen cans and glass containers with tightly bulged lids. It should be assumed that the contents might be toxic owing to *C. botulinum*. Unless precautions are taken, the contents may spew badly when the container is punctured. This spewing can be controlled by leaving the disinfectant towel in place, holding the towel firmly by gripping the container with one hand, and puncturing the container lid, through the cloth, with a sterile sharp instrument such as an ice pick. This will relieve the internal container pressure and control dangerous spewing. The lid may then be wiped clean with the disinfectant towel, and this towel replaced by another that has been soaking in disinfectant. Containers so treated are ready for opening and sampling in the same manner as containers the end surfaces of which have been sterilized by flaming.

For bacteriological examination, an opening in the lid, large enough to permit withdrawal of samples, should be made with a can opener that can be sterilized by flaming. During the examination, air contamination may be minimized by covering the opened can, when not withdrawing samples, with a disinfectant-moistened towel or a sterile petri dish cover. At the time the can is opened, the odor should be noted and recorded.

Bacteriological Examination

The first step in the bacteriological examination is removal of a sample for cultures. The choice of culture media depends greatly on the type of product and type of spoilage. This will be discussed later. To obtain samples for cultures of liquid and semiliquid products, wide-mouth cotton-plugged pipets are generally used. To obtain samples of solid foods, forceps, spoons, or narrow-blade spatulas are generally used. A sample of 10 to 20 gm. should be taken from each container and placed in a sterile cotton-plugged test tube. Subsamples from this are used to prepare cultures in appropriate media, as may be indicated by the microscopic examination.

The second step in the bacteriological examination is preparation of smears for microscopic examination. Often, microscopic examination will aid in selecting the media to be employed for culturing. Samples for making smears may be obtained with a sterile inoculating loop. A simple stain, such as methylene blue or crystal violet, is generally employed for staining the smears.

In sampling solid or semisolid foods, samples should be taken from near the center of the container unless spoilage of the product near the surface is obvious. If spoilage of these conduction-heating products is due to organisms surviving the heat process, initial spoilage is most likely to occur in a small area surrounding the geometrical center. If spoilage is due to organisms entering via leakage, initial spoilage is more often near one end or the other of the container. For solid and semisolid products in the initial stages of spoilage, it is often helpful to make smears, for microscopic examination, with product from different locations in the container.

pH Determination. Spoiled and normal cans from the same lot should be examined. Because small variations in pH (as small as 0.15) are often significant, a glass electrode pH meter should be used.

Microscopic Examination. Each stained smear should be carefully examined for evidence of:

1. *Leakage.* For products with normal pH values above 4.0, container leakage usually, but not necessarily, will be evidenced by a mixed flora. The presence of micrococci and/or yeasts is almost certain evidence of leakage.
2. *Underprocessing.* If spoilage is due to bacteria surviving the heat process, normally there will be no more than one or two types present—usually only one except in cases of gross underprocessing. In this respect the microscopic examination is only indicative because, morphologically, several species of spoilage bacteria are very similar. However, coupled with information regarding product history, container condition, pH, etc., the microscopic findings often give a very good indication of the specific causative organism and the type of media that should be used for culturing the container contents.

Culturing. Unless otherwise indicated by the microscopic picture, all products should be cultured in two types of media. Glucose–tryptone broth containing bromcresol purple indicator and liver or heart broth are most frequently employed. Composition and preparation of these media will be described later.

A minimum of eight tubes of each medium should be used for culturing the contents of each container. It is advisable to use rather large culture tubes—for example, 22 × 175 mm.—each containing about 20 ml. of medium. To one half of the tubes of each medium, 3 or 4 ml. of product are added per tube; to the other half, 0.3 to 0.4 ml. of product is added. For products that cannot be pipetted, these quantities may be estimated. Immediately prior to inoculation, the liver or heart broth should be exhausted by steaming or heating in a water bath, and cooled to 45°F. or less. Immediately after inoculation the liver or heart broth tubes should be stratified with a sealing mixture. Though sterile Vaseline or 2% agar containing 0.05% sodium thioglycolate may be used, a more satisfactory seal is obtained with a mixture of 1 part paraffin, 1 part Vaseline, and 4 parts mineral oil. To prepare the latter mixture, the ingredients are heated together and mixed in an open container. The heating should be sufficient to drive off any entrapped moisture. The mixture is then distributed in appropriate containers and sterilized in the dry-air oven at about 163°C. (325°F.) for 2 hours. To avoid entrapping air bubbles, when stratifying the tubes the mixture should be at a temperature of 80°C. (176°F.) or higher.

The primary reason for using the rather large volume of medium per tube is to dilute and buffer acidity and other possible growth inhibitors in the sample. The small sample is included because frequently the larger sample will inimically alter the medium for bacterial growth.

Four tubes of each of the inoculated media (two containing the large sample and two containing the small sample) are incubated at 27° to 32°C. (80° to 90°F.). The other four tubes of each medium are incubated at 50° to 55°C. (122° to 131°F.). For products having a normal pH below 4.0, only the lower incubation temperature need be employed. Normally cultures at the higher temperatures are incubated for 2 or 3 days. Those at the lower temperature may be incubated for a week or more.

Examination of Incubating Cultures. As soon as growth appears in both tubes of any set (often in less than 24 hours), the contents of one of the tubes should be examined microscopically. Also, at this time a record should be made of growth appearance, presence or absence of gas, odor, etc. For microscopic examination, smears should be made and stained with Gram's stain. The following are particularly important to observe:

1. Morphology
2. Gram characteristic
3. Presence or absence of spores
4. Purity of culture

Incubation of the other tube of each set should be continued as prescribed above. After these prescribed incubation periods, the cultures should again be carefully examined both macroscopically and microscopically. If thermal resistance determinations are contemplated, all positive cultures should be placed under refrigeration subsequent to incubation and examination, unless the thermal resistance determinations are to be carried out immediately.

It is often advisable to determine if the cultured organisms will grow in the food being investigated. A positive result here indicates, though it does not assure, that the organisms were the cause of spoilage and not contaminants entering during examination. Food sterilized in test tubes may be used, but it is better to use normal containers of food that have failed to spoil under incubation. To inoculate the container, the end is first sterilized as explained above. It is then punctured with a sterile sharp instrument, such as an ice pick. With a sterile syringe, 1 or 2 ml. of the culture is introduced deep into the container contents. The small hole is then closed with solder. The inoculated containers are incubated at the temperature indicated by the above studies. The containers should be incubated for at least two weeks if spoilage does not occur earlier. Bacterial growth may be slower in the food than in the culture media.

Interpretation of Observations

General. If gas is produced anaerobically in the meat broth, if growth in the glucose–tryptone (GT) broth is without a pellicle, and if the organism is a rod form, it probably is a non-sporeformer. Because some facultative anaerobes do not sporulate readily in these media, the culture should be streaked on an appropriate nutrient agar to check its ability to sporulate. Appropriate agars will be described later. Growth with pellicle in the GT broth indicates spore-bearing aerobic bacilli. If spores are not present in organisms from the pellicle, streaking on an appropriate agar is again indicated.

Low-Acid Foods (pH above 4.5). Only bacterial spores should survive heat processes given low-acid foods. If viable cocci, non-spore-bearing rods, yeasts, or molds are observed, it is virtually certain that they gained entrance via container leakage. If only nonviable, vegetative forms are observed, any one of three things may be indicated: incipient spoilage before processing, autosterilization subsequent to spoilage by organisms that gained entrance via leakage, or autosterilization subsequent to spoilage by organisms that survived the process. Fermented products, such as pickles or kraut, are exceptions— some nonviable organisms may be expected in smears from these products.

GASEOUS SPOILAGE. Pure cultures of spore-bearing bacteria from swollen containers of low-acid foods almost invariably indicate underprocessing, though leakage cannot be ruled out completely. However, unless heat resistance determinations indicate otherwise, underprocessing may be assumed. If the isolated organism is a thermophilic anaerobe, it most likely is a member of the *C. thermosaccharolyticum* group. If it is a mesophilic anaerobe, it most likely is a member of the *C. botulinum*–*C. sporogenes* group.

Sometimes mixed cultures of spore-bearing mesophilic anaerobes are obtained from swollen containers of underprocessed acid food, and even of spore-bearing thermophilic anaerobes and spore-bearing facultative anaerobes of the *B. stearothermophilus* group. The latter would not contribute to the swelling of the container, but its viable spores sometimes reside in spoiled product. This situation of mixed flora in an underprocessed product can come about when the heat process is inadequate to free the food of the organism or organisms of lower resistance. When a mixed flora is observed, the organism of greatest importance depends on the relative ability of the organisms to grow in the product, on relative heat resistance, and on the temperature conditions to which the product is likely to be subjected subsequent to heat processing.

If no gas-forming bacteria are obtained in culture, but only flat-sour types from swollen containers, the flat-sour organisms may have been the cause of spoilage. The swelling may have been due to hydrogen produced by the interaction of acid produced by the organisms and the metal of the containers. In such cases, gas analyses and examination of the metal for corrosion are helpful to a diagnosis. The possibility that the organisms cultured might produce gas in the product and not in the culture media should not be overlooked. This may be checked by inoculating sterile test tubes of product and incubating them.

Swelled containers may be observed where there appears to be no evidence of microbial spoilage or hydrogen production by the interaction of acid and metal. In this case, spoilage by thermophilic anaerobes should be suspected. Thermophilic anaerobes sometimes grow in a product to the extent of producing a swell, then die (autosterilization). The spoilage often produces a change in appearance of the product, and sometimes the spoiled product has a butyric acid odor. Careful examination usually reveals ghosts of some of the bacterial cells. These may appear as intact granulated cells or just rows of dots (granules) without cell walls.

FLAT-SOUR SPOILAGE. This type of spoilage is evidenced by a lowering of pH (as little as 0.1 to more than 1 unit), and usually a slight to pronounced sour off-odor. In underprocessed low-acid products it is generally caused by thermophilic facultative anaerobes of the *B. stearothermophilus* group. If a spore-bearing thermophile is obtained, underprocessing is definitely indicated. If the organisms cultured are spore-bearing mesophiles, leakage may be indicated. Thermal resistance determinations should be made to establish the significance of such organisms.

Frequently, cultures from flat-sour products will be negative. This is more likely, the lower the pH, and the longer the products have been stored subsequent to heat processing. Sterility is the result of autosterilization after spoilage has occurred. Careful examination of stained smears often reveals ghost cells.

SULFIDE SPOILAGE. This unusual type of spoilage is generally due to underprocessing. It is caused by hydrogen sulfide producing thermophilic anaerobes of the *C. nigrificans* group. Spoiled containers are usually flat or only softly swollen, and their contents are generally darkened, owing to the formation of iron sulfide. The odor of hydrogen sulfide is usually evident. If a spore-bearing thermophilic anaerobe is cultured from such spoiled products, it should be subcultured in sulfite agar for further identification. *Clostridium nigrificans* will blacken this agar.

Another unusual type of spoilage is blackening of canned beets caused by the facultative anaerobe *B. betanigrificans*. The spoiled containers remain flat, and the pH of the contents is usually higher than normal. If a spore-bearing thermophilic facultative anaerobe is isolated from blackened beets, underprocessing is indicated.

Acid Foods (pH 4.0 to 4.5). As in low-acid foods, only bacterial spores should survive heat processes normally given these foods. Because of acidity, most of the spore-bearing organisms so important in low-acid foods will not grow in these foods. Some less heat-resistant, acid-tolerant sporeformers will, however. If viable cocci, non spore bearing rods, yeasts, or molds are observed, they probably gained entrance via container leakage. However, if any of these are found, the heat process given the product should be reviewed before it is concluded that the spoilage was due to leakage. Non-spore-bearing forms should be destroyed in acid products receiving processes wherein the lowest food temperatures reached are above about 180°F.

HYDROGEN SWELLS. If there is no evidence of microbial spoilage, swelling was probably due to hydrogen produced by interaction of the food acids and the metal of the container.

SWELLS DUE TO MICROBIAL SPOILAGE. These are usually caused by butyric anaerobes of the *C. pasteurianum* type. Spoilage by these organisms is usually accompanied by the odor of butyric acid. They usually will produce sufficient gas to burst cans and blow lids from glass containers. They are mesophilic obligate anaerobes.

Thermophilic anaerobes of the *C. thermosaccharolyticum* type have occasionally caused spoilage of some acid products. The spoiled products usually have pH values near 4.5 or slightly above. Some products with pH values as high as 4.7 have been, and still are by a few processors, given processes designed for acid foods (pH of 4.0 to 4.5).

FLAT-SOUR SPOILAGE. This type of spoilage of acid foods is usually caused by *B. coagulans* (*B. thermoacidurans*). This organism is a facultative anaerobe which, though normally classified as a thermophile, will grow at ordinary temperatures. However, growth at the lower temperatures is usually slower. There is considerable strain variation in this regard.

High-Acid Foods (pH below 4.0).   Spoilage of these foods is usually caused by micrococci, non-spore-bearing rods, yeasts, or molds. The products are normally given heat processes adequate to free them of these low-resistance organisms. If any of them are observed as the cause of spoilage, underprocessing or leakage is indicated. It is often difficult to determine whether the organisms gained entrance via leakage or survived the heat process. Careful examination of container seams or closures and thorough knowledge of the heat processing are helpful to a diagnosis.

Stains

As was indicated above, the only stains normally used in the bacteriological examination of spoiled canned foods are simple stains and the gram stain. Dilute crystal violet is to be preferred as a simple stain.

Gram Stain. Solution A is prepared by dissolving 2 gm. of crystal violet (90% dye content) in 20 ml. of ethyl alcohol (95%).

Solution B is prepared by dissolving 0.8 gm. of ammonium oxalate in 80 ml. of distilled water.

The crystal violet stain is prepared by mixing solutions A and B.

Lugol's iodine solution is prepared by dissolving 1 gm. of iodine and 2 gm. of potassium iodide in 300 ml. of distilled water.

A safranin counterstain is prepared by dissolving 0.25 gm. of safranin-0 in 10 ml. of 95% ethyl alcohol and adding 100 ml. of distilled water.

STAINING PROCEDURE

1. Apply crystal violet solution to smears for 1 or 2 minutes.
2. Wash in tap water and remove excess water with blotting paper.
3. Apply Lugols' iodine solution for 1 or 2 minutes.
4. Wash in tap water and remove excess water with blotting paper.
5. Decolorize with 95% ethyl alcohol. Apply alcohol and leave for 10 seconds; then hold slide in slanting position and drip alcohol on the smear until the alcohol running off is clear. Wash in tap water immediately.
6. Counterstain with safranin solution for 20 to 30 seconds.
7. Wash in tap water and blot dry.

Crystal Violet Stain. Make up ammonium oxalate crystal violet solution as for the gram stain.

STAINING PROCEDURE

1. Apply above solution for 1 or 2 minutes.
2. Wash in tap water and blot dry.

Media

Glucose–Tryptone Broth

Tryptone	10 gm.
Glucose	5 gm.
0.4% alcoholic solution of bromcresol purple	5 ml.
Distilled water	1000 ml.

(Add 20 gm. of agar to obtain solid medium.)

Sterilize by autoclaving for 20 minutes at 121°C. (250°F.)

Liver Broth. One pound of fresh beef liver from which the fat has been removed is ground, mixed with 1000 ml. of distilled water and the mixture boiled slowly for 1 hour. The pH is then adjusted to 7.6, and the liver particles are removed by straining the material through cheesecloth. The volume of the broth is made up to 1 liter, and the following are ingredients added:

Tryptone	10 gm.
K₂HPO₄	1 gm.
Soluble starch	1 gm.

In each 22 × 175-ml. tube, 15 ml. of broth is placed, then liver particles, previously removed, to a depth of about 1 inch. Adding the broth first minimizes entrapment of air. The broth is sterilized by autoclaving for 20 minutes at 121°C. (250°F.)

Beef Heart Broth.   A heart infusion is prepared exactly as described above for preparing a liver infusion. The heart particles are removed, and volume of the broth is adjusted. To the broth are added:

Isoelectric casein	5.0 gm.
Gelatin	10.0 gm.
Glucose	0.5 gm.
K₂HPO₄	4.0 gm.
Sodium citrate	3.0 gm.

The broth is distributed in tubes, and the meat particles are added as was done in preparing the liver broth. The broth is sterilized by autoclaving for 20 minutes at 121°C. (250°F.)

Sulfite Agar

Tryptone	10 gm.
Sodium sulfite	1 gm.
Agar	20 gm.
Distilled water	1000 gm.

The broth is heated to liquefy the agar and is then distributed in tubes. To each tube is added a piece of iron wire or a nail that has previously been cleaned with dilute HCl. The broth is sterilized by autoclaving for 20 minutes at 121°C. (250°F.) This medium should not be kept for more than one week. It is useful for detection of hydrogen sulfide-producing organisms.

Thermoacidurans Agar (*Difco*)

Yeast extract	5 gm.
Proteose peptone	5 gm.
Dextrose	5 gm.
K_2HPO_4	4 gm.
Agar	20 gm.
Distilled water	1000 gm.

This medium is useful for the detection and cultivation of *B. coagulans* (*B. thermoacidurans*). Better sporulation will generally be obtained if 10 mg. of $MnSO_4$ is added per liter of medium.

Nutrient Agar plus Manganese

Beef extract	3 gm.
Tryptone	5 gm.
$MnSO_4$	10 mg.
Agar	20 gm.
Distilled water	1000 gm.

This medium is useful for cultivation of *B. stearothermophilus* and related organisms. It generally promotes good sporulation. The pH of the prepared medium should be adjusted, if necessary, to 6.8 to 7.0.

Pork Infusion Agar. One pound of fresh pork from which the fat has been removed is ground, mixed with 1000 ml. of distilled water, and the mixture is boiled slowly for 1 hour. The pork particles are removed by filtering through two layers of cheesecloth, between which is a pad of glass wool. The filtrate is cooled in the refrigerator, after which the fat is skimmed off and the volume is made up to 1 liter with distilled water. The broth is placed in a straight-walled beaker, and the following ingredients are added:

Peptone	5 gm.
Tryptone	1.5 gm.
K_2HPO_4	1.25 gm.
Soluble starch	1 gm.
Glucose	1 gm.

The medium is heated with stirring to dissolve these ingredients. The pH is adjusted to 7.6. Next, 15 gm. of agar is added, and the medium is autoclaved for 25 minutes at 110°C. (230°F.) to liquefy the agar and promote formation of precipitate. While still hot the medium is placed in a refrigerator at about 4.4°C. (40°F.) until solidified. The solidified medium is then removed from the beaker by loosening it from the wall with a spatula and inverting the beaker. The layer containing any settled precipitate is trimmed off and discarded. The remaining agar is cut into small cubes which are returned to the beaker, and the medium is reliquefied by autoclaving for 25 minutes at 110°C. (230°F.). It is then dispensed in tubes and sterilized.

This medium is one of the better for subculturing heated spores of the putrefactive anaerobes to obtain spore counts.

SELECTED REFERENCES

American Public Health Association (1958). "Recommended Methods for the Microbiological Examination of Foods," 207 pp. American Public Health Association, New York.

Frazier, W. C. (1958). "Food Microbiology," 472 pp. McGraw-Hill, New York, Toronto, London.

Hersom, A. C., and Hulland, E. D. (1964). "Canned Foods—An Introduction to Their Microbiology" (Baumgartner), 5th ed., 291 pp. Chem. Publ. Co., New York.

Townsend, C. T., Somers, I. I., Lamb, F. C., and Olson, Norman A. (1956). "A Laboratory Manual for the Canning Industry," 2nd ed., Chapter 4. National Canners Association Research Laboratories, Washington, D. C.

CHAPTER 4

Organisms of Greatest Importance in Food Pasteurization

Heat treatments given canned foods are generally thought of as sterilization processes, though those given high-acid canned foods and even those given some acid canned foods are commonly referred to as pasteurization processes. Whether the term *sterilization* or *pasteurization* is used to label a heat treatment designed to reduce the microbial population of a food, the basic purpose of the heat treatment is the same—that is, to free the food of microorganisms that may endanger the health of food consumers or cause economically important spoilage of the food in storage and distribution. In other words, the purpose is to sterilize the food with respect to certain types of microorganisms, regardless of how the heat treatment applied is labeled. Therefore, the chief difference between pasteurization and sterilization is in concept and common usage. *Pasteurization* is the term that is most often applied to relatively mild heat treatments given foods that because of their nature will not support growth of the more heat-resistant organisms or that are refrigerated, frozen, concentrated or dehydrated to prevent significant growth of the more heat-resistant organisms. *Sterilization* is the term that is most often applied to more severe heat treatments, e.g., those given most low-acid canned food items, that are designed to free foods of virtually all microorganisms regardless of their heat resistance. Obviously, there can be no sharp line of demarcation between the two terms in either concept or usage.

Organisms of greatest importance in the spoilage, and therefore heat processing, of canned foods were discussed in Chapter 2. For the purpose of discussion these were grouped in accordance with pH of the foods in which they would grow and cause spoilage, and with respect to their oxygen and growth-temperature requirements. In considering organisms of greatest importance in pasteurization of foods, a similar grouping is both difficult and impractical. For example, most of these organisms are mesophilic and most of them are facultatively anaerobic. Though acidity may be a significant influencing factor on the heat resistance of many of these, available kinetic data are inadequate for meaningful generalization in this regard. A more logical grouping of organisms for discussion of pasteurization processes seems to be (1) pathogenic microorganisms, (2) toxin-producing microorganisms, and (3) spoilage microorganisms.

The list of foods that are given pasteurization heat treatments has been steadily growing since the pioneering discoveries of Louis Pasteur during the period 1860 to 1870. The principle of one of Pasteur's early discoveries—that is, that abnormal fermentations of wine, beer, and vinegar can be stopped by heating the bottles to between 55°C. (131°F.) and 60°C. (140°F.)—is now applied in treatment of a multitude of different food products.

It is not within the scope of this text to discuss in great detail the great variety of pasteurization processes applied to the many different food items. Rather, the objective is to point out the basic considerations involved in establishing adequate pasteurization processes. Of first consideration in this regard are the microorganisms of chief concern and the kinetics of death of these organisms when subjected to moist heat. Consideration for the most part will be confined to foods in their natural state, with regard to moisture content when a pasteurization treatment is applied, that is, to foods with no more than about 20% naturally occurring soluble solids content.

Pathogenic Microorganisms

Consideration here will be confined to a number of microorganisms that may, not uncommonly, contaminate foods that when ingested with foods have the ability to pass through the intact mucosa of the alimentary tract, grow in the body tissues, and produce damage therein causing disease syndromes. Infectious diseases so produced are often referred to as *food infections* in contrast to afflictions caused by microorganisms, e.g., *Clostridium botulinum* and most coagulase-positive staphylococci, that normally are not pathogenic when ingested. When growing in foods prior to ingestion, however, they produce toxins that are absorbed from the alimentary tract when the food is ingested and thereby cause intoxication syndromes. The latter are usually referred to as *food poisonings* or *food intoxications*. Often infectious and intoxication disease syndromes are very similar, e.g., in the case of *Salmonella* infections and *Staphylococcus* food poisonings.

Mycobacterium tuberculosis.   This is a non-spore-forming, gram-positive, acid-fast, rod-shaped bacterium that causes tuberculosis, a disease that may appear in a diversity of forms, such as the more common pulmonary tuberculosis or the less common tubercular meningitis and tuberculosis of the bone. Tubercle bacilli readily pass through the intact mucosa of the alimentary tract and from there may spread to involve practically every organ of the body. Historically, tuberculosis stands out as one of the most important scourges of civilized communities.

From the standpoint of food pasteurization, *M. tuberculosis* is of chief concern in milk because it is infectious to cattle and frequently occurs in the milk of infected animals; in fact, milk in times past was the most common vehicle of infection, although other foods, particularly meat from infected animals, and sputum may be sources of infection also. The first real defense against transmission of tuberculosis through milk from cattle to humans was pasteurization, a heat treatment based on the early work of Louis Pasteur mentioned above. Pasteurization of milk as practiced, to ensure inactivation of *M. tuberculosis,* was clearly defined in the *Milk Ordinance and Code,* recommended by the United States Health Service (1953) as follows: "The terms 'pasteurization,' 'pasteurized,' and similar terms shall be taken to refer to the process of heating every particle of milk or milk products to at least 143°F., and holding at such temperature continuously for at least 30 minutes, or to at least 161°F., and holding at such temperature continuously for at least 15 seconds, in approved and properly operated equipment: *provided* that nothing contained in this definition shall be construed as barring any other process which has been demonstrated to be equally efficient and which is approved by the State health laboratory." "Boiled down" this code specifies, for milk pasteurization, any heat treatment that is equivalent in lethality, with respect to the destruction of *M. tuberculosis,* to the time–temperature relationships of 30 minutes at 143°F. and 15 seconds at 161°F. Other equivalents are discussed in Chapter 15.

At the time the above standards were developed, $M.$ $tuberculosis$ was regarded as the most heat resistant of the pathogenic organisms likely to be present in milk. This consideration was based on extensive literature covering heat resistance studies made throughout the world; in fact, at the time there were more than one hundred reports on the thermal resistance of $M.$ $tuberculosis$ and other pathogens that were thought to be of importance in the pasteurization of milk. Suffice it to say here that appraised thermal resistance of $M.$ $tuberculosis$ on which the standards were based may be described as follows: $D_{150} = 0.3$ to 0.4 minutes and $z = 8$ to 10 Fahrenheit degrees. These values may also be considered as representing the resistance observed for the more resistant of various strains of $M.$ $tuberculosis$ so far studied (cf. Evans et $al.$, 1970).

It should be noted that, in order to compare resistance values directly, the resistance parameter z is equally as important as the parameter D. Whereas D denotes resistance, per se, to a given temperature, z denotes relative resistance of an organism to different lethal temperatures. For definition and further discussion of D and z parameters see Chapter 7. It logically follows that, knowing the D value characterizing resistance of an organism to one temperature and the z value denoting its relative resistance to other temperatures, the D value characterizing resistance of the organism to any other temperature may be readily computed.

$Brucellae.$ The brucellae are non-spore-forming, gram-negative, minute rod-shaped bacteria that cause brucellosis in animals and man. This disease is very infectious and causes contagious abortion (Bang's disease) in cattle and undulent fever in humans. Though the disease may be acute in man it is more generally chronic and characterized by recurrent undulating fever. There are three recognized species of importance: $Brucella$ $abortus$, $Brucella$ $melitensis$, and $Brucella$ $suis$, identified with brucellosis in cattle, goats, and swine, respectively. All three species may cause brucellosis in humans, although $B.$ $suis$ is probably most infectious for man. $Brucella$ $suis$ is also highly infectious for cattle and may be transmitted to man through the milk from cows infected with this species.

The chief source of brucellosis infection in man is milk from infected cows, however, the disease can be transmitted in meat from infected animals. Like $M.$ $tuberculosis$, $Brucella$ organisms pass readily through the intact mucosa of the alimentary tract. Also, as with tuberculosis, milk pasteurization is the oldest and most effective defense against transmission of brucellosis from cattle to humans. It should be noted that, in regard to both tuberculosis and brucellosis, eradication programs have markedly reduced the number of infected cattle and consequently have greatly reduced the magnitude of the transmission problem; however they have not been and most likely never will be effective enough to eliminate the necessity of milk pasteurization.

The brucellae are appreciably less resistant to heat than are the most resistant strains of *M. tuberculosis*. Best evidence from published work indicates that the resistance of the more resistant brucellae in milk may be described as follows: $D_{150} = 0.10$ to 0.20 minutes and $z = 8$ to 10 Fahrenheit degrees. Comparing resistance values, it becomes obvious that pasteurization processes designed to sterilize with respect to *M. tuberculosis* are ample for sterilizing with respect to *Brucella* spp.

Coxiella burnetti.   This is a very minute rickettsia-like organism that causes a febrile disease in man known as Q fever (Enright *et al.* 1953). Epidemiological studies have implicated cows, sheep, and goats as sources of this infectious organism with regard to humans. When these animals are infected, their milk contains the organism, which probably explains the mode of transmission from animal to man. The disease in naturally occuring outbreaks was first recognized in the United States in 1946. The infection in humans is characterized by a febrile condition that is often accompanied by pneumonitis. The mortality rate is low.

Enright *et al.* (1956) reported results of their extensive studies on the thermal resistance of *Coxiella burnetti* in milk. According to their results the average maximum resistance of the organism may be described as follows: $D_{150} = 0.50$ to 0.60 and $z = 8$ to 10 Fahrenheit degrees. Therefore, their results show the resistance of *Coxiella burnetti* to be somewhat more than that of *M. tuberculosis*.

Prior to the advent of Q fever in the United States and the studies on thermal resistance of *Coxiella burnetti*, minimum standards for milk pasteurization were based on the resistance of *M. tuberculosis*. As indicated in the above discussion of *M. tuberculosis*, the milk pasteurization standards were equivalent in lethality to 30 minutes heating at 143°F. In light of the findings of Enright *et al.* (1956) on the resistance of *Coxiella burnetti*, the standards for batch or vat pasteurization of milk were changed (July 19, 1956) to make them equivalent in lethality to 30 minutes heating at 145°F. In other words, batch pasteurization processes were made more severe so as to sterilize with respect to *Coxiella burnetti*. Different equivalent time–temperature relationships are discussed in Chapter 15.

Salmonellae. The salmonellae are non-spore-forming, gram-negative, rod-shaped bacteria that cause salmonellosis in animals and man. Salmonellosis is one of the most important communicable diseases in the United States today. The number of human cases per year has been conservatively estimated at 2,000,000 though the actual number of cases may be much higher. At present there are some 1300 different identifiable serotypes of solmonellae, though these are all now generally grouped under three recognized species according to the "three species concept" proposed and emended by Ewing (1963, 1966, 1967). The three species recognized are *Salmonella cholerae-suis*, *Salmonella typhi*, and *Salmonella enteriditis*. Of these, *S. cholerae-suis* is considered the type species and all salmonellae other than *S. cholerae-suis* and *S. typhi* are considered to be serotypes of *S. enteriditis*.

Members of the genus *Salmonella* are widely distributed in nature. Though their natural habitat is often considered to be the animal body, they will multiply either inside or outside the animal body if placed in a suitable environment. Among the wide variety of *Salmonella* serotypes there are types that affect most all, if not all, warm-blooded animals and at least some cold-blooded animals, such as snakes, lizards, and turtles. At the present time it must be assumed that all of the known *Salmonella* serotypes are potentially pathogenic to man, though serotype *typhimurium* has been responsible for far more infections in man than any other, and only 12 serotypes account for 78% of the documented human infections (National Academy of Sciences, 1969). Although typhoid fever appears to be specific for man, the other salmonella infections occur naturally not only in man but in numerous other animal species as well. There are a few salmonellae that are primarily adapted to particular domestic animal hosts, such as *S. cholerae-suis* and some serotypes of *S. enteritidis:* however, all of these can cause gastroenteritis in man. With respect to *S. enteritidis*, there are some 1300 host-unadapted serotypes that seemingly attack man and other animals with equal facility.

Typhoid and paratyphoid fevers are characterized by prolonged incubation periods of 10 to 20 days or more, a generalized disease with febrile manifestations, and a tendency to produce carriers and to become endemic. Carriers may be described as individuals who harbor (and transmit) pathogens even though they portray no clinical manifestations of disease themselves. They may be individuals who continue to harbor and transmit a causative pathogen after their own complete recovery from a specific disease or they may be individuals who never experienced the clinically detectable disease but simply harbor and transmit the causative organism of that disease without themselves being adversely affected. It should be noted that public health measures have succeeded in bringing the typhoid fevers under satisfactory control and that the causative organisms are now of minor significance with regard to food pasteurization.

As a group the salmonellae, compared to *M. tuberculosis, C. burnetti,* and many nonpathogenic vegetative forms, have been found to have appreciably lower resistance to heat, making their removal from certain foods (e.g., liquid eggs) possible by the application of relatively mild pasteurization processes. Ng (1966) studied the heat resistance of 300 salmonella cultures representing 75 different serotypes. He compared their resistance with that of a reference strain, serotype *typhimurium TMI,* in a trypticase soy medium at pH 6.8 heated at 57°C. (134.6°F.) The average resistance reported for most of the strains is equivalent to a D_{150} of about 0.014 minutes, that of the least resistant strain to a D_{150} of about 0.007 minutes, and that of the most resistant strain to a D_{150} of about 0.070 minutes. Winter *et al.* (1946) isolated a strain of serotype *senftenberg,* designated as 775W, that, comparatively, has a much higher resistance. Resistance of this strain reported by Ng (1956) is equivalent to a D_{150} of about 0.894 minutes. Davidson *et al.* (1966) reported isolating a *senftenberg* strain from meat that had a similar resistance.

Anellis *et al.* (1954) reported on the resistance of a number of *Salmonella* serotypes in liquid whole egg. In acidified egg (pH 5.5) the range of resistance observed for serotypes other than *senftenberg* was approximately equivalent to $D_{150} = 0.023$ to $D_{150} = 0.129$, the average being about $D_{150} = 0.075$. The resistance value they reported for *senftenberg* was equivalent to a D_{150} of about 1.00. They also found the resistance of all strains to decrease, progressively, with increase in egg pH from about 5.5 to about 8.5; e.g., the resistance value reported for serotype *senftenberg* in egg at pH 8.55 was equivalent to D_{150} of about 0.036 in contrast to the D_{150} value of about

1.00 for the organism in egg at pH 5.45. This pH effect on resistance of salmonellae has been observed by others and may account for the somewhat higher resistance observed by Anellis *et al.* for the organisms heated in acidified egg (pH 5.5) and those observed by Ng for the organisms heated in trypticase soy medium (pH 6.8).

Similar resistance values have been observed by others for salmonellae in various food products, including milk and poultry meats (cf. Osborne *et al.*, 1954; Bayne *et al.*, 1965; Evans *et al.*, 1970; and Marth, 1969).

Based on the studies discussed above and others it has been concluded generally that serotype *senftenberg* 775W represents a most unusual and rare mutant and that pasteurization schedules for foods should not necessarily be based on this one strain. It may be noted that if schedules were based on the resistance of this strain they would be in the order of 10 times more severe than schedules based on the more resistant salmonellae other than 775W. Obviously, the resistance of salmonellae other than *senftenberg* 775W was the basis for a regulation published in the Federal Register, January 1, 1957, which was to the effect that "Strained and filtered liquid whole egg shall be flash heated to not less than 140°F. and held at this temperature for not less than $3\frac{1}{2}$ minutes." Whereas this schedule should accomplish at least a 5 log cycle ($5D$) reduction in salmonellae of average resistance in acidified eggs, it probably would accomplish less than one log cycle ($1D$) reduction in serotype *senftenberg* 775W. More accurate estimates cannot be given because of the influence on resistance of such factors as pH (discussed above) and soluble solids content as affected by added sugar, salt, etc. It should be noted that there are innumerable pasteurization schedules presently being employed for reduction of salmonellae and other pathogenic and nonpathogenic organisms in a wide variety of food products. Needless to say it is impossible here to review all these and even impractical to review any of them in detail, or to so review the great volume of thermal resistance data on which the schedules may have been based. Suffice it to say that any given pasteurization schedule should be based on best available death kinetic data for the organisms of concern in the specific product to be pasteurized. See Chapter 15 for further discussion regarding the establishment of equivalent time–temperature relationships in pasteurization.

Shigella dysenteriae.   This species is closely related to salmonellae and is often included with them in discussions. It causes the enteric disease, shigellosis, which symptomatically is so similar to salmonellosis that differential diagnosis usually depends on isolation of the causative organism and species identification. Its thermal resistance has been found to be intermediate compared with the wide range of resistance reported for various strains and serotypes of salmonellac. For example, Evans *et al.* (1970) reported a thermal resistance for the most resistant of five strains of this species, heated in milk, that may be characterized by a D_{150} equal to about 0.05 minutes and a z equal to about 8.4F°. Therefore, from the standpoint of food pasteurization this species may be considered along with the salmonellae.

Toxin-Producing Microorganisms

Micrococcus pyogenes var. *aureus—Coagulase-Positive Staphylococci.* This is a gram-positive micrococcus the cells of which often appear in clusters in culture. It is widespread in nature and is well known as a pathogen. It causes a number of infections in man and animals, notably, pimples, boils, carbuncles, internal abscesses, osteomyelitis, and septicemia. Though highly infectious under certain circumstances, it commonly resides on the skin of the most healthy individuals. It not uncommonly causes mastitis in cows and contaminates the milk of infected cows. Because of its prevalence in minor and major infections and in residence on the skin of healthy individuals, and because it readily multiplies outside the animal body when placed in a suitable environment, its chance of contaminating a wide variety of foods is indeed great.

Though many strains of this species are highly pathogenic under certain circumstances, this characteristic of the organism must be considered of minor importance from the standpoint of food contamination. Normally, they are noninfectious when ingested by healthy individuals. They are of chief concern as food contaminants because of their ability to grow in a wide variety of foods and produce an exotoxin that induces acute gastroenteritis when ingested. As a group they are mesophilic and facultatively anaerobic. They can grow and produce toxin in foods having pH values slightly above 4.00 and upward. Virtually all strains of *Micrococcus pyogenes* var. *aureus* are coagulase positive, that is, they produce the enzyme coagulase that coagulates human (and rabbit) blood plasma. It is a distinguishing characteristic of the group. In fact, in discussion of food poisoning, these micrococci are generally referred to as simply coagulase-positive staphylococci.

The gastroenteritis induced by the toxin of coagulase-positive staphylococci is very similar, symptomatically, to that associated with salmonellosis, which makes quick definitive diagnosis very difficult. One important distinction is the time elapsing from ingestion until the onset of symptoms, principally nausea, abdominal pain, diarrhea, and vomiting. In the case of staphylococcus food poisoning, symptoms usually appear from 1 to 6 hours after ingestion of the affected food. As pointed out above, the time from ingestion to the onset of symptoms of salmonellosis is generally 6 to 24 hours. However, distinction in this respect is not always clear.

The mortality rate for staphylococcus food poisoning is very low. The affliction is usually very incapacitating but normally of short duration (1 to 2 days). Recovery is generally complete, though secondary and serious infections may result occasionally. Chief concern, aside from temporary physical discomfort, is economic in the nature of working time lost and expense of medical care in some cases.

As indicated above the coagulase-positive staphylococci, though of some concern as food contaminants because of their pathogenicity, are of chief concern because of their ability to grow and produce toxin in a wide variety of foods. As is the case with other vegetative forms, the heat resistance of these organisms is very low compared to the heat resistance of spores of the toxin-producing organism, *C. botulinum*. It should be noted, however, that whereas botulinum toxin is quite heat labile, being destroyed readily by temperatures of 190°F. and above, the toxin produced by the staphylococci is very heat stable, withstanding temperatures of 250°F. and higher for long periods of time. Therefore, coagulase-positive staphylococci must be destroyed in a food, or prevented from growing therein, by imposing conditions, e.g., low temperatures, that are inimical to their growth. Consequently, they are of great importance in the pasteurization of foods.

As with salmonellae, review of the rather extensive literature regarding heat resistance of coagulase-positive staphylococci shall not be attempted. Reference is made to only a few more carefully conducted studies that define the maximum resistance that may be expected of these organisms. Heinemann (1957) determined the heat resistance of the most resistant of three strains of *Micrococcus pyogenes* var. *aureus* cultured in skim milk. As nearly as can be approximated from values reported the resistance of this strain was characterized by a D_{150} of about 0.25 minutes and a z of 9.2 Fahrenheit degrees. Evans *et al.* (1970) reported on the resistance of the two most resistant of ten strains of the same species. Cells for their studies were produced on brain–heart infusion agar and suspended in whole milk for resistance determinations. They assumed the z value was the same as that reported by Heinemann (1957). On this basis the equivalent D_{150} value would be 2.22 minutes. These studies and others concerning the resistance of the coagulase-positive staphylococci in different food products indicate a range of resistance for these organisms in the order of $D_{150} = 0.20$ to 2.20 minutes and $z = 8$ to 10 Fahrenheit degrees.

It may be noted that the maximum resistance observed for members of this group is somewhat higher than that observed for *Salmonella enteriditis* serotype Senftenberg and considerably higher than that observed for other pathogens of importance in pasteurization. When it is recognized that the coagulase-positive staphylococci are normally not pathogenic when ingested but must grow in foods to produce toxin in order to create a significant health hazard as far as foods are concerned, it seems reasonable to believe that pasteurization schedules, designed to free foods of the other pathogens of lesser resistance, may be satisfactory with respect to control of these organisms in many foods; however, in certain cases, e.g., pasteurized ham, pasteurization schedules should be based on the resistance of these more resistant organisms.

Clostridium botulinum Type E. This is a gram-positive, sporulating, anaerobic, motile bacillus that produces a powerful neuroparalytic exotoxin. It is not pathogenic in the sense of causing infection in man. However, its toxin when ingested causes an often fatal intoxication known as botulism.

The incubation period in botulism is usually 8 to 24 hours. Dolman (1960) described the course of events somewhat as follows: Abdominal symptoms of nausea, vomiting, pain, and distension may develop within a few hours after ingestion of the toxin. These are generally followed by persistent constipation and the onset of a peculiar paralytic syndrome. When sufficient toxin has been absorbed from the alimentary tract to reach sites adjacent to cranial and peripheral motor nerve endings, it exerts an inhibitory effect upon the mechanism of acetylcholine synthesis, or release, thereby interfering with the transmission of neural impulses to muscles and glands. Frequently, the cranial nerves supplying the muscles of the eyes and throat are affected first, resulting in dim, fuzzy, or double vision, and difficulty in swallowing. This is usually accompanied, or followed, by generalized weakness, and in severe cases by respiratory and cardiac failure. Salivation is often impaired causing dryness of the mouth. There is no rash nor fever, and consciousness is seldom impaired even in the terminal stages. In fatal cases, death usually occurs between the second and fifth day but may occur as early as 20 to 24 hours, or as late as 10 to 14 days after ingestion of affected food. (With minor exceptions, this quite accurately describes the events in botulism caused by the other types as well.)

There are six known Types of *C. botulinum*, designated alphabetically from A to F. Human botulism is nearly always caused by Types A, B, and E. A few cases have reportedly been caused by Types C and F. The spores of Types A and B are the most heat resistant, their resistance being characterized by D_{250} values in the range of 0.1 to 0.2 minutes. They are of great concern in the sterilization of canned low-acid foods, as discussed in Chapter 2. Comparatively, the heat resistance of Type E spores is very low, being described by D_{180} values of the order of 0.1 to 3.00 minutes (cf. Gonsalves, 1966).

Whereas the natural habitat of *C. botulinum* Types A and B is for the most part terrestrial, that of Type E is essentially aquatic (marine and inland lake waters). The natural habitat of the rather recently described and much less prevalent Type F is apparently aquatic. Limited studies indicate that the heat resistance of spores of the latter is about the same as that of spores of Type E. As a matter of interest, it should be noted that Type C has been incriminated in a few cases of human botulism and that the heat resistance of its spores is intermediate between that of Types E and F and that of Types A and B (Segner and Schmidt, 1971).

Because the natural habitat of Type E is aquatic, it is of primary importance in the pasteurization of fish and fish products, e.g., smoked fish. These are products that for quality reasons cannot be sterilized (severely heated) and therefore are pasteurized (mildly heated); refrigeration is employed to inhibit significant growth of microorganisms surviving the pasteurization treatment, until the products reach the consumer. Herein lies the significance of *C. botulinum* Type E as an organism of great importance in the pasteurization of these products. Whereas *C. botulinum* Types A and B fail to grow below a temperature of about 10°C. (50°F.) some strains of Type E will grow at temperatures near 4.4°C. (40°F.) (Schmidt, 1964; Graikoski and Kempe, 1965). Consequently, normal refrigeration systems during storage and distribution of these products are generally adequate to prevent growth of *C. botulinum* Types A and B; however, they may at times be inadequate to prevent growth of Type E. Furthermore, these products are much more likely to be contaminated with Type E than with Types A or B.

In view of the above, the obvious procedure to assure reasonable safety in regard to fish and fish products that are mildly heated but refrigerated during storage and distribution is to apply heat processes adequate to sterilize them with respect to *C. botulinum* Type E. These are generally considered pasteurization processes. In view of the resistance values noted above for spores of this organism, such processes should be at least the equivalent of 30 minutes heating at 82.2°C. (180°F.). This should accomplish about a 10 log cycle reduction in the initial population of the more resistant Type E spores. It should be noted that regulatory agencies in some states in the Great Lakes Region now require smoked fish to be heated to an internal temperature of at least 82.2°C. (180°F.) and held at this temperature for at least 30 minutes, or the equivalent thereof. Further information regarding equivalent processes is given in Chapter 15.

Spoilage Microorganisms

Nonsporulating Bacteria. Reference here is to saprophytic bacteria that normally are neither pathogenic nor toxin producing. Virtually all such organisms, and there is a multitude of different species and strains, may if given the opportunity grow and cause spoilage of foods of one sort or another. However, among the more important genera from the standpoint of food spoilage are *Pseudomonas, Achromobacter, Lactobacillus, Leuconostoc, Proteus, Micrococcus,* and *Aerobacter.*

Unfortunately, there is a grave paucity of definitive heat resistance kinetic data available for most of these organisms. However, judging from the effectiveness of pasteurization processes employed for treating a wide variety of foods for the purpose of extending their keeping quality (under refrigeration) it may be deduced that the average maximum heat resistance of most of these organisms may be characterized by D_{150} values in the range of 1.00 to 3.00 minutes and z values in the range of 8 to 10 Fahrenheit degrees. For example, the heat resistance results obtained by Collier and Townsend (1954) indicated that the heat resistance of the more resistant *Lactobacillus* isolates studied was characterized by a D_{150} value of about 1.00 minute and a z value of about 10 when the organisms were suspended in tomato juice or tomato paste when heated. When the same organisms were heated in neutral phosphate buffer the D_{150} value characterizing their resistance was about 3.00 minutes. It should be noted that pasteurization schedules, for a wide variety of food products, based on resistance values of this order have been very effective in extending the keeping quality of the products under refrigeration.

Yeasts and Molds.   Again, there is a grave paucity of definitive heat resistance kinetic data for yeasts and molds that are of importance in the spoilage of foods, and therefore in pasteurization. However, judging from the success of pasteurization schedules, based on resistance values of the order of those discussed above for vegetative forms of bacteria, in controlling spoilage by these organisms, it must be concluded that their heat resistance is of the same order as that of the vegetative forms of bacteria. Notable exceptions are the *Byssochlamys* and *Penicillium* species discussed in Chapter 2.

SELECTED REFERENCES

Annelis, A., Lubas, J., and Rayman, M. M. (1954). Heat resistance in liquid eggs of some strains of the genus *Salmonella*. *Food Res.,* **19,** 377.

Bayne, H. G., Garbaldi, J. A., and Lineweaver, H. (1965). Heat resistance of *Salmonella typhimurium* and *Salmonella senftenberg* 775W in chicken meat. *Poultry Sci.,* **44,** 1281.

Collier, C. P., and Townsend, C. T. (1954). "Container sterilization for acid products by hot fill-hold-cool procedures." National Canners Association, Information Letter No. 1472.

Davidson, C. N., Boothroyd, M., and Georgala, D. L. (1966). Thermal resistance of *Salmonella senftenberg. Nature,* **212,** 1060.

Dolman, C. E. (1960). Type E botulism: A hazard of the North. *Arctic,* **13,** 230.

Enright, J. B., Sodler, W. W., and Thomas, R. C. (1956). Observations on the thermal inactivation of the organism of Q-fever in milk. *J. Milk and Food Technol.,* **19,** 313.

Enright, J. B., Thomas, R. C., and Mullet, P. A. (1953). Q fever and its relation to dairy products. *J. Milk and Food Technol.,* **16,** 263.

Evans, D. A., Hankinson, D. J., and Litsky, Warren. (1970). Heat resistance of certain pathogenic bacteria in milk using a commercial plate heat exchanger. *J. Dairy Sci.,* **53,** 1659.

Ewing, W. H. (1963). An outline of nomenclature for the family *Enterobacteriaceae. Int. Bull. Bacteriol. Nomencl. Taxon.,* **13,** 95.

Ewing, W. H. (1966). "Enterobacteriaceae: Taxonomy and Nomenclature," NCDC Publ., Atlanta, Georgia.

Ewing, W. H. (1967). "Revised Definitions for the Family Enterobacteriaceae, Its Tribes and Genera," NCDA Publ., Atlanta, Georgia.

Foster, E. M., Nelson, E. F., Speck, M. L., Doetsch, R. N., and Olson, J. C. (1957). "Dairy Microbiology." 492 pp. Prentice-Hall, Englewood Cliffs, New Jersey.

Gonsalves, A. A. (1966). The effect of sodium chloride on the heat resistance of *C. botulinum* Type E. Masters Thesis. 77 pp. University of Massachusetts.

Graikoski, J. T., and Kempe, L. L. (1964). Heat resistance of *Clostridium botulinum* Type E Spores. Bacteriol. Proc., Am. Soc. Microbiol., p. 3.

Heinemann, B. (1957). Growth and thermal destruction of *Micrococcus pyogenes* var. *aureus* in heated and raw milk. *J. Dairy Sci.*, **40**, 1585.

Marth, E. H. (1969). Salmonellae and salmonellosis associated with milk and milk products. A review. *J. Dairy Sci.*, **52**, 283.

Milk Ordinance and Code: Public Health Service Bull. 220, (USPHS) (1953).

National Academy of Sciences. (1969). "An Evaluation of the Salmonella Problem," 207 pp. Natl. Acad. of Sciences, Washington, D. C.

Ng, H. (1966). Heat sensitivity of 300 *Salmonella* isolates, pp. 39–41 in *U. S. Dept. Agr., Agr. Res. Ser.* ARS 74-37.

Osborne, W. W., Straka, R. P., and Lineweaver, H. (1954). Heat resistance of strains of *Salmonella* in liquid whole egg, egg yolk, and egg white. *Food Res.*, **19**, 451.

Riemann, H. (1969). "Food-Borne Infections and Intoxications" (Hans Riemann, ed.), 677 pp. Academic Press, New York, N. Y.

Schmidt, C. F. (1964). Spores of *C. botulinum*. Formation, resistance, germination, pp. 69–81. Botulism: Proc. of a Symposium. U. S. Public Health Service Publication No. 999-FP-1. Cincinnati, Ohio.

Segner, W. P., and Schmidt, C. F. (1971). Heat resistance of spores of marine and terrestrial strains of *Clostridium botulinum* Type C. *Appl. Microbiol.*, **22**, 1030.

Winter, A. R., Stewart, G. F., McFarlane, V. H., and Solowey, M. (1946). Pasteurization of liquid egg products. III. Destruction of *Salmonella* in liquid whole egg. *Am. J. Pub. Health,* **36**, 451.

CHAPTER 5

Contamination and Its Control

As will become obvious from later discussions relating to the death of bacteria subjected to moist heat, there is a direct relationship between the number of bacteria in a product and the degree of heating necessary to sterilize the product. Other things being equal, the greater the number of bacteria, the more severe is the heating required. Because of the deleterious effect of heat on many organoleptic and nutritive properties of foods, it is imperative that heat treatments be very little more severe than those just adequate to free the foods of undesirable organisms. It follows, therefore, that bacterial contamination of foods to be heat processed should be kept at the minimum.

The chief sources of contamination of foods to be canned may be considered under four main headings: (1) Raw materials. (2) Food-handling equipment. (3) Containers. (4) Food handling.

Raw Materials

Nonformulated Items—Fruit and Vegetable Products. This category includes fruit and vegetable products to which are added only minor ingredients, such as salt, sugar, and sometimes starch. With these products, the primary source of contamination is soil. Soils frequently contain as many as 10^8 bacteria per gram. Of these, as many as 10^6 may be in the spore state. Therefore, preliminary to the canning of fruits and vegetables it is essential to remove adherent soil and dust from them by thorough washing. For cleaning, particularly vegetables, the use of wetting agents in the wash water is useful.

Studies have shown that spoilage of these products, given heat processes that were normally adequate, in many cases could be traced to failure to remove soil and dust laden with heat-resistant spores. Proper washing requires a generous supply of running water. Because of bacterial build-up, the use of static or recirculating water supplies is poor practice.

Minor ingredients, particularly sugar and starch, may be important sources of contamination. Many products that require some rather low-temperature heat processing in the course of their manufacture are potential sources of thermophilic bacteria. About 1930, the National Canners Association established standards for spores of thermophiles in sugar and starch. These standards are now applied to a variety of ingredients other than sugar and starch. They are as follows:

TOTAL THERMOPHILIC SPORE COUNT. For five samples examined, there shall be a maximum of not more than 150 spores and an average of not more than 125 spores per 10 gm. of product.

THERMOPHILIC ANAEROBIC SPORES. These shall be present in not more than three (60%) of five samples and in any one sample to the extent of not more than four (65 + %) tubes. (According to the recommended procedure, each tube would contain two-thirds gram of product.)

SULFIDE SPOILAGE SPORES. These shall be present in not more than two (40%) of five samples and in any one sample to the extent of not more than five spores per 10 gm.

For methods for determining thermophilic spore counts, the reader is referred to the National Canners Association's "Laboratory Manual for the Canning Industry."

Formulated Items. Many canned food items are formulated from a number of different ingredients. In addition to sugar and starch, many of these ingredients may contain thermophilic spores. Among them are mushrooms and mushroom powders, spices, various types of flour, condiments, cocoa, molasses, meat extracts, yeast extracts, condensed milk, powdered milk, and powdered eggs.

Standards have not been established for ingredients other than sugar and starch. The same or slightly modified methods may be employed for determining counts in these products. The significance of counts obtained depends on the proportion of the total product represented by any ingredient. It is the total number of heat-resistant spores in the product that determines the severity of the sterilization process.

Control Measures. With regard to those ingredient materials that are heat-processed during some stage of their manufacture, control of thermophilic spore build-up requires rigid sanitation of processing equipment and careful control, and sometimes adjustment, of temperatures employed in the processing operations. In many cases it has been possible to increase processing temperatures to levels above those that permit growth of thermophilic bacteria (above 170°F., 77°C.), without adversely affecting the product. Even when these higher processing temperatures are employed, precautions must be taken to avoid areas in the equipment, such as recesses or surfaces above product level, wherein the higher temperatures may not be maintained at all times.

In the case of certain ingredients,—for example, dry spices—it is often impossible to reduce materially the bacterial spore load by altering the conditions of production. Gaseous fumigants such as ethylene and propylene oxides are often used to reduce spore loads in such products before they are included as ingredients of canned foods. Bacterial spores may be virtually eliminated by such treatments. Heat processing is frequently employed to reduce spore loads of certain materials, such as meat extracts, before they are used as ingredients of canned foods.

Food-Handling Equipment

Many types of equipment are employed in handling and processing various food products prior to canning. Much of this equipment is designed with mechanical considerations in mind and with little or no attention given to sanitary aspects. Attention to certain principles of design can often minimize or eliminate potential sources of bacterial contamination. The following are some of the more important principles.

1. Construction and layout of the various units should be as simple as possible. So-called "dead ends" and other pockets should be avoided. Spoilage organisms may grow in entrapped food in such recesses and serve as reservoirs of contamination to product passing through the equipment. Such recesses usually are hard-to-clean areas.

2. Food pipelines should be as short as possible and made up of internally flush-fitting lengths. Couplings should be of the sanitary quick-release type to facilitate easy dismantling and cleaning. Pipe "turns" should be made with round, smooth, easy-to-clean elbow joints. All valves should be of a sanitary type that does not provide recesses in which food may accumulate and become stagnant. Valves should be easily removed for cleaning. In-place diversion lines should not be used. No matter how these are valved, they provide "dead-end" pockets for accumulation of product.

3. Wood should be avoided in canning equipment. Porous wood surfaces soon become soaked with food juices in which bacteria may grow and feed back into fresh food products passing through the equipment. Wood surfaces are difficult to clean and virtually impossible to disinfect.

4. Whenever possible, "stainless" metal should be used in canning equipment. Surfaces should be smooth and free of recesses where food may accumulate. All food-handling equipment should be so constructed that it may be easily cleaned and disinfected in place or readily dismantled for cleaning and disinfecting.

5. All equipment should be raised from the floor or otherwise constructed so that floors may be kept clean. There should be no floor areas inaccessible to cleaning.

6. There should be no pipes, conveyors, or other difficult-to-clean equipment directly above open food-handling equipment. Drippage because of pipe leakage or condensate formation can be a hazardous source of contamination, as can falling debris from moving equipment such as a conveyor.

7. Water for washing foods and food-handling equipment should be of the highest quality. It is good practice to use water with a residual chlorine concentration of 5 to 7 p.p.m.

8. All equipment should be periodically cleaned and sanitized. The frequency of cleanup depends on a number of factors, including type of food being processed, temperature of the food being processed, and temperature of the food-handling environment. The cleanup procedure also depends on a number of factors, but basically it consists of the following:

   a. Gross food residues are removed with warm water. Very hot water is generally not used at this stage because it often causes food to coagulate and to adhere to the equipment. Where necessary, the equipment is dismantled for cleaning.

b. After the initial cleaning, remaining food residues are removed by brushing and the use of a suitable detergent. With regard to detergent selection, a publication prepared by the Association of Food Industry Sanitarians, Inc., should be consulted.

c. Chlorine or quaternary ammonium disinfection is sometimes carried out as a third step. If so, the detergent solution should first be flushed from the equipment with water. Chlorine solutions ranging in concentrations from 20 to 200 p.p.m. are employed. These strong chlorine solutions are corrosive to some metals and usually should be flushed from the equipment after 10 or 15 minutes.

d. To remove disinfectant solutions, or detergent solutions if the disinfection step is not carried out, hot water is employed. The hot water should be applied for sufficient time to heat the equipment to the extent that it will dry rapidly after the flushing. It is important that the equipment dry rapidly, because any residual moisture will promote bacterial growth.

Containers

During transport and storage, food containers may become heavily contaminated with bacteria-laden dust. Many such bacteria will be in the spore state. A high proportion of the vegetative cells will be sporeforming types. Most of the contamination can be removed from containers by jet-spraying them with hot water.

Food Handling

Under ideal growth conditions, the number of bacteria in a food may double every 20 or 30 minutes. Application of principles inherent in bacterial growth curves is extremely important to the prevention of bacterial build-up in food products during their handling prior to heat processing.

Food packing has, through years of progressive development, grown in scope and complexity until now it consists of a multitude of varied and technical processing operations. The processing operations to which any given food is subjected are so interrelated with respect to their influence on the quality of the finished product that the control necessary during any one operation depends both on how the food was handled in previous operations and on the control possible in subsequent operations.

The large present-day food plants are departmentalized to the extent that a food product may be handled by a number of departments during its preparation. Under such a system, standardization of operations is essential to the production of food items of uniformly high quality. For example, the thermal process required to sterilize a canned food item depends greatly on how the food is handled during preparation. If the process is sufficiently severe to sterilize a food regardless of how it has been handled in previous operations, the quality of the finished product will seldom be either high or uniform. The rate of thawing indicated for a food stored in the frozen state will necessarily depend to some extent on how the food was handled prior to its being frozen. The temperature of refrigerated storage and the length of time a food may be safely stored at a given temperature are determined to a great extent by how the food is handled prior to storage. These are only a few of the examples that might be cited to emphasize the interrelationship of processing operations; but they serve to illustrate the necessity for standardization of operations if food products of uniformly high quality are to be produced.

Education as the result of advances in the science of nutrition and food technology has placed present-day emphasis on quality. Competition with respect to both organoleptic and nutritive qualities of processed food is so keen that the food packer needs to use every facility at his command to produce uniformly high-quality food products.

How can the bacteriologist best serve the food packer in his efforts to produce higher-quality food products? First, by applying available information concerning principles that govern bacterial growth behavior in foods, the bacteriologist can materially aid in standardization of practically every food processing operation. Second, by well-designed research he can obtain valuable information on which to base improvements in present-day methods of processing.

Food packing may well be considered as representing a battle between man and microbe. Therefore, application of knowledge concerning the growth behavior of bacteria should constitute the initial step in the packer's effort to "outwit" the microbe. Bacterial decomposition of food or food components is brought about by the growth of bacteria in the food. An understanding of the factors that influence bacterial growth is thus essential to successful food plant control.

The Bacterial Growth Curve

Under any given set of conditions, growth of bacteria can be represented by a curve that assumes a rather unique form. Such a curve, representing increase in number of bacteria with respect to time, is known as a growth curve. For the purpose of this discussion, a curve (Fig. 1) has been constructed by plotting the logarithms of assumed numbers of bacteria, per unit volume of food, in the direction of ordinates against time in the direction of abscissae. The curve shown in Fig. 1 is hypothetical only in the sense that it does not represent actual data. It does represent in a general way the growth behavior of bacteria subsequent to their being transferred from an unfavorable environment to one favorable for their growth.

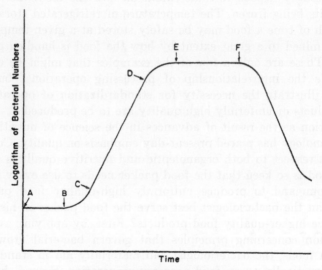

FIG. 1. Bacterial growth curve.

Bacteria subjected to certain adverse conditions will become virtually inactive. When such organisms are again subjected to favorable conditions for growth, it is only after an appreciable lapse of time that growth occurs in the form of cell reproduction. This period of growth lag is referred to as the lag phase (A to B, Fig. 1). At the end of the lag phase, bacteria start multiplying; and, very soon thereafter, the increase in number of bacterial cells is directly related to time. This is the logarithmic or rapid growth phase and is represented by a straight line (C to D, Fig. 1).

The number of bacteria in any medium, probably because of the exhaustion of nutrients essential for growth and because of the accumulation in the medium of metabolic products inimical to growth, will eventually reach a maximum, and the number present will remain virtually constant at this maximum for a period of time. This period, represented by the flat portion of the growth curve (E to F, Fig. 1), is known as the stationary growth phase. After a certain length of time, death of the cells begins. The resulting period of decline and death is represented by the receding portion of the curve. From the standpoint of food plant control, chief interest lies in factors influencing the length of the lag phase and the rate of growth of bacteria in the rapid growth phase. Therefore, only the lag phase and the rapid growth phase will be discussed further here.

It should be emphasized that a curve representing the growth behavior of a given species of bacteria under one set of environmental conditions will never be identical in form to a curve representing the behavior of the same species under any other set of conditions. For example, with changing conditions, the length of the lag phase may vary from less than an hour to many hours or even days. Furthermore, the rate at which bacteria multiply in the phase of rapid growth depends on many different factors. The factors commonly considered as exerting the greatest influence on the character of these growth phases are: (1) number of bacteria in product initially; (2) age of bacteria present; (3) frequency with which bacteria are transferred to new media; (4) growth phase of bacteria at the time they are transferred to new media; (5) kind of bacteria present in product; (6) temperature of bacterial environment; and (7) nature of substrate supporting bacterial growth.

The significance of these factors with reference to bacterial development in foods will now be considered. Such considerations will be based on the proposition that the ideal of food plant operation, toward which the food packer should be persistently working, is to accomplish all processing steps during the lag phase of microbial growth.

Number of Bacteria. Generally speaking, the greater the number of organisms present, the shorter will be the lag phase of growth, and the more rapidly will bacterial multiplication occur in the rapid growth phase. Hence, every practical effort should be made to hold to the minimum the number of organisms finding their way into a product. This involves a rigid program of food plant sanitation, in which disinfection plays a major role. For example, if food-handling equipment is not properly cleaned, bacteria may grow in the small amounts of food remaining on it and heavily contaminate food subsequently handled by the equipment. Extensive research in the dairy, meat, and other food industries has established that improperly cleaned food-handling equipment constitutes one of the most potent sources of contamination. Contaminated water supplies, contaminated ice, and minor product ingredients of poor quality are often important sources of contamination. Elimination of the major sources of contamination, which may be located by flow sheet methods of bacteriological analysis, is the first step toward making it possible to carry out all processing operations within the lag phase of bacterial growth.

Age of Bacteria.   The older organisms become, the longer will be their lag phase of growth. It is obviously impossible to determine the exact age of organisms present in a food product. However, knowing the sources of contamination will often allow for fairly accurate conjecture as to the behavior of bacteria in a food product. For example, consider a number of organisms entering food with dust contamination, in comparison with the same number entering with small amounts of food that have remained for several hours on improperly cleaned equipment. The organisms entering with the dust will ordinarily be relatively old and will show considerable growth lag. The organisms in the food left on the equipment, on the other hand, will very likely be quite young and after entering the fresh food may demonstrate little or no growth lag. Hence, it is far more important to keep idle equipment free of small amounts of food than it is to keep it dust-free.

Frequency of Transfer and Growth Phase of Bacteria Transferred. Frequent transfer of organisms to new media will tend to keep them in the phase of rapid growth. If bacteria are in the log phase of growth when transferred, they will probably continue to grow in the fresh product at a rapid rate. Bacteria in small amounts of food remaining on improperly cleaned equipment are frequently in the rapid growth phase when fresh food is again handled by the equipment, and, when entering the fresh food, instead of showing a lag in growth, these bacteria often continue to grow at a rapid rate. If the temperature of food handled by equipment located in a warm room never exceeds 40°F., it might be thought that any bacteria entering such food would immediately stop growing. However, this is seldom the case if the bacteria entering the food are in the rapid growth phase—bacteria will continue rapid growth at temperatures far below the minimum temperature that will initiate their rapid growth. Hence, all food-handling equipment located where the temperature is currently above 50°F. should receive special attention in cleaning procedures. If the temperature of the food is favorable for bacterial growth, it is often desirable to clean certain food-handling equipment periodically during the day to prevent bacteria, in food adhering to the equipment surfaces, from reaching the phase of rapid growth. Equipment having wood or other porous or rough surfaces that contact the food must receive careful attention to prevent its becoming a hazardous source of bacteria in the rapid growth phase. Recesses in food-handling equipment, in which food may accumulate and be overlooked in routine cleaning procedures, should be eliminated as far as possible.

Kind of Bacteria. Knowledge concerning the types of bacteria that may be present is often extremely valuable. Different species of bacteria vary considerably in their growth behavior when subjected to any given set of environmental conditions. Bacteria will show the shortest lag in growth under conditions most favorable for their growth. A knowledge of the kind of bacteria present will often allow conditions to be adjusted so that an essential operation may be completed before appreciable growth occurs in the product. For example, if it is known that thermophilic bacteria are present in a product to be canned and heat-processed, and that these bacteria will not be killed by the thermal process to be given the product, it is generally possible to cool the product, subsequent to thermal processing, rapidly enough to avoid initiating active growth of the bacteria surviving the process. In fact, this is a common practice for handling certain foods known to be contaminated with thermophilic bacteria. It is a classical example of operating within the lag phase of bacterial growth. The kinds of bacteria present will depend primarily on the sources of contamination; and a study of the sources of contamination will often lead to the possible elimination of bacteria that make certain operations difficult.

Temperature. Temperature control provides one of the greatest weapons against active bacterial growth. The length of the lag phase and the rate at which bacteria multiply in the rapid growth phase are influenced markedly by the food temperature. By temperature adjustment alone it is often possible to lengthen the lag phase of bacterial growth sufficiently to permit completion of a processing operation before active bacterial growth occurs in the product. The lag phase of growth of bacteria in many perishable foods may be lengthened by as much as 100% by maintaining the temperature of such food at 10°C. (50°F.) instead of 18°C. (65°F.) during certain processing operations; and, by lowering the temperature of food-holding coolers from 10°C. (50°F.) to 4.5°C. (40°F.), it is often possible to lengthen the lag phase of surface contaminants of foods stored in the coolers as much as 200%.

Adequate temperature control during food-cooling operations is essential to the prevention of spoilage unless the food is sterilized prior to cooling. Cooling nonsterile foods in bulk is a common industrial practice—for example, cooling of pasteurized milk, of meat immediately subsequent to animal slaughter, of foods to be preserved

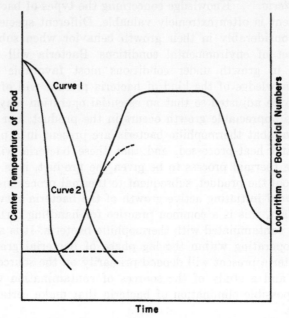

FIG. 2. Curves showing effect that rate of food cooling may have on growth of bacteria in food.

in the frozen state, of many canned food items subsequent to thermal processing, and of many different foods during various processing steps. The rate of cooling should be such that the temperature of the product is lowered rapidly enough to prevent any appreciable increase in the number of bacteria present. If temperatures favorable to bacterial growth prevail sufficiently long to initiate rapid bacterial growth, significant spoilage or build-up in spore load may occur before this growth can be stopped. Lower temperatures are required to stop growth of bacteria in the rapid growth phase than are required to keep bacteria in an inactive state of growth. This can best be illustrated by superimposing food-cooling curves on the growth curve shown in Fig. 1.

Figure 2 graphically illustrates what may be expected on cooling a food product at each of two different rates. It will be noted that cooling curve 2 intersects the growth curve near the end of the section of the curve representing the lag phase of bacterial growth. The broken line projected from the point of intersection of the two curves represents subsequent bacterial growth to be expected. Cooling curve 1, on the other hand, intersects the portion of the growth curve representing the rapid growth phase. The broken line projected from this point of intersection again represents subsequent bacterial growth to be expected. These effects have been observed many times in commercial practice, and it has been found that sufficient study of the growth behavior of bacteria commonly present in a given food product will dictate, within very close limits, the rapidity with which the product should be cooled.

Nature of Medium.   Composition of the substrate medium will influence markedly the length of the lag phase of bacterial growth and the rate of growth of bacteria in the rapid growth phase. Water content, pH of food, salt content, and many other factors are important in this respect. Certain salts have long been used as inhibitors of bacterial growth. Various organic acids have been widely used as preservatives. The effectiveness of these inhibitors increases directly with an increase in their concentration in food products.

From the standpoint of food plant control, water contents should receive special consideration. Other things being equal, the more free water present in, and on the surface of a food product, the shorter will be the lag phase of bacterial growth, and the greater will be the growth of bacteria in the rapid growth phase. Free water influences bacterial growth to a far greater extent than does bound water—free water being defined as that which is present in excess of that bound physically and chemically with other substances in the food. There is some free water naturally present in most fresh foods; but, with reference to food plant control, the concern is not primarily with the naturally occurring free water but with water added in washing procedures, with water that may collect on the surface of foods during handling operations, and with water added as an ingredient of a food product. The lower the free water content is kept, the greater will be the possibilities for bacteriological control.

To prevent free water from collecting by condensation on the surface of food products, certain relationships between temperature and relative humidity must be maintained. If food is kept at one given temperature, there are certain requirements for the temperature and relative humidity of an environment into which the food may be transferred without permitting water of condensation to collect on its surface. These relationships of temperature and humidity may be readily calculated for any circumstances encountered. They should receive careful consideration in the design and operation of the food plant.

Practical Considerations

The quality of a processed food product as measured in terms of organoleptic and nutritive properties depends chiefly on three general factors: (1) properties of the food prior to its being processed; (2) microbial action in the food during all processing operations performed in manufacture of the food product; and (3) nature of conditions imposed on the food to control microbial action in it during and subsequent to manufacture of the food product. Any program of food plant control should, therefore, be designed and executed with the objective of controlling these three general factors. The following suggestions for a program of food plant control, based on the foregoing discussions, are offered in the hope that they may prove helpful to the food packer in his future efforts toward product quality improvement:

1. Set as the objective the execution of all processing operations within the lag phase of microbial growth.

2. Institute routine bacteriological flow sheet analyses to cover every processing operation during manufacture of every food product.

3. Establish bacteriological standards for foods and for product ingredients used in the manufacture of each food product.

4. Eliminate all major sources of contamination as located by flow sheet analyses.

5. Correct conditions that flow sheet analyses have established as allowing significant increases in bacterial numbers during any processing operation.

6. Adjust the temperature of food during such operations, and the time food is held at such temperatures, according to requirements established by growth curves of bacteria in food, obtained under controlled conditions of food incubation.

7. Eliminate addition of free water to food wherever possible—especially water collecting on product through condensation. The latter may be eliminated by proper adjustment of temperature and humidity relationships.

8. If the entire lag phase is being consumed in a number of processing operations, leaving no operating time for succeeding operations, apply the same principles for adjusting conditions of all operations as were applied for adjusting those of a single operation. Strive to adjust conditions so that all operations may collectively be accomplished during the lag phase of bacterial growth.

9. If the thermal process used for pasteurization or sterilization is more severe than that considered necessary for safety from the standpoint of consumer health and prevention of economically important spoilage, determine, by methods to be discussed later, the process necessary to accomplish the desired extent of heat application and adjust the time and temperature of thermal processing accordingly.

SELECTED REFERENCES

Brownlee, D. S., Guse, U. C., and Murdock, D. I. (1947). In-plant chlorination of cannery water supply. *Food Packer*, **28**(1), 52.

Felice, B. A. (1953). Chlorination as applied to sanitation in a fruit cannery. *Western Canner and Packer*, **45**(4), 25, 55.

Griffin, A. A. (1946). Break-point chlorination practices. Technical Publication No. 213. Wallace and Tiernan, Newark, New Jersey.

Hammer, B. W. (1938). "Dairy Bacteriology," 2nd ed. Wiley, New York.

Hall, J. E., and Blundell, C. C. (1946). The use of break-point chlorination and sterilized water in canning and freezing plants. *Natl. Canners Assoc. Information Letter* No. 1073.

Harris, J. J. (1946). Chlorination in the food plant. Parts I and II. *Canner*, **103** (9), 18, 48; **103**(10), 14, 26.

Hersom, A. C., and Hulland, E. D. (1964). "Canned Foods—An Introduction to Their Microbiology," Baumgartner 5th ed., Chem. Publ. Co., New York.

Jensen, L. B. (1945). "Microbiology of Meats," 2nd ed., Garrard Press, Champaign, Illinois.

Mercer, W. A., and Somers, I. I. (1957). Chlorine in food plant sanitation. *Advan. Food Res.*, **7**, 129.

Ritchell, E. C. (1947). Chlorination of cannery water supply. *Natl. Canners Assoc. Information Letter* No. 1200.

Somers, I. I. (1951). Studies on in-plant chlorination. *Food Technol.*, **5**, 46.

Tanner, F. W. (1944). "The Microbiology of Foods," 2nd ed. Garrard Press, Champaign, Illinois.

The Association of Food Industry Sanitarians, Inc. (1952). "Sanitation for the Food Preservation Industries." McGraw-Hill, New York.

Townsend, C. T., Somers, I. I., Lamb, F. C., and Olson, N. A. (1956). "A Laboratory Manual for the Canning Industry," 2nd ed. National Canners Association Research Laboratories. Washington, D. C.

Zinsser, H., and Bayne-Jones, S. (1939). "Textbook of Bacteriology," 8th ed., Chapter III. Appleton-Century, New York.

6

Producing, Harvesting, and Cleaning Spores for Thermal Resistance Determinations

General Considerations

The heat resistance of bacterial spores of a given species is dependent on the nature of the medium in which the spores are produced and on the temperature at which they are produced. It is generally advisable to provide conditions that will produce spores of maximum heat resistance. Best spore crops of obligate anaerobes are usually obtained at the lowest incubation temperature at which vigorous vegetative growth occurs. With *Bacillus* species, spores of highest resistance are generally obtained at the highest temperature at which vigorous vegetative growth occurs (Williams and Robertson, 1954).

Cytology of Spore Formation. During the incubation of cultures for spore production, continuous microscopic examination of the cultures is extremely important, to determine when spore harvest is indicated. Incubation should be long enough to obtain mature spores from a high percentage of the cell population. However, with some species, incubation should not be extended much beyond this. Regermination—that is, germination of some spores produced in the culture —may occur. For this reason a brief discussion of cytological changes during sporulation seems appropriate.

The cytology of sporogenesis has been studied by a number of workers employing phase and electron microscopy. For the most part, this discussion is based on the important studies of Young and Fitz-

James (1959a, b) and the excellent review by Murrell (1961). Spore formation begins when the nuclear material of the vegetative cell, in the form of two condensed chromatin bodies, fuses and forms an axial thread. A portion of this chromatinic material then migrates to one end of the cell, the sporangium. Next, a thin septum, growing centripetally from opposite sides of the cell and close to the one end, confines the chromatin and a small amount of cytoplasm. (At this point the process of sporogenesis is believed to become irreversible; that is, the cell is committed at this stage, which may be reached in 30 to 35 minutes after the last cell division.) The septum continues to grow from its slightly thickened base, inside the cell wall, round inside the end of the cell until it completely encloses the chromatin body. This spore body now becomes free in the sporangium. At this stage it stains densely with dilute cold stains. During maturation, a cortical region and at least two coat layers, and, in some organisms, an exosporium, are formed around the spore body enclosed by the initial septum. This initial spore septum or membrane becomes the cell wall of the germ cell on spore germination. With formation of the cortex and the first spore coat, the spore becomes refractile. From the completion of the septum to the development of refractility, the spore body appears as a densely stained body in the light microscope and is known as a forespore.

When the spore becomes refractile, it also becomes impermeable to dilute cold stains and, compared with vegetative cells, highly resistant to heat. When virtually all spores in a crop become refractile, and impermeable to dilute cold stains, immediate cessation of incubation may be indicated. Therefore, during incubation for the production of spores, microscopic examination should be made at least daily.

Dipicolinic Acid Formation and Uptake of Calcium.   Dipicolinic acid (DPA) does not occur in vegetative cells. It first appears at the onset of sporulation. Its association with the formation of the cortex (Hashimoto *et al.*, 1960) suggests the cortex as the location of DPA in the spore. This is supported by the fact that it is almost completely released in germination exudates (Powell and Strange, 1953) during spore germination and by cytological evidence that the cortex disappears during the first few minutes of the germination process (Mayall and Robinow, 1957).

Coincident with DPA formation is the incorporation of calcium. Powell and Strange (1956) and Vinter (1956a, b) observed that calcium uptake followed the synthesis of DPA, the calcium content of spores reaching several times that of vegetative cells. These findings have since been confirmed by many other workers. It is not yet known whether calcium is required in reactions in the synthesis of DPA or is needed to combine with the end product. Lund (1959) observed a molar ratio of 1 to 1 for calcium and DPA in spores and exudates from germinated spores. However, Perry and Foster (1955) and Slepecky and Foster (1959) found that less calcium than DPA was present.

Both DPA and calcium are thought to be concerned in resistance of spores to heat. The formation of DPA and the incorporation of calcium are coincident or nearly so with the increase in heat resistance. Moreover, the release of DPA and some calcium during germination is coincident with loss of heat resistance.

Factors Affecting Sporulation

Temperature.   Observations on the effect of temperature on sporulation have been almost entirely qualitative in nature. These observations, however, indicate that the temperature of incubation affects both the rate and the amount of spore formation. They also indicate that, generally, sporogenesis occurs most rapidly at or near the optimum growth temperature. The amount of sporulation is reduced by unfavorable temperatures.

pH.   Accurate studies of the effect of pH on sporogenesis have been prevented by difficulties in controlling pH during growth. However, it has been observed by most workers in the field that, with some species, pH values not affecting vegetative growth may partially or completely prevent spore formation. Some species, such as *C. pasteurianum* and *B. coagulans,* have low optimum pH levels (5.4 to 5.7) for sporulation. Others, such as *B. stearothermophilus, C. thermosaccharolyticum, C. botulinum, C. sporogenes,* and closely related organisms, have rather high optimum pH levels (usually about 7.00 or slightly above). It should be noted that optimum pH levels vary not only with species but with strains within the same species and to some extent with composition of the medium.

Oxygen.   *Bacillus* species require oxygen for the production of sporangia. Whether it is required in the final stages of sporogenesis is not yet clear.

Little information has been published on the effect of oxygen on sporogenesis of *Clostridium* species. What little is known, plus the experience of workers in the field, indicates that, though some sporulation may occur at low oxygen pressures, best sporulation is obtained in the complete absence of oxygen.

Minerals.   Certain ions, notably potassium and manganese, have been found essential for maximum production of sporangia or for the process of sporogenesis in a number of species of bacteria. Further study may show that they are required by all spore-bearing species. Other ions, such as Ca, Mg, Fe, Zn, Co, PO_4, and SO_4, are variously required by different species and strains. The amount of these ions required varies widely among species and strains. For further details regarding specific requirements, see Murrell (1961). Some of these ions are required in much greater amounts for sporulation than for vegetative growth. Many rather complex media adequate for good vegetative growth may contain too low a concentration of certain of them for spore formation to occur. It should be noted that a deficiency of some ions may limit growth but induce spore formation.

Carbon Source.   Numerous carbon sources have been reported to affect sporulation differently than they affect vegetative growth. Certain sugars may reduce or prevent sporulation at one concentration but increase sporulation at lower concentrations. Depletion of carbon sources in the growth media initiates sporulation of some species. The effect of different carbon sources varies so greatly among different species and strains that meaningful and definitive assessments are impossible at this time. In testing for sporulation response by fermentable sugars, it is important to observe pH changes. In some cases sufficient acid may be produced during vegetative development to reduce the pH to unfavorable levels for sporulation.

Nitrogen Source.   Nitrogen requirements of some groups of organisms vary from simple inorganic sources to complex mixtures of amino acids and possibly peptides. As was pointed out by Murrell (1961) the nitrogen nutrition in such groups of organisms is difficult to assess. As he further pointed out, it is probable that cells that can grow on a simple medium will synthesize a pool of intracellular substrates and reserve compounds as complicated as those of cells growing in a complex medium.

Significant observations on the nitrogen status during formation of sporangia may be briefly summarized as follows:

1. Nitrogen depletion as a cause of sporulation initiation has been observed for organisms with either simple or complex nutritional requirements.
2. In complex amino acid media, sporulation may be induced by a deficiency of one or more amino acids.
3. Certain amino acids may have specific roles in spore formation.

Most of the work to date has been done with the bacilli. However, evidence now accumulating indicates that the above observations may apply equally to some if not all of the clostridia.

Miscellaneous Factors. Some investigators have observed increased sporulation for certain organisms growing in association with other organisms, or on the addition of filtrates of cultures of other organisms. Little is known of the factors involved. Low concentrations of some fatty acids may increase sporulation in some organisms, and inhibit sporulation in others. Also, low concentrations of certain fatty acids have been observed to increase heat resistance of spores of some organisms. Low concentrations of others have been observed to decrease heat resistance of spores of some organisms. Some amino acids inhibit spore formation by some organisms.

So little is as yet known of specific effects of these miscellaneous factors that definitive assessment is impossible. It has been found that the inhibitory effects on sporulation of some substances, whether in the medium initially or produced in the medium during vegetative growth, may frequently be counteracted by the addition of adsorbents such as soluble starch or activated carbon.

Producing and Harvesting Spores

As is evidenced by the foregoing discussions, nutrient requirements for good sporulation are often different from those for good vegetative growth. In some media that support excellent vegetative growth, sporulation may be very poor. The media listed in Table 1 for producing spores of the different food spoilage bacteria generally support good sporulation as well as good vegetative growth. For composition and preparation of the different media, see Chapter 3. The temperatures and times of incubation given have been found to produce spores of highest heat resistance.

Obligate Aerobes and Facultative Anaerobes.   Spores of these organisms are produced on the surface of appropriate agar media. A generally applicable procedure will be described.

Tube slants of agar are inoculated with an actively growing culture of the test organism and incubated for 2 to 4 days at the appropriate temperature. When microscopic examination shows a high percentage of mature spores, the slants are removed from incubation, and the

TABLE 1

Media, incubation times, and temperatures for spore production

		Incubation	
Medium		Temperature	Time
Obligate aerobes			
B. subtilis and related species	Nutrient agar	27°–30°C.	2–4 days
Facultative anaerobes			
B. stearothermophilus	Nutrient agar	50°–55°C.	2–3 days
B. coagulans	Thermoacidurans agar	50°–55°C.	2–4 days
Obligate anaerobes			
C. botulinum (Types A and B)	Beef heart casein broth	28°–30°C.	1–2 weeks
C. sporogenes and related species (including P.A. 3679)	Liver or beef heart broth	28°–30°C.	1–2 weeks
C. thermosaccharolyticum	Liver broth with iron strips	50°–55°C.	1–2 weeks
C. pasteurianum	Liver–tomato broth with 10% soil added	28°–30°C.	1–2 weeks

surface growth is suspended in sterile distilled water. To suspend the growth, 3 ml. of water is added to each tube, and the growth is suspended by scraping the surface with a sterile inoculating loop. Suspensions from a number of slant tubes are collected in a sterile screw-cap dilution bottle. The bottle of suspension is then heated for 20 minutes at 80°C. (176°F.) to destroy vegetative cells and to activate the spores for germination. This heated suspension is used to inoculate bottle slants of agar for mass spore production.

Screw-cap bottles (150 ml.) are convenient for making bottle slants. Larger or smaller bottles may be used if preferred. When 150-

ml. bottles are used, 50 ml. of agar is placed in each and sterilized by autoclaving for 20 minutes at 15 pounds. After sterilization the bottles are slanted so as to obtain maximum agar surface.

Each bottle is inoculated with 2 ml. of heated spore suspension, prepared as described above. Each bottle is then tilted back and forth until the inoculum covers the entire surface. The inoculated bottles are incubated at the appropriate temperature until growth and sporulation are virtually complete. (During incubation, the bottles are slanted at an angle of about 45° so that excess inoculum and condensation collect toward the bottom of the bottles.) After incubation, 2 ml. of sterile distilled water is added to each bottle, and the growth is scraped from the agar surface with an inoculating loop. Suspensions from all bottles are then combined in an appropriate sterile container. Two milliliters of sterile distilled water is again added to each bottle, and the remaining growth is washed from the surfaces and combined with the first washings. This crude spore suspension is then stored under refrigeration preparatory to cleaning the spores.

Obligate Anaerobes.   Tube cultures in which good sporulation has occurred are used for inoculating larger volumes of media for mass spore production. Again, 150-ml. screw-cap dilution bottles are convenient for mass spore production. About the same ratio of meat particles to broth is employed as was used in the tubes. One milliliter of tube culture is used to inoculate each bottle. Just prior to inoculation, air is exhausted from the bottles of medium by autoclaving them for 15 minutes at 5 pounds. After autoclaving, the bottles are placed in a water bath thermostatically controlled at 80°C. (176°F.). Time is allowed for the temperature of the medium to approximate that of the bath before inoculation. After inoculation, the bottles are left in the bath for 20 minutes. (This heating activates the spores for rapid germination.) The inoculated bottles are incubated at the appropriate temperature until microscopic examination shows sporulation to be virtually complete. (With this procedure, neither the addition of thioglycolate nor stratification of the medium with sealing mixture is normally required to maintain anaerobiosis.)

After incubation, the larger particles of undigested meat are removed by filtering the cultures through two layers of sterile cheesecloth, between which is a pad of glass wool. Cultures from all bottles are combined in an appropriate vessel. This crude spore suspension is then stored under refrigeration preparatory to cleaning of the spores.

Cleaning of Spores

Because heat resistance of spores is influenced to varying degrees by substances in the medium in which the spores are produced, some cleaning is usually advisable. Several methods have been proposed for freeing spore suspensions of vegetative cells and extraneous materials. The method of choice depends to some extent on what studies are to be performed, but chiefly on the condition of the crude spore suspension and the bacterial species involved. If the suspension contains a large number of vegetative cells, the use of lysozyme may be indicated. No more than 0.3 mg. of lysozyme per milliliter of suspension is required; usually considerably less is adequate. If the pH is adjusted to about 11, lysis occurs almost immediately on the addition of lysozyme.

Some bacterial species, when they sporulate, build up strong lytic systems for vegetative cells. Vegetative cells of these will usually lyse under storage at 4°C. Storage for one or two weeks usually is adequate. Vegetative cells of obligate anaerobes usually can be lysed by bubbling air through the suspension. Aeration for 15 or 20 minutes usually suffices.

From the standpoint of heat resistance, a few vegetative cells are not objectionable. Consequently, lytic procedures are not generally carried out. However, solutes are extremely important. These can be reduced to low levels by repeated washing with distilled water. The crude spore suspension is distributed in sterile centrifuge tubes, and the spores are spun down by centrifugation at 1000 g for 5 minutes. After centrifugation, the supernatant is discarded, and the spores are resuspended in sterile distilled water and again centrifuged. This washing procedure is repeated twice or more. For spores of obligate aerobes and facultative anaerobes, three washings generally suffice. Spores of obligate anaerobes produced in meat broths should be washed at least five times.

After the final washing the spores are resuspended in an appropriate amount of sterile distilled water and stored at about 4°C. (39.2°F.) until used in thermal resistance determinations.

SELECTED REFERENCES

Halvorson, H. O. (1957). "Spores. A Symposium," 164 pp. American Institute of Biological Science, Washington, D. C.
Halvorson, H. O. (1961). "Spores II. A Symposium," 296 pp. Burgess, Minneapolis, Minnesota.

Hashimoto, I., Black, S. H., and Gerhardt, P. (1960). Development of fine structure, thermostability, and dipicolinate during sporogenesis in a *Bacillus*. *Can. J. Microbiol.*, **6**, 203.

Lund, A. J. (1959). Physiology of bacterial spores. *Rept. Hormel Inst. Univ. Minn.*, p. 14.

Mayall, B. H., and Robinow, C. F. (1957). Observations with the electron microscope on the organization of the cortex of resting and germinating spores of *B. megaterium*. *J. Appl. Bacteriol.*, **20**, 333.

Murrell, W. G. (1961). Spore formation and germination as a microbial reaction to the environment. *Symp. Soc. Gen. Microbiol.*, **11**, 100.

Perry, J. J., and Foster, J. W. (1955). Studies on the biosynthesis of dipicolinic acid in spores of *Bacillus cereus* var. *Mycoides*. *J. Bacteriol.*, **69**, 337.

Powell, J. F., and Strange, R. E. (1953). Biochemical changes occurring during the germination of bacterial spores. *Biochem. J.*, **54**, 205.

Powell, J. F., and Strange, R. E. (1956). Biochemical changes occurring during sporulation in *Bacillus* species. *Biochem. J.*, **63**, 661.

Slepecky, R., and Foster, J. W. (1959). Alterations in metal content of spores of *Bacillus megaterium* and the effect on some spore properties. *J. Bacteriol.*, **78**, 117.

Townsend, C. T., Somers, I. I., Lamb, F. C., and Olson, N. A. (1956). "A Laboratory Manual for the Canning Industry," 2nd ed. National Canners Association Research Laboratories, Washington, D. C.

Vinter, V. (1956a). Sporulation of bacilli. Consumption of calcium by the cells and decrease in the proteolytic activity of the medium during sporulation of *Bacillus megatherium*. *Folia Biol. (Prague)*, **2**, 216.

Vinter, V. (1956b). Sporulation of bacilli. III. Transference of calcium to cells and decrease in proteolytic activity in the medium in the process of sporulation of *Bacillus megatherium*. *Čsl. Mikrobiol.*, **2**, 80 [*Biol. Abstr.*, **32**, 16, 672 (1958)].

Williams, O. B., and Robertson, W. J. (1954). Studies on heat resistance. VI. Effect of temperature of incubation at which formed on heat resistance of aerobic thermophilic spores. *J. Bacteriol.*, **67**, 377.

Young, I. E., and Fitz-James, P. C. (1959a). Chemical and morphological studies of bacterial spore formation. I. The formation of spores in *Bacillus cereus*. *J. Biophys. Biochem. Cytol.*, **6**, 467.

Young, I. E., and Fitz-James, P. C. (1959b). Chemical and morphological studies of bacterial spore formation. II. Spore and parasporal protein formation in *Bacillus cereus* var. *Alesti*. *J. Biophys. Biochem. Cytol.*, **6**, 483.

CHAPTER 7

Death of Bacteria Subjected to Moist Heat

Temperature slightly above the maximum allowing bacterial multiplication cause death of vegetative bacterial cells. Bacterial spores, however, generally survive temperatures much higher than the maximum allowing multiplication of vegetative forms. Because bacterial spores are far more resistant to heat than are vegetative forms, they are of primary concern in the sterilization of most foods. However, as was discussed in Chapters 2 and 4, in the case of certain types of foods, sterilization involves, almost entirely, the destruction of vegetative forms.

Order of Death

From the practical standpoint, the bacteriologist's definition of death of bacteria is quite satisfactory; that is, a bacterium is dead when it has lost its ability to reproduce. Virtually all studies on death of bacteria have used failure of reproduction as the criterion of death. Results of such studies show that, when bacteria are subjected to moist heat, death expressed in terms of reduction of individuals is usually quite orderly. Generally, the number of viable cells reduces exponentially with time of exposure to a lethal temperature. Consequently, if logarithms of numbers of survivors are plotted against times of exposure, a straight line will be obtained (see Fig. 3). This is commonly referred to as a logarithmic order of death. According to Rahn (1945a), "From the earliest quantitative measurements by Chick

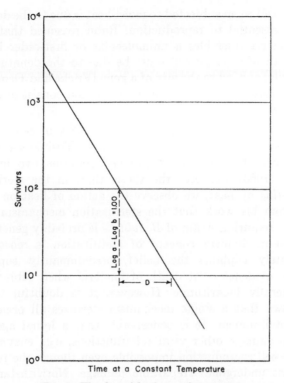

10^4

10^3

10^2

10^1

10^0

Survivors

Log a − Log b = 1.00

D

Time at a Constant Temperature

FIG. 3. The logarithmic survivor curve.

(1910) to the extensive investigations by Watkins and Winslow (1932), death of vegetative cells as well as death of spores has been found to be logarithmic." Though this may overstate the situation somewhat, literature during and after the period referred to by Rahn is replete with evidence in support of a logarithmic order of death of bacteria. A number of exceptions have been noted, but many of these may describe certain events other than bacterial death. Certain deviations and their possible causes will be discussed later. Suffice to say now that the overwhelming evidence that bacterial death itself is essentially logarithmic is most comforting when one considers the problem of evaluating sterilization processes.

Many explanations have been offered to account for the logarithmic order of death of bacteria. One of the most plausible explanations was given by Rahn (1929; 1945b)—namely, that loss of reproductive power of a bacterial cell when subjected to moist heat is due to the denaturation of one gene essential to reproduction. Rahn reasoned that, since the death of bacteria resembles a unimolecular or first-order bimolecular reaction, death of a single cell must be due to the denaturation of a single molecule; and, since the size of a gene (Fricke and Demerec, 1937) is that of a small protein molecule, a gene would consist of only one or two molecules.

Present knowledge regarding the role of DNA in cell reproduction is supportive of Rahn's theory. The work of Yoshikawa (1968) and others indicates that each microbial cell contains two molecules of DNA. Wax (1963) proposes the thesis that, in the sterilization of *Bacillus subtilis* by heat, we observe the action of heat on DNA, and concludes from his work that the sterilization mechanism for a certain thymine-requiring strain of *B. subtilis* is probably genetic damage.

This single-mechanism concept of sterilization is most attractive and adequately explains the belief, predominantly supported by experimental evidence, that death of bacteria when subjected to wet heat is generally logarithmic. However, it is doubtful that it can ever be shown that a single mechanism sterilizes all organisms in a sterilizing environment. It is conceivable that a lethal agent, as wet heat, can so damage other vital cell functions, e.g., enzyme activity, as to render cell reproduction impossible even though the reproductive gene has not undergone irreparable damage. Notwithstanding, bacterial death is generally logarithmic and it must be concluded at this time that genetic damage is primarily, if not solely, responsible for loss of reproductive ability.

What the true explanation is regarding the cause of the logarithmic order of death of bacteria does not alter evidence that it exists and that it should be fully considered in the evaluation of sterilizing processes for foods. In the words of Rahn (1945a), ". . . it permits us to compute *death rates* and to draw conclusions from them which are independent of any explanation. Death rates make it possible to compare the heat resistance of different species at the same temperature, or the heat resistance of one species at different temperatures. It also enables us to describe in quantitative terms the effect of environmental factors, such as concentration of the medium or its pH, upon sterilization."

Since death of bacteria is generally logarithmic, it is convenient to consider it strictly so and to account for slight deviations as they may occur. Considering it so allows it to be described mathematically in the same manner as a unimolecular or first-order bimolecular chemical reaction. In a unimolecular reaction only one substance reacts, and its rate of decomposition is directly proportional to its concentration—decomposition of phosphorus pentoxide is such a reaction. In a first-order bimolecular reaction one reactant is in such great excess that variation in its concentration is negligible, and rate of decomposition of the second reactant is directly proportional to its concentration—hydrolysis of sucrose is such a reaction when water is present in great excess.

Expressed mathematically,

$$-\frac{dC}{dt} = kC,$$

or

$$-\frac{dC}{C} = k \, dt \tag{1}$$

in which C = concentration of decomposing reactant.

k = proportionality factor.

$-dC/dt$ = rate at which concentration decreases.

Integrating equation 1 between limits, concentration C_1 at time t_1 and concentration C_2 at time t_2, we have

$$-\int_{C_1}^{C_2} \frac{dC}{C} = k \int_{t_1}^{t_2} dt$$

$$-\ln C_2 - (-\ln C_1) = k(t_2 - t_1)$$

and

$$k = \frac{\ln C_1 - \ln C_2}{t_2 - t_1} = \frac{2.303}{t_2 - t_1} \log \frac{C_1}{C_2}$$

This latter equation may be modified to give

$$k = \frac{2.303}{t} \log \frac{C_0}{C}$$

or

$$t = \frac{2.303}{k} \log \frac{C_0}{C} \tag{2}$$

in which C_0 = initial concentration of reactant.

C = concentration after reaction time t.

For the survivor curve (see Fig. 3), let a represent the initial number of cells (comparable to C_0 in equation 2) and b represent the number of surviving cells (comparable to C in equation 2) after heating time t. Then

$$t = \frac{2.303}{k} \log \frac{a}{b} \tag{3}$$

From Fig. 3 it is noted that the time required to destroy 90% of the cells is the time required for the curve to traverse one log cycle. If this time is represented by D (decimal reduction time) (Katzin *et al.*, 1943), the slope of the survivor curve may be expressed as

$$\frac{\log a - \log b}{D} = \frac{1}{D}$$

Substituting in the general equation of a straight line,

$$y = mx$$

we obtain

$$\log a - \log b = \frac{1}{D} t$$

or

$$t = D (\log a - \log b) = D \log \frac{a}{b} \tag{4}$$

in which t = time of heating.

D = time required to destroy 90% of cells.

a = initial number of cells.

b = number of cells after heating time t.

In comparing equations 2 and 4, it becomes obvious that

$$D = \frac{2.303}{k},$$

and that both D and k represent the slope of the survivor curve. That D is the more convenient term to use in comparing heat resistance will become obvious as we proceed.

Factors Causing Apparent Deviations from the Logarithmic Order

As was pointed out above, the order of death of bacteria is generally considered to be logarithmic and subject to mathematical description. However, deviations from a straight line have been noted when logarithm of survivors was plotted against time of heating at a constant lethal temperature. The question arises whether observed deviations represent deviations from the logarithmic order of death or whether such nonlinear curves are describing occurrences in addition to death of the bacteria. In conducting heat resistance determinations and interpreting the data obtained, it is extremely important to be aware of a number of factors known to cause deviations from the straight-line survivor curve. The influence of such factors is best understood by considering types of nonlinear curves that they may produce.

Heat Activation for Spore Germination. Some, and perhaps all, bacterial spores require activation for rapid germination. Heat is a very effective spore germination activator. The extent of heating required for rapid activation of any given spore population depends on a number of factors, some of which are:

1. Species or strain of bacteria.
2. Nature of medium in which spores are produced.
3. Nature of suspending medium in which spores are stored.
4. Temperature at which spores are stored.
5. Age of spores.
6. Nature of medium in which heated spores are subcultured.
7. Temperature of heating.

There may be still other factors as yet unknown or unsuspected. Suffice it to say that when any significant amount of heat is required for rapid germination, the observed so-called survivor curve will not be a straight line.

The curve in Fig. 4 is typical of those obtained when a significant amount of heat is required. This curve depicts the activation requirement in addition to spore death. The curve indicates that, during the first few minutes, many more spores were being activated for germination than were being destroyed. The curve also indicates that, by the end of 5 minutes of heating, most spores were activated for rapid germination. It further indicates that spore death per se, as judged by the ability of the heated spores to germinate and reproduce in the subculture medium supplied, was strictly logarithmic.

A curve similar to that shown in Fig. 4 may be produced by a combination of spore activation and spore death. Such a curve is shown in Fig. 5. Such a curve would be produced if, during the first few minutes of heating, some spores were being destroyed at the same rate as others were being activated for rapid germination. In other words, for a time there

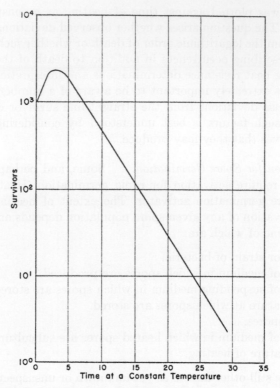

FIG. 4. Influence of heat required for activation on the character of the survivor curve.

is neither a rise nor a fall in the subculture count. Then, as activation nears completion, death rate exceeds activation rate, depicted by the downward bending of the curve. On completion of activation, death alone is depicted, which again is strictly logarithmic.

Different initial combinations of rates of activation and death during the early period of heating will produce different shapes of the so-called lag phase of the survivor curve. The exact character of this portion of the curve will depend on the degree of predominance of one rate over the other. The important thing to remember is that deviation from a straight line during the early period of heating because of the spore activation requirement in no way indicates a nonlogarithmic order of death.

It has been observed that, when a number of different temperatures are being employed to compare death rates, the lag caused by the heat

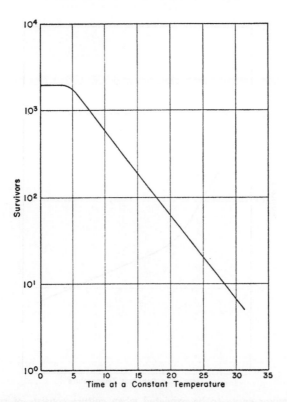

FIG. 5. Character of the survivor curve when during the first few minutes rate of activation equals rate of death.

activation requirement decreases as the temperature increases. Lags may be very apparent when relatively low temperatures are used. With much higher temperatures, there may be no detectable lag. This is believed to be due to the fact that time for spore activation decreases with increasing temperature. At the higher temperatures activation may occur so quickly that the time required is so short as to be undetectable by the techniques employed.

Mixed Flora.   When a culture contains two or more species or strains of different heat resistance, the survivor curve would not be a straight line. Figure 6 depicts the type of curve that would be expected if two species or strains made up the culture. The first steep part of the curve describes for the most part death of the organism of lower resistance,

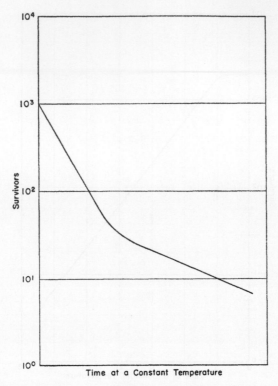

Fig. 6. Survivor curve to be expected when culture consists of two strains or species of different resistance.

and the second part describes death of the more-resistant one. Figure 7 is an actual survivor curve determined for spores of a mixture of two and possibly three anaerobic mesophiles. Figure 8 is an actual survivor curve determined for spores of a pure culture of Type E *C. botulinum* isolated from the mixed culture. For both determinations, the spores were suspended in distilled water. When curves resembling those in Figs. 6 and 7 are observed, mixed flora should be suspected.

Clumped Cells.　　Considering spores that do not require a significant amount of heat for rapid germination, or vegetative cells that require no heat activation for active multiplication, the type of survivor curve depicted in Fig. 5 will also be obtained if cells initially occur in clumps of two or more cells during heating and remain so when subcultured to obtain colony counts. In the beginning each colony will originate from

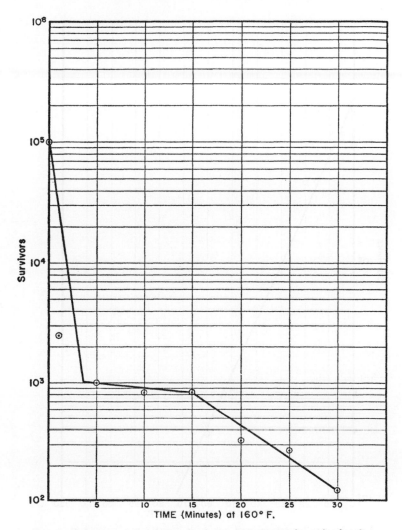

Fig. 7. Experimentally determined survivor curve for mixed culture.

two or more cells. There will be no change in colony count until at least some clumps have been reduced to one viable cell per clump. As the number of clumps being reduced to one viable cell per clump increases, the curve bends downward, becoming a straight line only when virtually all clumps have been reduced to no more than one viable cell per clump.

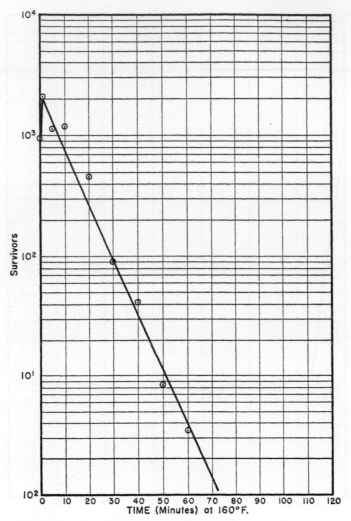

Fig. 8. Experimentally determined survivor curve for spores of *C. botulinum*, Type E, isolated from mixed culture the survivor curve for which is depicted in Fig. 7.

In this case, the deviation of the curve from a straight line during the early period of heating is caused by an inadequacy of technique, rather than by any peculiarity in the order of death. In other words, it is quite possible to obtain a curve of this type even though the order of death is strictly logarithmic.

Flocculation during Heating.   Solids in many liquiform food products flocculate during heating. The reaction is often referred to as "curdling." Gentle agitation usually promotes flocculation. Cream-style corn, cream of mushroom soup, and meat homogenates are good examples of such products. Flocs (clumps) so formed may be difficult to disperse (break up); they often withstand high dilution and vigorous agitation.

Salt solutions such as neutral phosphate buffer, frequently used as suspending media for bacteria in heat resistance determinations, will often flocculate the suspended bacteria into multicellular clumps during heating. Again, clumps so formed may be difficult to disperse by dilution and agitation.

In heat resistance methods in which samples are withdrawn periodically and cultured to obtain colony counts, flocculation during heating can cause serious errors in viable cell counts. The cells may be perfectly dispersed in the suspending medium prior to heating. Samples taken at this time and cultured will yield colonies originating for the most part from single cells. But, as flocculation proceeds with continued heating, more and more colonies in subculture will originate from multicellular clumps, until completion of the flocculation reaction.

Figure 9 shows a curve of the type expected when flocculation proceeds toward completion during the early period of heating. As clumping nears completion, the colony count will tend to level off during the time the number of surviving cells per clump is being reduced to one cell per clump. Only when virtually all clumps in which viable cells remain contain only one viable cell per clump will the rate of reduction in colony count assume that of a logarithmic order.

In Fig. 9, the first part (*a*) of the curve describes a mixture of flocculation and death. The second part (*b*) represents a reduction in number of viable cells per clump, death of some cells occurring singly, and possibly some further flocculation. The third part (*c*) represents the reduction, with time, of colony count when each colony originates from a single surviving cell—it is the only part that can be called a true survivor curve. If sampling were stopped at any time prior to the end of period *b*, the true curve would never be observed.

Deflocculation during Heating.   This is a rare occurrence, but it can happen, particularly when the suspending medium is being vigorously stirred during the thermal resistance determination. If the medium is flocculated to any extent before heating, and stirring is started, some of the flocs can be broken up by the agitation and cause a rise in the colony count. This has been observed for meat homogenates that were

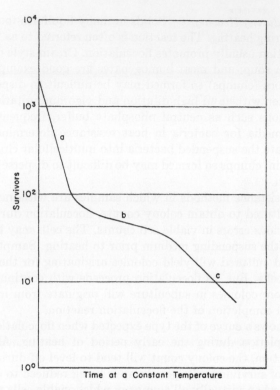

FIG. 9. Curve to be expected when clumping occurs during early part of heating.

heated gently before stirring was started. Deflocculation is usually most evident during the early period of heating, and its effect may be mistaken for a heat activation effect, since the survivor curves are similar (see Fig. 4).

Nature of Subculture Medium.   That the nature of the medium in which heated vegetative cells or spores are subcultured can significantly alter the character of the observed survivor curve has been demonstrated by a number of workers, among them Curran and Evans (1937), Olsen and Scott (1946, 1950), and Murrell *et al.* (1950).

As was pointed out above, a bacterial cell that fails to reproduce (divide and produce progeny) under conditions considered favorable for such activity, is generally considered by the bacteriologist as a dead cell—that is, failure of reproduction is the criterion of death. This is a severe definition, but it is used universally in interpreting data for construction of survivor curves. And, regardless of its severity, it is used for want of a better one—as yet, we have no choice but to use failure of reproduction as the criterion of death. It has been suggested (Rahn, 1945b) that it would be more appropriate to consider bacteria that fail to reproduce as sterile rather than dead. Of course, they might in some cases be only sterile, but if they cannot multiply they might as well be dead as far as influence on their environment is concerned. Therefore, choice of words in this case has little practical value. The real problem is how to tell for certain whether bacteria are dead (or sterile) or are lying dormant. To date, this problem has not been solved. In studying thermal resistance, it is present practice to provide those conditions considered most favorable for culturing heated cells; but there is never total assurance that all cells surviving the heat treatment will find these conditions to their liking and make their viability known through reproduction. Colony counting or positive tube techniques being the procedures employed for enumerating heated bacteria, it is obvious that viable cells that fail to reproduce under conditions provided them will be considered dead. By viable cells is meant cells that under another set of conditions would be capable of reproduction. This brings us to the obvious conclusion that we can never be certain that we are counting all survivors of a given heat treatment.

In any thermal resistance determination, it is essential to estimate the number of viable cells present in a sample prior to the application of heat, the effect of which is to be measured, and to know the number of viable cells present after the heat application. The observed time required to destroy a given percentage of initially viable cells depends greatly on the adequacy of the subculture medium to support growth of both the unheated and the heated cells.

In estimating numbers of bacterial spores, we commonly find one subculture medium yielding counts many times as great as those obtained with another. And very rarely can a medium be found that will yield counts by subculturing techniques as large as those obtained by direct microscopic examination. Thus, the question remains whether some of the spores enumerated microscopically were dead before the application of heat or whether conditions adequate for their germination and outgrowth were not provided—one condition being the nature of the subculture medium. In consequence, this leads to the obvious demand that, in thermal resistance determinations, the medium, among those available for trial, that yields the highest counts should be selected.

In choosing a medium for thermal resistance studies, selection should be made on the relative ability of various media to support growth of severely heated cells as well as on their ability to support growth of unheated or mildly heated cells. This is particularly true when the resistance of bacterial spores is being measured. Various types of media may give equal estimates of the number of viable unheated spores but divergent estimates of the numbers of spores surviving heat treatments. Certain types of media contain substances that are far more inhibitory to severely heated spores than they are to unheated or mildly heated spores.

Olsen and Scott (1946) found that addition of small amounts of starch to the subculture medium promoted better germination of and outgrowth from spores surviving heat treatments. These same workers (1950) reported serum albumin and charcoal to have a similar effect. Table 2 shows the striking effect of these additives in promoting outgrowth from heated spores of two strains of *Clostridium botulinum*.

TABLE 2

Destruction of C. botulinum spores at 100°C. Comparison of starch, charcoal, and serum albumin additions to pork infusion thioglycolate agar.
Viable counts per ml. after 14 days at 25°C.[a]

| | Strain 07 | | Strain 213B | |
Medium	0 min.	40 min.	0 min.	50 min.
Pork infusion thioglycolate agar	52,000	20	980,000	24
+0.1% starch	65,000	520	1,140,000	490
+0.1% charcoal	49,000	360	1,100,000	450
+0.1% serum albumin	52,000	340	1,020,000	320
+0.1% starch and 0.1% charcoal	49,000	530	1,080,000	410
+0.1% starch and 0.1% serum albumin	61,000	410	980,000	390

[a] From Olsen and Scott (1950).

As was concluded by these workers, the results show that all three substances produce similar effects, the combined effects of starch and charcoal, or starch and albumin, being no greater than those of starch alone. They consider it most likely, therefore, that each substance increases the germination of heated spores by virtue of its capacity to adsorb substances that inhibit spore germination. They also found that starch in 0.1% concentration increased the numbers of colonies developing in pork infusion medium inoculated with heated spores of the following organisms: *C. histolyticum, C. oedematiens, C. bifermentans, C.*

botulinum type *CB*, *C. tetani*, *C. sporogenes*, *C. thermosaccharolyticum*
N.C.A. No. 3814, *Clostridium* sp. N.C.A. No. 3679.

It is interesting to note that these workers found a number of starches to be equally effective—potato, maize, rice, and wheat starches. Also, it was undegraded starch that was effective, enzyme- or acid-hydrolyzed starch showing little or no effect. Glucose, maltose, and glucose-1-phosphate were without effect. Though they give no evidence regarding the nature of the inhibitory substance adsorbed from the medium, they postulate that unsaturated fatty acids are probably involved. It is known that inhibitory substances occur widely and have been shown to inhibit growth of several different types of bacteria (Kodicek and Worden, 1945; Ley and Mueller, 1946; Pollock, 1947; Murrell *et al.*, 1950). Wynne and Foster (1948), studying germination of mildly heated spores of *C. botulinum*, confirmed the effectiveness of starch for removing inhibitory substances from the medium, as was previously reported by Olsen and Scott (1946). Wynne and Foster (1948) demonstrated that germination of *C. botulinum* is inhibited by traces of long-chain unsaturated fatty acids and that such inhibition can be prevented by the incorporation of starch in the medium. Olsen and Scott (1950) pointed out that this finding establishes a reason for the effect of starch, but it still remains to be determined whether unsaturated fatty acids are the only inhibitors in media or whether other substances are involved. Roth and Halvorson (1952) found that oxidized fats and fatty acids exerted considerably greater inhibitory effect on germination of spores of several species than did unoxidized fats or fatty acids. The status of this subject was so well summarized by Olsen and Scott (1950) that it seems pertinent to quote them: "if the ideal conditions for germination of heated spores of a culture of *Cl. botulinum* were known, then a true estimate of the spores surviving a given heat treatment could be obtained by inoculating the spores into a medium providing the appropriate conditions. The accuracy of such estimates would be affected only by sampling errors. At present, however, these ideal conditions are not known and one can only compare estimates under various conditions and assume that the greatest estimates approximate most closely to the true value." Extend these statements to include organisms other than *C. botulinum*, and we have quite an accurate evaluation of the present state of affairs. However, further discussions will point out appropriate conditions, other than the proper medium, which are not always easy to provide. Also, important sampling errors are sometimes difficult to avoid.

Let us now consider what effect this one factor, starch in the subculture medium, may have on the form and slope of survivor curves. Again let us refer to results reported by Olsen and Scott (1950) and by Murrell *et al.* (1950). Values from the former publication have been plotted as Figs. 10 and 11 and values from the latter as Fig. 12. The effect of starch in the media on the slope of the survivor curves is indeed striking. Curves in Figs. 10 and 12 show the effect of heat of activation.

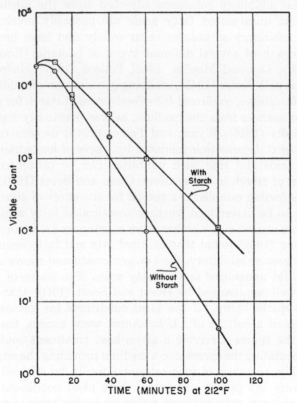

FIG. 10. Effect of 0.15% soluble starch in subculture medium on outgrowth of heated spores of *C. botulinum*—strain L12 (Olsen and Scott, 1950).

It is evident from the curves in Figs. 10, 11, and 12 that the more severe the heat treatment, the more sensitive survivors are to inhibitory substances in the medium. It is also evident, as was pointed out by the authors, that the highly erroneous data obtained with the control media, when plotted, approximate straight lines (after first portions), and argu-

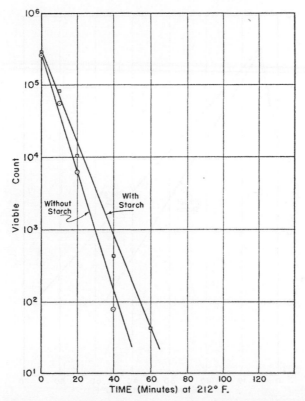

FIG. 11. Effect of 0.15% soluble starch in subculture medium on outgrowth of heated spores of *C. botulinum* strain 213B (Olsen and Scott, 1950).

ments based on conformity to these types of survivor curves cannot be accepted as an indication that the data are reliable. (The initial portions of these curves have been discussed above.) Again, it should be emphasized that there is no assurance that survivor curves for the spores subcultured in the starch-containing media are the true ones. However, they certainly approximate the true curves far more closely than do those plotted from data obtained with the control media.

Anaerobiosis. Generally speaking, other things being equal, the more severe the heat process that spores survive, the longer it takes for the spores to germinate. This effect is referred to as heat injury. Apparently there is a period of reparation required of spores that have survived severe heating. Therefore, when culturing heated spores of obligate

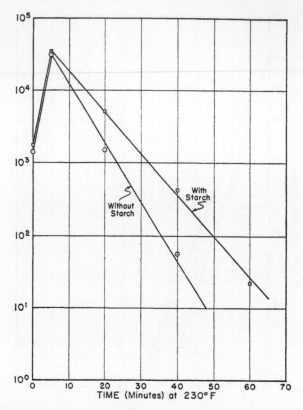

FIG. 12. Effect of 0.10% soluble starch in subculture medium upon outgrowth of heated *Bacillus* spores—strain 320 (Murrell, Olsen, and Scott, 1950).

anaerobes, it is imperative to maintain strict anaerobiosis in the medium for sufficiently long to allow heat-damaged spores to recover, germinate, and produce detectable outgrowth. If this is not done, askew survivor curves, or survivor curves of steeper slopes than those of the true curves, will be observed. Methods of obtaining and maintaining anaerobiosis in subculture are discussed in Chapter 3.

Thermal Destruction Curves

Thermal destruction (TD) curves, often referred to as thermal death time curves, reflect the relative resistance of bacteria to different lethal temperatures. They are most conveniently constructed by plotting the logarithm of D, or some multiple of D, in the direction of ordinates against exposure temperature in the direction of abscissae (see Fig. 13).

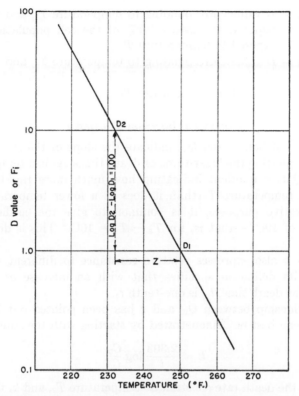

FIG. 13. Thermal destruction curve passing through 1 minute at 250°F.

Over the range of temperatures of chief concern in food sterilization, TD curves are essentially straight lines.

The term z, employed in process calculation methods to account for the relative resistance of a microorganism to the different temperatures, is equal numerically to the number of Fahrenheit degrees required for the TD curve to traverse one log cycle. Considering any one such portion of the curve, the slope of the curve may be expressed as follows:

$$\frac{\log D_2 - \log D_1}{z} = \frac{1}{z}$$

And, the general equation of the TD curve may be conveniently written as follows:

$$\log D_2 - \log D_1 = \frac{1}{z}(T_1 - T_2)$$

in which, $D_2 = D$ value corresponding to temperature T_2, and the time required to destroy 90% of the cell population when exposed to temperature T_2.

$D_1 = D$ value corresponding to temperature T_1, and the time required to destroy 90% of the cell population when exposed to temperature T_1.

Relationship between Q_{10} and z

The values of both z and Q_{10} indicate the slope of the straight line obtained by plotting the logarithms of death times against temperature. The term Q is a quotient indicating how much more rapidly death proceeds at temperature T_2 than it does at a lower temperature, T_1. For comparative purposes, it is common to give the coefficient for an increase of 10C.°—that is, for $T_2 - T_1 = 10$C.° This is designated as Q_{10}.

The term z also expresses relative resistance to different temperatures, and its definition specifies that with an increase of zF.°, or $0.555z$C.°, the death time, t_2, is one-tenth t_1.

The relationship between Q_{10} and z has been pointed out by Rahn (1945a). It can best be demonstrated by starting with the equation

$$k = \frac{2.303}{t} \log \frac{C_0}{C}$$

Let k_2 be the death rate constant at temperature T_2, and k_1 the death rate constant at temperature T_1, which is lower than T_2, and t_2 and t_1 the corresponding death times. Then

$$k_2 = \frac{2.303}{t_2} \log \frac{C_0}{C} \tag{5}$$

and

$$k_1 = \frac{2.303}{t_1} \log \frac{C_0}{C} \tag{6}$$

By dividing equation 5 by 6, we obtain

$$\frac{k_2}{k_1} = \frac{t_1}{t_2}$$

This ratio is the temperature coefficient for $n° = T_2 - T_1$. Then

$$Q_n = Q_1{}^n = \frac{t_1}{t_2}$$

where Q_1 = coefficient for 1°C. Then

$$Q_{10} = Q_1^{10}$$

By definition,

$$n = 0.555z°C., \text{ and } \frac{t_1}{t_2} = 10$$

Therefore,

$$Q_1{}^n = Q_1{}^{0.555z} = \frac{t_1}{t_2} = 10$$

and

$$0.555z \log Q_1 = \log 10 = 1$$

or

$$\log Q_1 = \frac{1}{0.555z}$$

Since $Q_{10} = Q_1^{10}$,

$$\log Q_{10} = 10 \log Q_1 = \frac{10}{0.555z} = \frac{18}{z}$$

and

$$z = \frac{18}{\log Q_{10}}$$

SELECTED REFERENCES

Chick, H. (1910). The process of disinfection by chemical agencies and hot water. *J. Hyg.*, **10**, 237.

Curran, H. R., and Evans, F. R. (1937). The importance of enrichment in the cultivation of bacterial spores previously exposed to lethal agencies. *J. Bacteriol.*, **34**, 179.

Curran, H. R., and Evans, F. R. (1945). Heat activation inducing germination in the spores of thermotolerant and thermophilic aerobic bacteria. *J. Bacteriol.*, **49**, 335.

Evans, F. R., and Curran, H. R. (1943). The accelerating effect of sublethal heat on spore germination in mesophilic aerobic bacteria. *J. Bacteriol.*, **46**, 513.

Fricke, H., and Demerec, M. (1937). The influence of wave-length on genetic effects of X-rays. *Proc. Natl. Acad. Sci. US.*, **23**, 230.

Katzin, L. I., Sandholzer, L. A., and Strong, M. E. (1943). Application of the decimal reduction time principle to a study of the resistance of coliform bacteria to pasteurization. *J. Bacteriol.*, **45**, 265.

Kodicek, E., and Worden, A. N. (1945). The effect of unsaturated fatty acids on *Lactobacillus helveticus* and other gram-positive micro-organisms. *Biochem. J.*, **39**, 78.

Ley, H. L., Jr., and Mueller, J. H. (1946). On the isolation from agar of an inhibitor for *Neisseria gonorrhoeae. J. Bacteriol.*, **52**, 453.

Murrell, W. G. (1961). Spore formation and germination as a microbial reaction to the environment. *Symp. Soc. Gen. Microbiol.*, **11**, 100.

Murrell, W. G., Olsen, A. M., and Scott, W. J. (1950). The enumeration of heated bacterial spores. II. Experiments with *Bacillus* species. *Australian J. Sci. Res.,* **3,** 234.

Olsen, A. M., and Scott, W. J. (1946). Influence of starch in media used for detection of heated bacterial spores. *Nature,* **157,** 337.

Olsen, A. M., and Scott, W. J. (1950). The enumeration of heated bacterial spores. I. Experiments with *Clostridium botulinum* and other species of *Clostridium. Australian J. Sci. Res.,* **3,** 219.

Pollock, M. R. (1947). The growth of *Haemophilus pertussis* on media without blood. *Brit. J. Exptl. Pathol.,* **28,** 295.

Rahn, O. (1929). The size of bacteria as the cause of the logarithmic order of death. *J. Gen. Physiol.,* **13,** 179.

Rahn, O. (1934). Chemistry of death. *Cold Spring Harbor Symp. Quant. Biol.,* **2,** 70.

Rahn, O. (1943). The problem of the logarithmic order of death in bacteria. *Biodynamica,* **4,** 81.

Rahn, O. (1945a). Physical methods of sterilization of microorganisms. *Bacteriol. Rev.,* **9,** 1.

Rahn, O. (1945b). Injury and death of bacteria. *Biodynamica Monograph* No. 3.

Reynolds, H., and Lichenstein, H. (1949). Germination of anaerobic spores induced by sublethal heat. *Bacteriol. Proc. (Soc. Am. Bacteriologists),* p. 9.

Roth, N. G., and Halvorson, H. O. (1952). The effect of oxidative rancidity in unsaturated fatty acids on the germination of bacterial spores. *J. Bacteriol.,* **63,** 429.

Secrist, J. L., and Stumbo, C. R. (1956). Application of spore resistance in the newer methods of process evaluation. *Food Technol.,* **10,** 543.

Schmidt, C. F. (1957). Thermal resistance of microorganisms. *In* "Antiseptics, Disinfectants, Fungicides and Chemical and Physical Sterilization" (G. F. Reddish, ed.), 2nd ed., pp. 831–884. Lea & Febiger, Philadelphia.

Stumbo, C. R. (1948). Bacteriological considerations relating to process evaluation. *Food Technol.,* **2,** 115.

Stumbo, C. R. (1949). Thermobacteriology as applied to food processing. *Advan. Food Res.,* **2,** 47.

Stumbo, C. R., Murphy, J. R., and Cochran, J. (1950). Nature of thermal death time curves for P.A.3679 and *Clostridium botulinum. Food Technol.,* **4,** 321.

Watkins, J. H., and Winslow, C.-E. A. (1932). Factors determining the rate of mortality of bacteria exposed to alkalinity and heat. *J. Bacteriol.,* **24,** 243.

Wax, R. G. (1963). Ph.D. thesis, Pennsylvania State University.

Wynne, E. S., and Foster, J. W. (1948). Physiological studies on spore germination with special reference to *Clostridium botulinum.* I. Development of a quantitative method. *J. Bacteriol.,* **55,** 61.

Yoshikawa, H. (1968). Chromosomes in *Bacillus subtilis* spores and their segregation during germination. *J. Bacteriol.,* **95,** 2282.

Youland, George C., and Stumbo, C. R. (1953). Resistance values reflecting the order of death of spores of *Bacillus coagulans* subjected to moist heat. *Food Technol.,* **7,** 286.

CHAPTER 8

Thermal Resistance of Bacteria

Methods of Measuring

The most common methods now in use for measuring thermal resistance of bacteria may be roughly classified as follows:

1. Thermal death time (TDT)—tube method (Bigelow and Esty, 1920).
2. Thermal death time (TDT)—can method (American Can Company, 1943).
3. "Tank" method (Williams *et al.*, 1937).
4. Flask method (Levine *et al.*, 1927).
5. Thermoresistometer method (Stumbo, 1948).
6. Unsealed TDT tube method (Schmidt, 1950).
7. Capillary tube method (Stern and Proctor, 1954).

All these methods, and some modifications of them, are being used at present to obtain data for thermal process calculation. Each has its advantages and disadvantages, some of which will be pointed out as each method is described.

TDT Tube Method. In this method, inoculated menstruum (water, buffer solution, culture medium, or food material) is distributed in small-diameter (7 to 10 mm) test tubes which are subsequently sealed near the mouth in the flame of a blast burner. The volume of product per tube usually is from 1 to 4 ml. The sealed tubes of inoculated product are generally heated in a thermostatically controlled bath of mineral oil, lard, vegetable oil, propylene glycol, butyl phthalate, or some other suitable medium. At predetermined intervals replicate tubes are removed and plunged into water at or below 70°F. After cooling, the tubes are usually opened aseptically and their contents transferred to tubes of sterile culture medium favorable for growth of the organism being studied. However, if the medium in which the bacteria were suspended for heating is favorable for growth of the test organism, the tubes are sometimes incubated directly without subculture.

If tubes heated for varying intervals of time are subcultured and the number of survivors is determined, a survivor curve may be constructed by plotting the logarithm of survivors against time of heating. By so doing, the D value may be taken directly from the graph. Another procedure commonly employed is to determine the number of survivors after two heating times, and calculate the value of D. These procedures and their relative merits will be described later in more detail.

Several devices have been employed for distributing inoculated menstrua in TDT tubes. For solutions and for homogenates that flow readily, a pipet or syringe is commonly used. For heavier products, where air entrapment may be a problem, the tubes should be filled from the bottom up. A syringe may be used for products that will pass readily through a syringe needle. For products such as ground meat, an Alemite grease gun equipped with a large-artery pumping needle has been used successfully (Stumbo et al., 1945).

The chief advantages of the TDT tube method, relative to some other methods, are:

1. It employs simple, inexpensive equipment available to most laboratories.
2. Bacterial growth in clear media and spoilage changes in some food products may be observed visually without opening the tubes.
3. Tubes may be easily opened for subculture with little danger of contamination.
4. Space required for incubation of unopened TDT tubes is small.

Chief disadvantages of the method are:

1. Filling, sealing, heating, and subculturing the tubes are very time-consuming operations, thereby making labor costs high.
2. In transferring contents for subculture there is always the possibility of leaving some survivors in the TDT tubes.
3. Generally only liquid products or homogenates can be used as menstrua.

4. Heating and cooling lags in the tube contents are appreciable and difficult to evaluate with respect to lethal value. Procedures for evaluating these heating and cooling lags have been proposed (Sognefest and Benjamin, 1944). However, it has been well established that these cannot give true corrections (see Stumbo, 1953). Corrections arrived at in this manner are for a single point only in the tube contents. Moreover, to arrive at them, a z value is needed, to account for relative resistance to the different temperatures during heating and cooling. To reach this z value we must first know the resistance of the organism to the different temperatures of concern.

5. Because of the heating and cooling lags, reasonable accuracy can be expected only at temperatures of about 240°F. and lower. The maximum temperature at which acceptable accuracy may be obtained depends on whether the product heats by convection or by conduction and on the resistance of the organism being studied. Because of the lags, when temperatures above a given level are used, most of the total lethal effect will be due to temperatures below that of the heating bath. With any organism, there is always a point above which a portion of the tube contents will not reach the temperature of the heating bath before all or most of the bacterial cells or spores have been killed. In brief, reasonable accuracy cannot be expected when an appreciable amount (more than 5 to 10%) of the total lethal effect is due to temperatures very much more than one degree below the temperature of the heating bath.

TDT Can Method. In this method, small cans (208 × 006) are used as containers of inoculated product. The cans of product are heated in small, accurately controlled steam retorts. This equipment was employed by Townsend *et al.* (1938) for heating products in both TDT tubes and TDT cans. The tubes or cans so heated are cooled by water in the retort, after which they are removed and subcultured or incubated unopened as in the TDT tube method discussed above.

Chief advantages of cans, compared to tubes, are:

1. Many products may be studied in their natural state under conditions closely simulating those employed in canning practice. For this reason inoculated packs employing large numbers of replicates may be carried out in the laboratory or pilot plant. The space required for incubation is small.

2. Spoilage by gas producers causes can swelling and is readily detected.
3. Cans can be filled and closed with far greater speed than is possible with glass tubes.

Chief disadvantages of the TDT can method are:

1. Heating and cooling lags are about the same for the cans as they are for glass tubes, thus similarly limiting the usefulness of the method.
2. Special can-closing equipment is required. However, this may be no more than an adaptor chuck on a regular closing machine.
3. A battery of small, accurately controlled steam retorts is required for extensive studies.
4. Only spoilage that produces can swelling can be detected visually. Cans must be opened and examined visually and by pH determination in order to detect flat-sour spoilage.
5. Opening cans and transferring contents for subculture without contamination is much more difficult with cans than with glass tubes.
6. When gas-producing organisms are studied, spoilage development must be followed closely in order to avoid bursting of the cans.
7. Spoiled cans must be examined carefully to rule out spoilage due to post-heating contamination through can leakage.

"Tank" Method.   The apparatus for this method was designed by Williams *et al.* (1937). It consists of a cylinder 4 inches in diameter and $4\frac{1}{2}$ inches deep (the tank, or chamber). This chamber is constructed of stainless steel and enclosed in a stainless-steel steam jacket. Maximum capacity of the chamber is about 900 ml. Through the bottom are four sampling tubes connected with sampling valves at the bottom of the tank. The stainless-steel cover is fitted with a $\frac{1}{8}$-inch tank exhaust petcock, a flange fitting for a rubber stopper through which a thermometer is inserted, and a packing gland for an agitator shaft. The jacket is fitted with steam inlet and outlet ports, and a drain petcock at the bottom. The motor-driven agitator consists of two four-bladed propellers mounted on a single shaft in such a manner as to produce opposing currents in the food during operation.

In making thermal resistance determinations, inoculated substrate (usually cold) is placed in the heating chamber. After the chamber has been closed and the agitator started, the product is brought to the desired temperature as rapidly as possible. Heating to the desired temperature generally requires a minimum of 2 or 3 minutes. After the temperature is reached, samples are withdrawn at selected time intervals for enumeration of survivors, usually by subculture and colony counts. With these data, survivor curves are constructed in the usual manner, and D values are taken from the graphs.

Chief advantages of the tank method are:

1. Liquid media or mixed solids and liquids may be studied. In the latter case, the solids are held in a wire basket.
2. Survivor counts may be readily obtained for selected time–temperature relationships.

Chief disadvantages of the method are:

1. The time required to bring the tank contents to temperature limits the method to temperatures no higher than 240° to 250°F. for studying the resistance of most bacteria.
2. Withdrawing measured samples, rapidly cooling them, and diluting them for subculture requires a great deal of skill.
3. Construction of the apparatus requires considerable precision tooling.
4. Heating and agitation may cause flocculation (curdling) of some food products. If this occurs, subculture counts, particularly during the early heating periods, may be of colonies produced from multi-cellular clumps rather than from single cells. This can cause serious deviations in survivor curves.
5. With heavy products, obtaining uniform temperature distribution in the mass of product in a relatively short time is difficult. This factor further limits the method to study of lower temperatures for longer times.

Flask Method. This method (see Fig. 14) is for studying resistance of bacteria to temperatures below the boiling point of water. It is of value in determining the resistance of nonsporulating organisms and sporeformers of low heat resistance (Type E *Clostridium botulinum*, butyric anaerobes, etc.). Generally, a three-neck flask (Woulff bottle) such as that used by Levine *et al.* (1927) is employed as the substrate container. Different sizes of these flasks are now available, with ground-glass fittings, including a ground-glass stirrer rod fitted in a ground-glass stopper.

Usually, a thermometer is introduced through one neck, and a small mechanical stirrer through another (the center one). The third is used

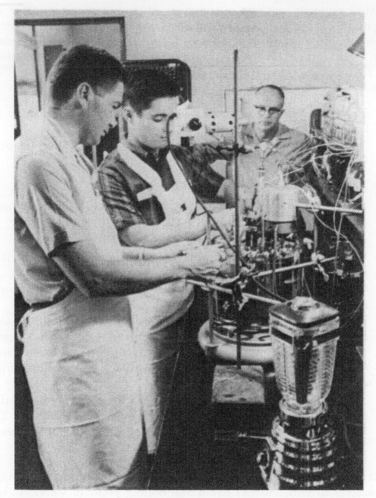

FIG. 14. The flask method being employed to determine the heat resistance of *C. botulinum,* Type E.

for introducing inoculum and withdrawing samples. After proper insertion of the thermometer and stirrer with shaft, the desired amount of substrate may be introduced and the entire assembly sterilized in the autoclave.

After sterilization, the assembly is placed in a heating bath which is thermostatically controlled at the desired temperature. Any of a number of materials may be used as a heating medium. Light or heavy mineral oil is very satisfactory. The assembly is located in the bath in such a way that the surface of the heating medium will be above the substrate surface at the wall of the flask when the flask contents are being stirred. (Special care must be taken to avoid splashing of the substrate on the walls of the flask above the level of the outside heating medium.)

After the assembly has been properly placed in the bath, the stirrer shaft is connected to an electric stirrer motor the speed of which may be controlled by a rheostat. This connection is most conveniently made with a short metal rod held in the chuck of the stirrer motor and a short rubber tube. The stirrer speed is then adjusted, and the contents of the flask allowed to heat up to the bath temperature.

When the substrate is up to bath temperature, a predetermined amount of inoculum is introduced into the substrate. The volume of inoculum should be small enough that its introduction does not cause an appreciable drop in the substrate temperature. In this manner virtually instantaneous heating of the inoculum to test temperature will be attained. Extremely high-population inocula should not be used. With many organisms, cells in high-population suspensions will clump badly when they are introduced into the hot menstruum. After inoculation, samples are withdrawn, at predetermined time intervals, by pipet or syringe and subcultured for counting of survivors. From the data thus obtained, survivor curves are constructed in the usual manner, and D values are taken from the graphs.

Chief advantages of the flask method are:

1. It is a convenient and rapid method for determining the resistance of "low-resistance" organisms.
2. It requires only readily available equipment.
3. When the method is properly employed, heating and cooling lags are negligible.

Chief disadvantages of the method are:

1. It is confined to use of temperatures below the boiling point of water.
2. Extreme care is required to avoid splashing of the flask contents above the level of the external heating medium and to avoid contamination of the walls of the flask above the substrate level when introducing the inoculum or withdrawing samples. If the walls of the flask are so contaminated, the bacteria temporarily

residing thereon may be exposed to temperatures far below that of the substrate. These bacteria subjected to less-severe heating may continually "feed" back into the main body of substrate and cause erroneous results. The usual result is a "tailing off" of the survivor curve.

3. With some substrates flocculation may occur, particularly during the early period of heating. If flocculation does occur and the samples are diluted for subculture, it is often helpful to add a nontoxic, noninhibitory surfactant to the diluent.

Thermoresistometer Method. The thermoresistometer was designed by Stumbo (1948) particularly for studying resistance of bacterial spores to temperatures in the higher range (above 240°F.). However, it is applicable for studying resistance to temperatures over the range 215° to 300°F. The maximum temperature for which accurate resistance values may be obtained depends on the resistance of the organism being studied—the higher the resistance, the longer will be the heating times.

The thermoresistometer is built by the RePP Industries Division of

FIG. 15. The thermoresistometer in operation.

the Virtis Company. Though the present model is superior in design to the original, functionally it is identical (see Fig. 15).

DESIGN. Three steam chambers, connected in line, are so arranged that they may be sterilized concurrently by steam at pressures up to 50 p.s.i.g. (see Fig. 16). A stainless-steel plate, *d*, called the carrier plate, traverses the three chambers. Steam pressure may be maintained in all chambers or in any one chamber independently because of a flat seal bearing arrangement (as *r*) around the plate between chambers *a* and *b* and another between chambers *b* and *c*.

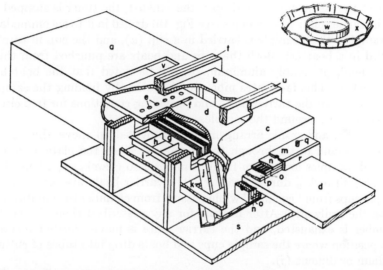

FIG. 16. Perspective drawing showing a longitudinal vertical cutout section of thermoresistometer (not drawn to scale).

Steam is supplied to the three chambers from a steam reservoir (not shown) having a volume of approximately 3.5 cubic feet. The reservoir is equipped with an automatic temperature controller, and, because its volume is large compared with volumes of the chambers, temperatures in the chambers may be controlled to within ±0.3°F. of the desired temperature.

Since each chamber is equipped with separate steam inlet and exhaust lines, each may be operated independently. The two lines to the center chamber, where samples are heated, have quick-acting valves operating from one lever. When the lever is in one position (up), the exhaust is closed and the inlet from the reservoir is open; when it is in the other position (down), the reverse is true. This is a "free exhaust" for releasing pressure in the center chamber. This chamber is also equipped with a second exhaust line having a condensate trap for releasing condensate and air during operation. Control systems for the first and last chambers are less elaborate, because steam is used only for sterilizing them prior to a resistance determination.

An electric timer graduated in increments of 0.001 minute and controlled by two microswitches is installed in the system. When samples enter the controlled steam in the center chamber, the timer is started; and when the lever, mentioned above, is moved to the down position, to close the steam supply and open the exhaust, the timer is stopped.

The carrier plate has six holes (see Fig. 16) drilled in it to accommodate six samples. Each sample is carried in a cup (w), and the cup in turn is carried in a boat (x). Both the cups and boats are punched from thin-gage metal, preferably aluminum. It will be noted that the boat has fluted edges. This facilitates more rapid heating by holding the sample cup away from the carrier plate and providing conditions for free circulation of steam around the sample.

A ratchet-and-crank arrangement is provided to move the carrier plate back and forth through the chambers. When the plate is moved forward, samples are drawn from chamber a to chamber b, where they rest on grillwork g during heating. (The carrier boats also prevent the sample cups from "cocking" while passing from chamber to chamber and along the grillwork.) After heating for any prescribed time, the center chamber is exhausted, and the carrier plate is moved farther forward to a position where the sample cups and boats drop into tubes of culture medium or diluent (j).

OPERATION. Removable covers are clamped on chambers a and c at v and k, and all three chambers are sterilized. After sterilization, the steam lines to chambers a and c are closed, and the steam from these two chambers is exhausted. The temperature of the center chamber is then adjusted to that desired, and the covers are removed from chambers a and c. Next, the carrier plate is positioned so that the six holes are directly under the opening in chamber a. These holes are closed at the bottom by a small bronze plate under the carrier plate. A boat (x) is placed in each of the six holes, and a cup containing 0.01 to 0.02 ml. of sample is placed in each boat. The samples are quickly drawn into the center chamber, where they are allowed to remain for any prescribed time. At the end of this prescribed time, the valve lever is thrown to the down position, closing the steam supply line from the reservoir, opening the ½-inch "free" exhaust line, and opening a microswitch which stops the timer. When the pressure in the center chamber drops to 1 pound or less, the carrier plate is moved forward until the samples drop into tubes of culture medium or diluent, positioned as shown in Figs. 15 and 16.

When the "end-point" technique is employed, tubes of medium containing treated samples are removed, closed, and placed under incubation. Usual techniques are employed in handling the tubes. Tube closures may be placed in sterile culture plates during the time the tubes are in the machine. While in the machine the tops of the tubes are in a light fog of steam as a result of slight leakage through the seal bearings. This is desirable, as it minimizes chances of contamination.

When it is desired to obtain several points for construction of survivor curves, all samples except those heated most severely are ejected into an appropriate diluent rather than a culture medium. Tubes containing the samples are thoroughly mixed and further diluted as indicated. The diluted samples are then subcultured to obtain survivor counts.

One may ask why the open samples do not spatter from the cups when the center chamber is suddenly exhausted. In the reason they do not lies the novelty of the thermoresistometer technique. The rate of exhaustion is controlled by the size of the exhaust pipe. The ½-inch pipe is used because it allows reasonably rapid exhaustion, but not so rapid as to cause spattering of the samples.

Chief advantages of the thermoresistometer method are:

1. Virtually instantaneous heating and cooling of samples (see Fig. 17). The maximum correction for heating and cooling lags is less than 0.3 second. Therefore, if no correction is applied, times as short as 6 seconds may be used with no greater than 5% error. This permits the determination of thermal resistance with good accuracy at relatively high temperatures (for *C. botulinum* Types A and B, 260°F.; for P.A.3679, 270° to 280°F.; for *B. stearothermophilus*, 280° to 290°F.).
2. Precise timing. Heating times may be readily reproduced to within 0.0005 minute.
3. Automatic subculturing of heated samples. When proper care is taken, chances of contamination are virtually nil. In one series of experiments, over 50,000 samples were treated in a thermoresistometer without detectable contamination of a single subculture.
4. Labor saving. It is not unusual for one technician to prepare and process as many as 300 samples in one 8-hour day.

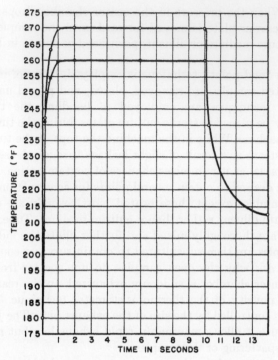

Fig. 17. Calculated heating and cooling curves representing temperatures in food sample, during thermal process, at point most remote from surface.

5. Simplicity of operation. No special skills beyond those required in any aseptic technique of culture transfer are required.

6. Low upkeep. Some machines have been in almost continuous operation for more than 15 years with almost no repairs being required, and these were only minor.

Chief disadvantages of the thermoresistometer method are:

1. Rather high initial cost. However, this is soon balanced by savings in labor costs and low costs of upkeep.

2. Substrates confined for the most part to liquid suspensions and homogenates.

3. Direct incubation of samples to observe spoilage is not possible. However, this can be overcome in many cases by using food prod-

ucts rather than formulated culture media for subculturing heated samples.

4. Method confined to study of temperatures above about 215°F.

Pflug and Esselen (1953) developed an instrument designed to accomplish essentially the same objectives as the thermoresistometer. It is more complicated in design and construction and subject to more mechanical failures than the thermoresistometer, as well as the same limitations.

Unsealed TDT Tube Method. This method, described by Schmidt (1950), consists in heating 13 × 100-mm. cotton-plugged tubes in steam and cooling them under pressure. This permits subculture by the addition of the medium to the suspension in the tube after heating, rather than the transfer of the suspension to the subculture medium with possible attendant loss of surviving organisms. The small retort system described for the TDT can method may be modified for use in this method (Schmidt *et al.*, 1955).

Chief advantages of this method are:

1. The subculture medium may be transferred to the heated spore suspension, rather than the reverse.
2. The laborious tasks of sealing and opening glass tubes are eliminated.

Chief disadvantages of the method are:

1. As in the TDT tube and TDT can methods, heating and cooling lags limit the temperatures that may be studied with accuracy.
2. Studies are limited to temperatures above 212°F. unless steam–air mixtures are employed. Temperature control is difficult with steam–air mixtures.

Capillary Tube Method. This method, described by Stern and Proctor (1954), employs samples contained in small sealed capillary tubes. Stern *et al.* (1952) described procedures for measuring rapid temperature changes in sealed tubes in an oil bath. Their measurements agreed well with values calculated by the method of Olson and Schultz (1942).

Chief advantages of this method are:

1. Rapid heating and cooling of samples is possible.
2. Heated samples may be stored for later subculture or incubated unopened for observation.

Chief disadvantages of the method are:

1. It involves the laborious tasks of sealing and opening glass tubes.
2. Introduction of the samples into the capillary tubes is a tedious operation. This is particularly true for heavy-bodied products.
3. There is the hazard of contamination during subculture.

Treatment of Thermal Resistance Data

There are two procedures commonly employed in treating thermal resistance data, depending on how the data were obtained.

1. Construction of survivor curves, by plotting the logarithm of survivors against time at constant temperatures, from which D values may be taken directly.
2. Assumption of the logarithmic order of death, and calculation of a D value from "initial" number and number surviving after some one heating time at each temperature studied.

The first procedure is quite satisfactory as long as reasonably straight lines are obtained. If straight lines are not obtained, it is often difficult to judge which portion of the curve is describing only bacterial death, or to draw a straight line best representing death alone (see Chapter 6).

A number of workers (Stumbo et al., 1950; Schmidt, 1957; and others) feel that the second procedure is the more satisfactory and practical in handling problems of thermal resistance and in calculation of sterilizing processes. In this method the "initial" number is usually the spore count taken after the application of sufficient heat to activate most spores for germination—this may be a few seconds or several minutes, depending on the spores under study and the temperature being applied. To obtain this initial count, the spores may be heated for activation in a medium in which flocculation or clumping does not occur. Or the suspension may be diluted prior to heating so that less than one activated spore will occur per unit volume of the suspending medium. This gives reasonable assurance that most colonies developing in subculture are from spores surviving singly, if each unit is subcultured individually. Also, it allows the application of "most probable" techniques to arrive at counts by considering the ratio of subcultures positive for growth to those negative for growth. These techniques, which are statistical in nature, will be discussed later.

The second point for use in calculation of the death rate parameter, D, is most easily arrived at in the following manner. A number of heating times are chosen that will reduce the number of survivors to very low levels compared with the number of spores initially present in the population. These times should be such that subcultures from replicate samples heated for the shortest time are all positive for growth, for the next longest time (or few times) are neither all positive nor all negative for growth, and for the next longest time, or times, are all negative for growth. When the number of spores in the population is so reduced by heat, most colonies developing in subculture will be from single spores; or most tubes positive for growth will be so by virtue of outgrowth from very few surviving spores.

In this "end-point" technique, if a solid medium is used for subculture, colonies developing in subculture may be counted directly to arrive at the number of survivors. If a liquid medium is used for subculture, the number of survivors may be estimated by considering only those heating times that resulted in some, but not all, of the subcultured replicate samples showing growth. A number of statistical procedures are available for estimating the most probable number of survivors from data of this type.

Stumbo et al. (1950) applied the equation of Halvorson and Ziegler (1932) to estimate the number of survivors. The equation, as applied, follows:

$$\bar{x} = \frac{2.303}{a} \log \frac{n}{q}$$

where $\bar{x}$ = most probable number of surviving spores per milliliter of heated material.

a = volume, in milliliters, of each heated sample.

n = total number of replicate samples subjected to one time–temperature combination.

q = number of samples subjected to this time–temperature combination which were negative for growth on subculture.

When this technique is used to estimate the "initial" number and the number surviving time, t, D may be calculated as follows:

$$D = \frac{t}{\log a - \log b}$$

The following is an example of treatment of data from an actual thermal resistance determination by the thermoresistometer technique. A suspension of P.A. 3679 spores was diluted such that less than one spore per 0.01 ml. would survive a heat activation treatment of 0.464 minute at 220°F. (that predetermined to give maximum count).

Ninety 0.01-ml. aliquots of this diluted suspension were subjected to 220°F. for 0.464 minute in the thermoresistometer and ejected into 90 tubes of pork infusion medium (one sample lost). After incubation for 3 weeks at 85°F., 20 of the 89 tubes were positive for growth. The number per milliliter of diluted suspension was computed as follows:

$$\bar{x} = \frac{2.303}{a} \log \frac{n}{q}$$

$$\bar{x} = \frac{2.303}{0.01} \log \frac{89}{69}$$

$$\bar{x} = \textbf{25.6}$$

This was then multiplied by the dilution factor of 10^5 to give an initial count of 2,560,000 in the original suspension.

A protocol of time–temperature combinations was then set up for studying the resistance of the spores to 220°, 230°, 240°, 250°, 260°, and 270°F. Each time first set down was corrected for the time required for activation, but the corrections could have been ignored (they were of the order of 0.1% of the heating time employed).

To each of the thermoresistometer cups required was added 0.01 ml. of the spore suspension and approximately 0.01 ml. of puréed peas. The purée and suspension in each cup were then mixed with a sharpened inoculating needle. The samples so prepared were then subjected in replicates of six to the time–temperature combinations detailed in the protocol. Times at different temperatures that resulted in sterilization of some but not all of the samples, as determined by subculture and incubation, were employed to compute D values. The $\bar{x}$ equation was employed to estimate the most probable number of spores surviving per six replicate samples. For example, of six samples heated for 90.4 minutes at 230°F., four were sterilized, and

$$\bar{x} = 2.303 \log \frac{n}{q} \times 6$$

$$\bar{x} = 13.8 \log \frac{6}{4}$$

$$\bar{x} = 13.8 \times 0.176 = \textbf{2.43}$$

Since the six samples before heating contained 153,600 (6 × 25,600),

$$D_{230} = \frac{90.4}{\log 153,600 - \log 2.43} = \textbf{18.8}$$

TABLE 3
Thermal resistance data

220°F.				230°F.				240°F.			
t	No. of + tubes	b	D	t	No. of + tubes	b	D	t	No. of + tubes	b	D
92.8	6	—	—	25.8	6	—	—	7.1	6	—	—
139.3	6	—	—	38.8	6	—	—	10.7	6	—	—
185.7	6	—	—	51.7	6	—	—	14.3	6	—	—
232.1	6	—	—	64.6	6	—	—	17.8	6	—	—
278.5	5	10.8	67.0	77.5	6	—	—	21.4	4	6.60	4.90
324.9	5	10.8	78.3	90.4	2	2.43	18.8	25.0	2	2.45	5.20
371.4	3	4.16	81.4	103.4	0	—	—	28.5	1	1.09	5.54
417.8	0	—	—	116.4	0	—	—	32.0	0	—	—

250°F.				260°F.				270°F.			
t	No. of + tubes	b	D	t	No. of + tubes	b	D	t	No. of + tubes	b	D
2.00	6	—	—	0.52	6	—	—	0.157	6	—	—
3.00	6	—	—	0.85	6	—	—	0.236	6	—	—
4.00	6	—	—	1.13	6	—	—	0.313	6	—	—
5.00	6	—	—	1.41	5	10.8	0.340	0.393	3	4.16	0.086
6.00	4	6.60	1.38	1.70	2	2.43	0.350	0.472	0	—	—
7.00	4	6.60	1.61	1.98	0	—	—	0.548	0	—	—
8.00	1	1.09	1.56	2.26	0	—	—	0.628	0	—	—
9.00	0	—	—	2.54	0	—	—	0.708	0	—	—

Thermal resistance data for all temperatures employed are summarized in Table 3.

For constructing a TD curve, the D values, where there were more than one, were averaged for each temperature employed. The TD curve obtained is shown in Fig. 18. The z value taken from this graph is 17.3.

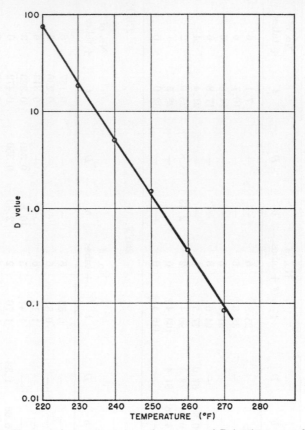

Fig. 18. Thermal destruction curve for spores of P.A.3679 suspended in puréed peas when heated.

This method was employed (Stumbo *et al.*, 1950) to obtain TD curves for spores of P.A. 3679 and *C. botulinum* in a number of substrates, including nine food products. All curves obtained were straight lines. The z values for P.A. 3679 spores ranged from 16.6 to 20.5, and those for *C. botulinum*, obtained for the spores in only four of the substrates, ranged from 14.7 to 16.3. Many subsequent studies by the author and others using this technique with a wide variety of sporeforming organisms in many different substrates have yielded straight-line TD curves. For this reason, it is felt that the technique is quite reliable for obtaining parameters to account for bacterial resistance in calculation of sterilizing processes.

Schmidt (1954) presented a method for treating thermal resistance data based on a procedure for bioassay suggested by Reed (1936). As in the method just discussed, only the initial number, a, and the "end-point" number, b, of spores were employed to determine D values. The difference in this method and that of Stumbo *et al.* is in the statistical procedure used to determine the most probable number (b) of spores surviving. Schmidt's method employs probability paper to determine the time at which 50% of the subculture tubes will be sterile. This is considered the LD_{50} point. It corresponds to 0.69 surviving organism per sample. For example, if six replicate samples are used,

$$b = \bar{x} = 2.303 \log \frac{6}{3} = 0.69$$

Values of D are then calculated by the equation

$$D = \frac{LD_{50}}{\log a - \log 0.69}$$

As was pointed out by Schmidt (1957), D values obtained in this manner and those obtained by the method of Stumbo *et al.* are generally in good agreement. Schmidt compared a number of D values calculated by the two methods. This comparison is given in Table 4.

Other methods have been suggested (Lewis, 1956) for calculating D values. These methods may be statistically superior to those just discussed, but they are extremely laborious and time-consuming. In view of present precision in thermal resistance determination, it is doubtful if the use of these statistical procedures of higher precision is warranted. This is certainly indicated by the close correlation between D values calculated by the method of Stumbo *et al.* (1950) and probability methods (Schmidt, 1954; Reynolds *et al.*, 1952; Esselen and Pflug, 1956).

As was indicated earlier, this "end-point" technique is preferred over the multiple-point technique because it tends to eliminate the description by TD curves of occurrences other than bacterial death—such as flocculation or deflocculation during heating and during dilution of heated samples. Moreover, it requires much less work to obtain heat resistance parameters.

TABLE 4

Comparison of D values

| | | | D_{250} calculated by: | | |
Organism	Substrate	TDT procedure	Stumbo *et al.* method	Schmidt method
P.A. 3679	Cream-style corn	Can	2.47	2.53
P.A. 3679	Whole-kernel corn (1)	Can	1.52	1.54
P.A. 3679	Whole-kernel corn (2)	Can	1.82	1.85
P.A. 3679	Phosphate buffer	Tube	1.31	1.32
F.S. 5010	Cream-style corn	Can	1.14	1.09
F.S. 5010	Whole-kernel corn	Can	1.35	1.38
F.S. 1518	Phosphate buffer	Tube	3.01	3.04
F.S. 617	Whole milk	Can	0.84	0.79
F.S. 617	Evaporated milk	Tube	1.05	1.05

Comparative Heat Resistance of Most Important Bacteria in Spoilage of Heat-Processed Foods

In considering bacteria important in spoilage of canned foods, the bacteria may be grouped in accordance with the nature of the foods in which they grow and cause spoilage. Though there are a number of food factors that may influence bacterial resistance to heat and growth of bacteria, the most important is perhaps acidity, generally expressed as pH. For this reason, in discussions of bacterial spoilage, foods are generally classified as low-acid, acid, and high-acid with pH ranges of above 4.5, 4.0 to 4.5, and below 4.0, respectively (see Chapter 2).

Table 5 lists the more important bacteria, or groups of bacteria, and their approximate range of heat resistance, in terms of D and z, in the different classes of canned foods.

Factors Influencing Thermal Resistance of Bacteria

Many factors influence the resistance of bacteria to heat. Many of the published results are difficult to interpret for various reasons: A number of different techniques have been and are being used to measure thermal resistance. Media for growing organisms for thermal resistance determinations are many and varied, as are the menstrua in which the bacteria are suspended during heating. A large number of different types of bacteria have been studied. More often than not, these have not been well described. For these reasons, only generalizations regarding factors reported to influence resistance can be made. At this time, it is impossible to say how any one factor quantitatively affects resistance.

TABLE 5

Comparative heat resistance of bacteria—canned foods

	Approximate range of heat resistance	
Bacterial groups	D	z
Low-acid and semi-acid foods (pH above 4.5)		
Thermophiles (spores)	D_{250}	
Flat-sour group (*B. stearothermophilus*)	4.0–5.0	14–22
Gaseous-spoilage group (*C. thermosaccharolyticum*)	3.0–4.0[a]	16–22
Sulfide stinkers (*C. nigrificans*)	2.0–3.0	16–22
Mesophiles (spores)		
Putrefactive anaerobes		
C. botulinum (types A and B)	0.10–0.20	14–18
C. sporogenes group (including P.A. 3679)	0.10–1.5	14–18
Acid foods (pH 4.0–4.5)		
Thermophiles (spores)		
B. coagulans (facultatively mesophilic)	0.01–0.07	14–18
Mesophiles (spores)	D_{212}	
B. polymyxa and *B. macerans*	0.10–0.50	12–16
Butyric anaerobes (*C. pastiurianum*)	0.10–0.50	12–16
High-acid foods (pH 4.00 and below)		
Mesophilic non-spore-bearing bacteria	D_{150}	
Lactobacillus spp., *Leuconostoc* spp.,		
and yeasts and molds	0.50–1.00	8–10

[a] Recently an organism of this group was described by Xezones *et al.* (1965) as having a D_{250} value in excess of 50 minutes. At present this must be considered as a very rare exception.

In considering bacteria most important in the pasteurization of foods, the bacteria are more conveniently grouped in accordance with their significance in consumer health and as causes of economically important food spoilage during refrigerator storage and distribution (see Chapter 4).

Table 6 accordingly lists the bacteria, or groups of bacteria, of greatest importance in the pasteurization of foods along with their approximate range of resistance (in terms of D and z).

For the purpose of discussion the classification of influencing factors given by Schmidt (1957) will be adopted, as follows:

1. "Inherent resistance.
2. "Environmental influences active during the growth and formation of cells or spores.
3. "Environmental influences active during the time of heating of the cells or spores."

TABLE 6

Comparative heat resistance of bacteria—pasteurized foods

Bacterial groups	Approximate range of heat resistance	
	D	z
Pathogenic and toxin-producing microorganisms	D_{150}	
Mycobacterium tuberculosis	0.20–0.30	8–10
Brucella spp.	0.10–0.20	8–10
Coxiella burnetti	0.50–0.60	8–10
Salmonella spp. (other than Senftenberg)	0.02–0.25	8–10
Salmonella senftenberg	0.80–1.00	8–12
Staphylococcus aureus	0.2–2.00	8–12
Streptococcus pyogenes	0.2–2.00	8–12
	D_{180}	
Clostridium botulinum Type E (spores)	0.10–3.00	9–16
Spoilage microorganisms	D_{150}	
Non-spore-bearing bacteria, yeasts, and molds	0.50–3.00	8–12

Inherent resistance varies not only with species but also with different strains of the same species. This has been observed by virtually all workers in the field. Different strains of the same species grown in the same medium and heated in the same menstruum may show widely different resistances.

Environmental factors active during growth of bacteria have been observed to exert widely variable influences on resistance of different species and strains of bacteria. The same has been observed for environmental factors active during the heating of bacterial suspensions.

Age.   From the published literature it is impossible to assess the influence of age on resistance. It is no doubt related to the environmental factors active during growth of the bacteria and during storage of cell suspensions. However, some workers, under the conditions of their experiments, could find no correlation between age and resistance of spores of some species (Williams, 1936; Gonsalves *et al.*, 1965; and others). Corran (1934) reported that aging up to 1 year increased resistance slightly. Vinton (1964) reported reduced resistance of P.A. 3679 spores stored for 20 years at 4.4°C. (40°F.).

It has been observed by some that heat resistance of certain non-sporulating bacteria increases as they progress in the log phase of growth and then decreases with continued incubation or with storage at refrigerator temperatures. Therefore, at this time, no general conclusion regarding influence of age on resistance can be drawn. However, in thermal resistance work, the possibility that resistance may change with age should not be disregarded.

Growth Temperature.   The temperature at which spores are produced has been shown by various workers to influence heat resistance (Williams, 1929; Curran, 1934; Theophilus and Hammer, 1938; Sugiyama, 1951; Williams and Robertson, 1954; El-Bisi and Ordal, 1956; and others). Generally speaking, it has been found that for species of thermophilic bacteria and for some species of mesophilic aerobic bacilli highest resistance is obtained for spores produced at the highest temperatures at which good vegetative growth occurs. Though exceptions have been reported, most resistant spores of mesophilic obligate anaerobes are usually produced at or near the lowest temperature at which good vegetative growth occurs. With respect to nonsporulating bacteria, it has generally been found that highest resistance is observed for cells produced at or near their optimum growth temperature.

Nature of Medium in Which Spores are Produced.   A number of studies suggest that resistance to heat in spores that occur naturally may be different from that shown by spores produced in laboratory media (Curran, 1935; Vinton *et al.*, 1947; Cameron *et al.*, 1936; and others). Many nutrient conditions have been found to increase or decrease resistance when compared to one nutrient condition taken as a standard. However, because of variability with respect to species and strains, no generally applicable statement with regard to the effect of any one nutrient condition can be made. Sugiyama (1951) found that, for *C. botulinum*, reduction in Fe^{++} and Ca^{++} concentrations in the sporulating medium below certain levels reduced heat resistance of the spores. Amaha and Ordal (1957) showed that reduction of Mn^{++} and Ca^{++} concentrations below certain levels in the sporulating medium reduced heat resistance of spores. They showed that Mn^{++} and Ca^{++} in the sporulating medium increased the resistance of *B. coagulans* spores. El-Bisi and Ordal (1956) found that a high phosphate level (1% KH_2PO_4) markedly decreased the resistance of *B. coagulans* spores. Many organic constituents of sporulating media influence spore resistance to heat. Williams (1929) observed wide variations in thermal resistance of spores of *B. subtilis* produced in media of different compositions. His studies show that different peptones or different brands of peptone may produce spores of different resistance. A casein digest medium was shown to produce spores of relatively high resistance.

Because the nature of sporulating media may influence resistance markedly, one should be cautious about accepting resistance values obtained for spores obtained in artificial media as representative of those found for spores occurring naturally in food products. For this reason, spores produced for thermal resistance values to be used in process determination should be grown in media that will yield spores of highest resistance.

Effect of Nature of Recovery Medium on Apparent Spore Resistance. For a discussion of this topic, see Chapter 6.

Effect of Nature of Medium in Which Bacteria Are Suspended When Heated. The chemical environment of the bacterial cell at the time it is subjected to heat has a marked effect on its resistance (see reviews by Stumbo, 1949; Schmidt, 1957; Hersom and Hulland, 1964). Many factors are important. Some of the major ones are:

1. pH of medium.
2. NaCl concentration.
3. Concentration of sugars and other carbohydrates.
4. Concentration of fats.
5. Agents used in curing meats, particularly sodium nitrite.
6. Water content.

It should be noted that variations in heat resistance that cannot be explained by variations in the above factors have been observed quite frequently by different investigators. Because so many of the studies relative to the factors named have been qualitative in nature, only general statements concerning the influence of these factors can be made. Increased acidity usually lowers the resistance of bacterial spores to heat. In many cases the effect is very pronounced. Too little is known concerning the influence of acidity on the resistance of vegetative forms to support a definitive generalization. As pointed out in Chapter 4, it has been shown that resistance of salmonellae in whole egg is greatest at a pH of about 4.5 and decreases as the pH is increased to about 8.00. However, it cannot be assumed that this phenomenon is common to all vegetative forms or even to salmonellae in other types of media. Low concentrations of NaCl (up to about 4%) tend to increase the resistance of many organisms, whereas higher concentrations tend to decrease resistance. The resistance of some spores may be as much as doubled by the addition of 2% NaCl to some food products. High sugar concentrations tend to increase resistance. High fat concentrations have in some instances been shown to increase resistance. The importance of this factor is minor for most foods. Curing agents, other than salt, in the concentrations used have only minor effects on thermal resistance of bacteria in meats. The agents nitrate and nitrite in water and culture media may have pronounced effects. Water content is of minor importance in most foods that have not been concentrated or dehydrated. Both spores and vegetative forms are generally more resistant in dehydrated foods. Bacteria in general are more resistant to heat when dry than when moist. Because most canned foods have relatively high water contents, variations are not expected to affect thermal resistance of bacteria to any great extent in them.

Miscellaneous Factors.   Kosker *et al.* (1951) found that allylisothio-cyanate in a concentration of 10 p.p.m. markedly reduced the thermal resistance of spores of *Aspergillus niger*. LaBaw and Desrosier (1953) found that some naturally occurring constituents of edible green plant extracts reduced the thermal resistance of spores of food spoilage organisms. The same workers (1954) reported that certain synthetic plant auxins decreased or increased the resistance of *B. coagulans* spores. Schmidt (1957) found the resistance of spores of two thermophilic flat-sour organisms, suspended in cream-style corn, to be unaffected by 200 p.p.m. of indole-3-acetic acid or β-naphthoxyacetic acid, two compounds found active by LaBaw and Desrosier (1954). El-Bisi *et al.* (1955) reported that four of five fungicides tested, when added to tomato juice, significantly reduced the resistance of *B. coagulans* spores. Frank (1955) observed significant differences in resistance of *B. coagulans* in different cationic environments. The cation solutions were prepared to simulate the cationic conditions of tomato juice and evaporated milk. No doubt many other factors, yet undescribed, influence thermal resistance of bacteria.

The marked effect of the chemical environment of bacteria, at the time heat is applied, on their thermal resistance makes it imperative that whenever possible in making thermal resistance determinations the bacteria should be suspended in the food for which a sterilization or pasteurization process is to be established.

SELECTED REFERENCES

Amaha, M., and Ordal, Z. J. (1957). Effect of divalent cations in the sporulation medium on the thermal death rate of *Bacillus coagulans* var. *thermoacidurans. J. Bacteriol.*, **74**, 596.

American Can Company. 1943. "The Canned Food Reference Manual," p. 248. American Can Co., New York.

Bigelow, W. D., and Esty, J. R. (1920). Thermal death point in relation to time of typical thermophilic organisms. *J. Infect. Diseases*, **27**, 602.

Cameron, E. J., Esty, J. R., and Williams, C. C. (1936). The cause of "black beets": an example of oligodynamic action as a contributory cause of spoilage. *Food Res.*, **1**, 73.

Curran, H. R. (1934). The influence of some environmental factors upon thermal resistance of bacterial spores. *J. Bacteriol.*, **27**, 26.

Curran, H. R. (1935). The influence of some environmental factors upon the thermal resistance of bacterial spores. *J. Infect. Diseases*, **56**, 196.

El-Bisi, H. M., and Ordal, Z. J. (1956). The effect of sporulation temperature on the thermal resistance of *Bacillus coagulans* var. *thermoacidurans. Food Res.*, **20**, 554.

El-Bisi, H. M., Ordal, Z. J., and Nelson, A. I. (1955). The effect of certain fungicides on the thermal death rate of spores of *Bacillus coagulans* var. *thermoacidurans. Food Res.*, **20**, 554.

Esselen, W. B., and Pflug, I. J. (1956). Thermal resistance of putrefactive anaerobe No. 3679 in vegetables in the temperature range of 250–290°F. *Food Technol.*, **10**, 557.

Frank, H. A. (1955). The influence of cationic environments on the thermal resistance of *Bacillus coagulans. Food Res.*, **20**, 315.

Gonsalves, A. A., Parliment, M. W., and Stumbo, C. R. (1965). Unpublished data.

Halvorson, H. O., and Zeigler, N. R. (1932). Application of statistics in bacteriology. *J. Bacteriol.*, **25**, 101.

Hersom, A. C., and Hulland, E. D. (1964). "Canned Foods—An Introduction to Their Microbiology," 5th ed. Chem. Publ., New York.

Kosker, O., Esselen, W. B., and Fellers, C. R. (1951). Effect of allylisothiocyanate and related substances on the thermal resistance of *Aspergillus niger, Saccharomyces ellipsoideus,* and *Bacillus thermoacidurans. Food Res.*, **16**, 510.

La Baw, G. D., and Derosier, N. W. (1953). Antibacterial activity of edible plant extracts. *Food Res.*, **18**, 186.

La Baw, G. D., and Derosier, N. W. (1954). The effect of synthetic plant auxins on the heat resistance of bacterial spores. *Food Res.*, **19**, 98.

Lewis, J. C. (1956). The estimation of decimal reduction times. *Appl. Microbiol.*, **4**, 211.

Levine, M., Buchanan, J. H., and Lease, G. (1927). Effect of concentration and temperature on germicidal efficiency of sodium hydroxide. *Iowa State Coll. J. Sci.*, **1**, 379.

Olson, F. C. W., and Shultz, O. T. (1942). Temperatures in solids during heating or cooling. *Ind. Eng. Chem.*, **34**, 874.

Pflug, I. J., and Esselen, W. B. (1953). Development and application of apparatus for study of thermal resistance of bacterial spores and thiamine at temperatures above 250°F. *Food Technol.*, **7**, 237.

Reed, L. J. (1936). Statistical treatment of biological effects of radiation. *In* "Biological Effects of Radiation" (B. M. Duggar, ed.), Vol. 1, 227. McGraw-Hill, New York.

Reynolds, H., Kaplan, A. M., Spencer, F. B., and Lichtenstein, H. (1952). Thermal destruction of Cameron's putrefactive anaerobe 3679 in food substrates. *Food Res.,* **17,** 153.

Schmidt, C. F. (1950). A method for determination of the thermal resistance of bacterial spores. *J. Bacteriol.,* **59,** 433.

Schmidt, C. F. (1954). Thermal resistance of microorganisms. *In* "Antiseptics, Disinfectants, Fungicides and Sterilization" (G. F. Reddish, ed.), 1st ed., pp. 720–759. Lea & Febiger, Philadelphia.

Schmidt, C. F. (1957). Thermal resistance of microorganisms. *In* "Antiseptics, Disinfectants, Fungicides and Sterilization" (G. F. Reddish, ed.), 2nd ed., pp. 831–884. Lea & Febiger, Philadelphia.

Schmidt, C. F., Bock, J. H., and Moberg, J. A. (1955). Thermal resistance determinations in steam using thermal death time retorts. *Food Res.,* **20,** 606.

Sognefest, P., and Benjamin, H. A. (1944). Heating lag in thermal death-time cans and tubes. *Food Res.,* **9,** 234.

Stern, J. A., Herlin, M. A., and Proctor, B. E. (1952). An electric method for continuous determination of rapid temperature changes in thermal death time studies. *Food Res.,* **17,** 460.

Stern, J. A., and Proctor, B. E. (1954). A micro-method and apparatus for the multiple determination of rates of destruction of bacteria and bacterial spores subjected to heat. *Food Technol.,* **8,** 139.

Stumbo, C. R. (1948). A technique for studying resistance of bacterial spores to temperatures in the higher range. *Food Technol.,* **2,** 228.

Stumbo, C. R. (1949). Thermobacteriology as applied to food processing. *Advan. Food Res.,* **2,** 47.

Stumbo, C. R. (1953). New procedures for evaluating thermal processes for foods in cylindrical containers. *Food Technol.,* **7,** 309.

Stumbo, C. R., Gross, C. E., and Vinton, C. (1945). Bacteriological studies relating to thermal processing of canned meats. *Food Res.,* **10,** 260.

Stumbo, C. R., Murphy, J. R., and Cochran, J. (1950). Nature of thermal death time curves for P.A.3679 and *Clostridium botulinum. Food Technol.,* **4,** 321.

Sugiyama, H. (1951). Studies of factors affecting the heat resistance of spores of *Clostridium botulinum. J. Bacteriol.,* **62,** 81.

Theophilus, O. R., and Hammer, B. W. (1938). Influence of growth temperature on the thermal resistance of some bacteria from evaporated milk. *Iowa State Coll. Agr. Expt. Sta. Res. Bull.,* 244.

Townsend, C. T., Esty, J. R., and Baselt, F. C. (1938). Heat-resistance studies on spores of putrefactive anaerobes in relation to determination of safe processes for canned foods. *Food Res.,* **3,** 323.

Williams, C. C., Merrill, C. M., and Cameron, E. J. (1937). Apparatus for determination of spore-destruction rates. *Food Res.,* **2,** 369.

Williams, F. T. (1936). Attempts to increase the heat resistance of bacterial spores. *J. Bacteriol.,* **32,** 589.

Williams, O. B. (1929). The heat resistance of bacterial spores. *J. Infect. Diseases,* **44,** 421.

Williams, O. B., and Robertson, W. J. (1954). Studies on heat resistance. VI. Effect of temperature of incubation at which formed on heat resistance of aerobic thermophilic spores. *J. Bacteriol.*, **67**, 377.

Vinton, C. (1964). Viability and heat resistance of anaerobic spores held 20 years at 40°F. *J. Food Sci.*, **29**, 337.

Vinton, C., Martin, S., and Gross, C. E. (1947). Bacteriological studies relating to thermal processing of canned meats. VII. Effect of substrate upon thermal resistance of spores. *Food Res.*, **12**, 173.

Xezones, H., Segmiller, J. L., and Hutchings, I. J. (1965). Processing requirements for a heat tolerant anaerobe. *Food Technol.*, **19**, 1001.

THERMAL PROCESS EVALUATION

CHAPTER 9

Important Terms and Equations

Symbols and Definitions

Unless otherwise indicated in the text, the following terms are used as defined:

a Initial population. Number of spores or vegetative cells of a given organism, per some given unit of volume before lethal heat is applied.

b Final population. Number of spores or vegetative cells per the same unit of volume after heating at a constant temperature for some designated time.

B Thermal process time, uncorrected for time required to bring the retort to processing temperature.

D Time required at any temperature to destroy 90% of the spores or vegetative cells of a given organism. Numerically, equal to the number of minutes required for the survivor curve to traverse one log cycle. Mathematically, equal to the reciprocal of the slope of the survivor curve.

D_r Time required at any designated reference temperature to destroy 90% of the spores or vegetative cells of a given organism.

F The equivalent, in minutes at some given reference temperature, of all heat considered, with respect to its capacity to destroy spores or vegetative cells of a particular organism.

F_c F value of all lethal heat received by the geometrical center of a container of food during process.

F_λ F value of all lethal heat received by any point in the container other than the geometrical center.

123

F_s Integrated lethal or degradative capacity of heat received by all points in a container during process. It is a measure of the capacity of a heat process to reduce the amount of any heat vulnerable factor that is destroyed exponentially with time of heating at a constant temperature.

F_i Time at any other temperature equivalent to 1 minute at some designated reference temperature.

f Time, in minutes, required for straight-line portion of semi log heating or cooling curve to traverse one log cycle.

f_h f of heating curve when it can be represented by one straight line. Also, f of first straight-line portion of a broken heating curve.

f_2 f of second straight-line portion of a broken heating curve.

f_c f of straight-line portion of semilog cooling curve.

g Difference, in degrees Fahrenheit, between retort temperature and the maximum temperature reached by the food at the point of concern.

g_c g when the point of concern is the geometrical center of the container.

g_λ g when the point of concern is any point in the container other than the geometrical center.

g_{bh} Difference between retort temperature and food temperature at the time of the break in a heating curve.

g_{h2} g value at the end of heating when the heating curve is broken.

I_h Difference between retort temperature and food temperature when heating is started.

I_c Difference between cooling water temperature and food temperature when cooling is started.

j A lag factor.

j_{ch} j of the heating curve for the geometrical center of the container. A factor that, when multiplied by I_h, locates the intersection of an extension of the straight-line portion of the semi–log heating curve and a vertical line representing beginning of heating.

j_{cc} j of the cooling curve for the geometrical center of the container. A factor that, when multiplied by I_c, locates the intersection of an extension of the straight-line portion of the cooling curve and a vertical line representing beginning of cooling.

$j_{\lambda h}$ j of the heating curve for iso-j region enclosing 0.19 of container volume.

$j_{\lambda c}$ j of the cooling curve for iso-j region enclosing 0.19 of container volume.

k Thermal diffusivity.

l Time, in minutes, required to bring retort to processing temperature.

L Lethal rate. Reciprocal of time, at any lethal temperature, equivalent to 1 minute at some designated reference temperature, or $1/F_i$.

n Total number of samples subjected to one time–temperature combination in a thermal resistance determination.

P_t Operator's process time. Time, in minutes, from the instant retort reaches processing temperature to the instant steam is turned off and cooling is started.

q Number of samples out of n samples that were sterilized by one time–temperature combination employed in a thermal resistance determination.

r Radius of container (inside dimension), except in broken heating equations, where it is a proportionality factor.

t Time.

T Temperature.

T_λ Temperature of any iso-j region.

T_{ih} Initial food temperature when heating is started.

T_{ic} Initial food temperature when cooling is started.

T_{pih} Pseudo-initial food temperature when heating is started. Temperature indicated by the intersection of the extension of the straight-line portion of the heating curve and a vertical line representing beginning of heating.

T_{pic} Pseudo-initial food temperature when cooling is started. Temperature indicated by the intersection of the extension of the straight-line portion of the cooling curve and a vertical line representing start of cooling.

T_r Retort or process temperature.

T_s Mass average can temperature at some given time during heating or cooling.

T_{ch} Temperature at the geometrical center of the can at some given time during heating.

T_{cc} Temperature at the geometrical center of the can at some given time during cooling.

T_x Reference temperature. The lethality equivalent of 1 minute at T_x is taken as 1 unit of lethality.

T_w Temperature of cooling water.

U The equivalent, in minutes at retort temperature, of all lethal heat received by some designated point in the container during process.

U_c U value of heat received by the geometrical center of a container.

U_λ U value of heat received by any point in the container other than the geometrical center.

U_{bh} U in f_h/U_{bh} corresponding to g_{bh}.

U_{h2} U in f_h/U_{h2} corresponding to g_{h2}.

v Fraction of can volume enclosed by an iso-j region.

$\bar{x}$ Most probable number of spores or vegetative cells of a given organism surviving some given time–temperature combination employed in a thermal resistance determination.

y_c Difference between temperature of the geometrical center of a container of food and temperature of the cooling water after some designated cooling time.

z Number of Fahrenheit degrees required for the thermal destruction curve to traverse one log cycle. Mathematically, equal to the reciprocal of the slope of the TD curve.

Equations

Thermal Resistance

$$\bar{x} = 2.303 \log \frac{n}{q}$$

$$t = D(\log a - \log b)$$

$$F_s = D_r(\log a - \log b)$$

$$\log D_2 - \log D_1 = \frac{1}{z}(T_1 - T_2)$$

$$L = \log^{-1} \frac{T - 250}{z}$$

Improved General Method

$$F = \frac{mA}{F_i d}$$

(See Chapter 11 for definitions of terms in this equation.)

Mathematical Methods

$$B = f_h(\log j_{ch} I_h - \log g_c)$$

$$B = f_h \log j_{ch} I_h + (f_2 - f_h) \log g_{bh} - f_2 \log g_{h2}$$

$$F_i = \log^{-1} \frac{250 - T_r}{z}$$

$$U_c = F_c F_i$$

$$U_\lambda = F_\lambda F_i$$

$$g_c = T_r - T_{ic}$$

$$g_\lambda = 0.5 g_c$$

$$j_{ch} = \frac{T_r - T_{pih}}{T_r - T_{ih}}$$

$$j_{cc} = \frac{T_w - T_{pic}}{T_w - T_{ic}}$$

$$j_{\lambda h} = 0.5 j_{ch}$$

$$j_{\lambda c} = 0.5 j_{cc}$$

$$I_h = T_r - T_{ih}$$

$$I_c = T_{ic} - T_w$$

$$F_c = U_c / F_i$$

$$F_c = \frac{f_h}{(f_h / U_c) F_i}$$

$$F_c = \frac{f_2}{(f_h / U_{h2}) F_i} - \frac{r(f_2 - f_h)}{(f_h / U_{bh}) F_i}$$

$$F_\lambda = U_\lambda / F_i$$

$$F_s = F_c + D_r \left(1.084 + \log \frac{F_\lambda - F_c}{D_r} \right)$$

$$F_s = F_c + D_r \log \frac{D_r + 10.93(F_\lambda - F_c)}{D_r}$$

$$F_s = D_r (\log a - \log b)$$

$$F_c \text{ desired} = F_{c_1} + \frac{F_{c_2} - F_{c_1}}{F_{s_2} - F_{s_1}} (F_s \text{ desired} - F_{s_1})$$

$$F_{sb} = F_{sa} + D_r \log \frac{V_b}{V_a}$$

$$P_t = B - 0.4l$$

$$k = \frac{0.398}{[(1/a^2) + (0.427/b^2)]f}$$

$$f'_h = \frac{[(1/a^2) + (0.427/b^2)]f_h}{(1/a'^2) + (0.427/b'^2)}$$

$$\frac{f_h}{f'_h} = \frac{ry}{r + y} \cdot \frac{r' + y'}{r'y'}$$

$$T_{sc} = T_w + 0.27(T_{cc} - T_w)$$

$$T_{sh} = T_r + 0.27(T_{ch} - T_r)$$

$$g_c = j_{ch}I_h \log^{-1}\left(-\frac{t}{f_h}\right)$$

$$y_c = j_{cc}I_c \log^{-1}\left(-\frac{t}{f_c}\right)$$

10

Basic Considerations

Probability of Survival

As was discussed in Chapter 7, bacteria die in an orderly fashion when subjected to moist heat. The order of death is believed to be logarithmic, or essentially so. A general equation of the survivor curve may be written as follows:

$$t = D(\log a - \log b)$$

in which t = time, in minutes, of heating at a constant lethal temperature.

D = time, in minutes, to kill 90% of the cells in the population.

a = initial number of viable cells in the population.

b = number of viable cells in the population after time t.

The population considered here is per some unit volume. Unit volume may be 1 ml., or it may be the volume of food in a container of a given size, or whatever volume—as long as volumes containing a and b are equal.

Whatever is considered as unit volume, in view of the order of bacterial death, there is no time–temperature combination that will sterilize an infinite number of such volumes. Therefore, in any finite number of such volumes, there is always a chance of survival in any one volume. For example, let us assume that there is one spore of C. botulinum in each of 10^{12} containers of a low-acid food. Let us assume further that these spores are characterized by a D_{250} of 0.21. A heat process is applied for which the sum of all lethal effects of all time-temperature combinations is equivalent to 2.52 minutes at 250°F. If all containers are given the same process, in how many should spores be expected to survive?

Since

$$t = D(\log a - \log b)$$
$$2.52 = 0.21(\log 10^{12} - \log b)$$
$$12 = 12 - \log b$$
$$\log b = 0$$
$$b = 1$$

This tells us that one container of the 10^{12} should not be sterilized by the process. It will be noted that here the unit of volume was taken as the volume of 10^{12} containers.

Suppose we had taken the unit of volume as the volume of one container.

Then

$$2.52 = 0.21(\log 1 - \log b)$$
$$\log b = -12$$
$$b = 10^{-12}$$

In this case b expresses probability of survival—that is, there is one chance in 10^{12} that any particular container of the 10^{12} containers should not be sterilized by the process. This is the same as saying that one of every 10^{12} containers should not be sterilized by the process.

The Unit of Lethality

In order to compare the relative sterilizing capacities of heat processes, it is essential to have a unit of lethality. For convenience, as will become apparent later, the unit chosen for food pasteurization and sterilization processes was 1 minute of heating at some given reference temperature—usually 250°F. In the case of severe (sterilization) processes—and total process lethality was designated by the symbol F. For example, if a process is assigned an F value of 3, it means that the sum of all lethal effects of all time-temperature combinations in the process is equivalent to the lethal effect of 3 minutes of heating at the designated reference temperature, on the assumption of instantaneous heating to this temperature and instantaneous cooling from this temperature.

Free of subscripts, F is a general term. It may refer to the sum of lethal effects at some one point in a container of food, or it may refer to the sum of lethal effects at an infinite number of points throughout a container of food. For example, if it refers to lethal effects at the geometrical center, it becomes F_c; if it refers to lethal effects at any point away from the geometrical center, it becomes F_λ; and if it refers to the sum of all lethal effects at the infinite number of points in the whole container, it becomes F_s. In any event, it is the sum of all lethal effects considered, expressed in minutes at some given reference temperature.

However, this is not enough. Because microorganisms vary in their relative resistance to different temperatures, to be meaningful, F must be identified so as to indicate this. As was brought out in Chapter 7, the term z accounts, in process evaluation procedures, for relative resistance to different temperatures. Therefore, the value of F determined for any process depends on the value of z employed in the calculation. The value of z employed is that characterizing the thermal destruction curve of the organism the process is designed to destroy. In other words, a process designed to destroy organisms characterized by a z of 18 would not cause the same percentage destruction of organisms characterized by a z of 12. Therefore, F correctly defined is the equivalent, in minutes at some given reference temperature, of all lethal heat in a process with respect to the destruction of an organism characterized by some given z value. Common designations are as follows: $F^{z=12}$, $F^{z=14}$, $F^{z=16}$ or simply F^{12}, F^{14}, F^{16}, etc. Because a z value of 18 is so commonly observed or assumed for spores, F values calculated with a z of 18 are often designated F_o or F_{so}.

The reference temperature is usually designated in subscript, e.g., F_{150}, F_{212}, F_{250}, etc.

Determination of Process Lethality Requirement

The criterion of process adequacy is the extent to which the bacterial population is reduced. There are two types of populations of concern. First is the population of organisms of public health significance. For canned foods having pH values above 4.5, $C.$ $botulinum$ is the organism of concern. It has been arbitrarily established that the minimum heat process should be such as to reduce any population of the most-resistant botulinum spores to 10^{-12} (the $12D$ concept). The most-resistant spores of botulinum so far studied are characterized by a D_{250} value of about 0.21. The minimum process requirement in terms of its equivalent in minutes at 250°F. may be computed as follows:

$$t = F_s = D_r(\log a - \log b)$$
$$F_s = 0.21(\log 1 - \log 10^{-12})$$
$$F_s = 0.21 \times 12 = 2.52 \text{ or } 2.5$$

Because no one has been able to get Types A and B (the most-resistant types) of botulinum to grow and produce toxin in products having pH values below about 4.6, the organism is considered to have no public health significance in the more-acid products. It is probably safe to assume that resistant botulinum spores contaminate low-acid foods at a rate, on the average, of no more than one spore per container. If this is the case, the above minimum process should reduce the chance of survival to one spore in one of one million million containers. This may seem at first to be such a ridiculously remote chance of survival that we might well be content with a much greater chance. However, when it is realized that in the United States alone many billions of cans of low-acid foods are produced annually, the $12D$ concept appears to be wholly justified.

Most low-acid foods are now processed beyond the minimum botulinum cook because of the occurrence of spoilage bacteria of greater heat resistance. Because these bacteria are not of public health significance, minimum processes to control them are arrived at through economic considerations. Most food processors frown on spoilage rates in commerce greater than one container per thousand. For mesophilic spore formers more resistant than botulinum, this can generally be readily accomplished by reducing the population of such organisms by 10^{-5}. Because the average maximum D_{250} value characterizing resistance of spores of these organisms is about 1.00, adequate heat processes to control them may be computed as follows:

$$F_s = 1.0(\log 1 - \log 10^{-5})$$
$$F_s = 5.00$$

Where thermophilic spoilage is a problem, more severe processes are often necessary. Generally, foods that show no more than about 1% spoilage on thermophilic incubation subsequent to heat processing will show less than 0.001% in United States commerce. Spores of the more-resistant thermophilic species have an average maximum D_r value of about 4.00. Since the extent of contamination with these organisms varies greatly with product, no generally applicable rule can be laid down for setting minimum processes. The average maximum contamination load must be determined to obtain an a value for the above equation. Suppose this were found to be one spore per gram. Then for a container of 20-ounce capacity,

$$F_s = 4.0(\log 570 - \log 0.01)$$
$$F_s = 19.0$$

This is of the order of many processes currently employed to control thermophilic spoilage. The F_c value of such a process would be about 14.0 to 16.0.

It should be pointed out that, except for the minimum botulinum cook, the above base processes are only estimates for average conditions. The exact processes for any given product must be tailor-made in view of the types of contaminating bacteria, the population of most-resistant types, the spoilage risk accepted, and the nature of the food product from the standpoint of its supporting growth of the different types of contaminating bacteria.

The Effective Mean F Value

Because of product and process variables, F values will vary somewhat. It is often the tendency to take the average of a number of experimentally determined F values as representative of a situation with respect to product sterilization. This is incorrect and, if relied on where variations are appreciable, may result in the use of inadequate processes. A few simple calculations will demonstrate the point. Suppose that two identical containers of product are given processes of appreciably different severity. The F_s value of the process given one container is 4.00, the F_s value of the process given the other is 2.00. The product in each container, prior to processing, contained 1000 spores characterized by $D_r = 1.00$. If we took the average F value as representative of the sterilizing capacity of both processes, how many spores would we predict survived in the two containers? Prior to processing there were 2000 spores in the two containers—that is, $a = 2000$. Then, since

$$F_s = D_r(\log a - \log b),$$
$$\text{Average } F_s = 3 = 1(\log 2000 - \log b)$$

and

$$b = \mathbf{2.00} \text{ spores in the two containers.}$$

This is incorrect. Let us determine how many actually should have survived. First, in the container given a process of $F_s = 4.00$: In this case,

$$4 = 1(\log 1000 - \log b)$$
$$b = 0.1 = 1 \text{ spore per 10 such containers.}$$

For the other container,

$$2 = 1(\log 1000 - \log b)$$
$$b = 10 = 10 \text{ spores per each of such containers.}$$

Then in the two containers, $b = 10.1$. What then is the mean effective F_s value?

$$a = 2000$$
$$b = 10.1$$
$$\bar{F}_s = 1(\log 2000 - \log 10.1)$$
$$\bar{F}_s = 2.297, \text{ a value considerably below the average.}$$

This simple example demonstrates the soundness of the normal procedure in practice of taking the lowest observed F value, rather than the average, as representative. If equal numbers of containers are receiving processes of different F values, the mean effective F value will always be nearer the low than the high value.

Heating Characteristics of Canned Foods

Still Cook.   There are no canned foods that heat purely by conduction or purely by convection. However, those heavy-consistency products that exhibit, except for initial lags, straight-line semilogarithmic heating curves are referred to as conduction-heating products. There is virtually no movement of these products in the container during heating or cooling. Likewise, light-consistency products that exhibit straight-line semilogarithmic heating curves are referred to as convection-heating products. During heating and cooling, these products are in continuous motion, owing to convection currents set up by temperature differences between product and heating medium.

Because of the lack of product movement in conduction-heating products, there is always, during heating or cooling, a temperature gradient from the geometrical center to the wall of the container. During heating the gradient is an ascending one from center to wall. During cooling it is a descending one from center to wall. For this reason the geometrical center is often referred to as the slowest heating and slowest cooling point.

Because of product movement in convection-heating products, temperatures throughout the product are reasonably uniform during heating and cooling. Owing to adhesive forces there is always, during heating, a thin layer of higher temperature next to the container wall. During cooling the reverse is true. Also, the slowest heating and cooling point is generally, though on the vertical axis, somewhat below the geometrical center. However, temperature at the geometrical center is thought to approximate closely the effective mean for the container.

Between these two extremes are products that exhibit broken heating curves. The more common of these are those that heat for some time by convection; then, because of starch or some other thickening agent "setting up," they thicken and continue to heat by conduction. Generally these products cool by conduction alone. Less common are products that for some time heat by conduction, then for the remainder of the period heat by convection. These are generally products with both liquid and solid phases, containing solid pieces which settle and pack in the lower two-thirds or so of the container. After a time, when the convection forces become strong enough, the pieces disperse and start moving with the liquid phase, causing convection heating from then on. These products generally cool by convection.

Agitated Cook.   Many products that heat by conduction in still cook will heat by "convection" during agitated cook. Actually this is not true convection, but rather speeded heating due to mechanical agitation causing product movement. Because heating and cooling curves exhibited closely resemble those for products that heat by convection, it is often referred to as mechanical convection. The important thing, from the standpoint of process evaluation, is that the semilogarithmic heating and cooling curves are generally straight lines with slope values comparable to those for curves exhibited by convection-heating products.

There are two general types of in-can agitation, axial and end-over-end. Perhaps the most important single factor in causing product movement is movement of the head-space "bubble" through the product as the cans turn over and over. For this reason close head-space control is vital to the success of the agitated cook process.

Turbulent Flow.   As in agitated cook, many products that heat by conduction during still cook will heat by "convection" when pumped through tubular heat exchangers, particularly if the exchangers are so constructed as to promote turbulence. Heating and cooling generally exhibit straight-line semilogarithmic curves with slope values comparable to those for curves exhibited by convection-heating products.

Method of Choice in Process Evaluation

For products exhibiting broken heating curves, processes may be most accurately evaluated by the general method, a graphical integration procedure. Though mathematical procedures have been proposed (Ball, 1928; Ball and Olson, 1957), they are complex and time-consuming. Data for one set of conditions usually cannot reliably be converted to apply to another set of conditions.

Though the general method is satisfactory for evaluating processes for convection-heating products, the mathematical methods to be presented here are more versatile, less time-consuming, and comparable in accuracy of results.

For products heating by conduction it is necessary, for accurate evaluation, to integrate lethal effects at an infinite number of points throughout the container. This can be done graphically, but it is a very tedious and time-consuming operation. There are now simple mathematical procedures for accomplishing the integration with a high degree of accuracy. These will be presented later.

Heat Penetration Measurement

Copper–constantan thermocouples are employed to measure the rate of heat penetration in canned foods. At the present time, thermocouple wire is of two calibrations—1921 and 1938, the latter being the more common. Potentiometers are calibrated accordingly; only wire that corresponds in calibration to that of the potentiometer should be used. Two types of thermocouple assemblies are commonly used. One consists of continuous lead wires with the thermocouple junction protruding from a ¼-inch molded plastic rod of appropriate length (usually 4 or 5 inches). Because of heat conduction through thermocouple wire, wire larger than 24 gage should not be used. A satisfactory continuous couple may be made by molding a ¼-inch plastic rod around duplex thermocouple wire in fiberglass insulation.

The continuous couple has the disadvantage that it cannot be placed in the can before closing because the projection of the thermocouple rod interferes with the closing. This disadvantage can be overcome by using an Ecklund-type receptacle adaptor. The regular gasketed receptacle is placed through a cutout in the can, and the receptacle is closed with a flush-fitting dummy nut. The can is then filled and closed. After being filled, the can is positioned with the receptacle up, the dummy nut is removed, and the thermocouple with the receptacle adaptor is inserted.

Nonprojecting, plug-in couples are available (Ecklund, 1949). These are convenient to use and do not interfere with the closing. The couples are provided with a head into which thermocouple leads may be plugged. The couples are introduced into the can through a hole made with a countersinking punch. They are held in place by a threaded receptacle entering the can from the outside and a lock nut on the inside. A gasket is inserted between the lock nut and the can wall to prevent leakage. With glass containers, they may be introduced through the lid. These couples are usually made with 30-gage wire. Wire up to 20 gage may be used for leads to the potentiometer. The lead wires usually enter the retort through a hole in the wall provided with a packing gland. The packing gland may be of a number of different designs, but it should be such that packing placed around the wires can be sufficiently compressed to prevent appreciable leakage.

Lead wires from the thermocouples are attached to a copper–constantan calibrated temperature-indicating or temperature-recording potentiometer. This may be either of two types: hand balancing or electronic automatic balancing. The latter is preferable because of accuracy of readings at frequent intervals, freedom from vibration effects, and saving of labor if it is the temperature-recording type.

Regardless of the type of potentiometer used, readings should be taken at frequent intervals. Only in this way can sufficient temperature data be obtained to construct reliable heat penetration curves.

Lead wires should, whenever possible, be continuous, without splicing, from thermocouple to potentiometer. If splicing is necessary, only copper-to-copper and constantan-to-constantan connections should be made. The use of unlike metals introduces supplementary junctions which may cause serious errors in temperature readings.

Plotting Heat Penetration Data

When the improved general method is employed in process evaluation, time–temperature data are plotted directly on lethal-rate paper to obtain a lethality curve, the area beneath which will be proportional to the lethality of the process.

When mathematical methods are employed, the time–temperature data are most conveniently plotted on semilogarithmic paper. To obtain the heating curve, the difference between retort temperature and food temperature is plotted on the log scale against time on the linear scale. This is conveniently accomplished by rotating the semilog paper through

180° and labeling the top line one degree below retort temperature, then plotting temperatures directly (see Fig. 19). To obtain the cooling curve, the difference between food temperature and cooling-water temperature is plotted on the log scale against time on the linear scale. In this case the semilog paper is kept in its normal position, the bottom line is labeled one degree above cooling water temperature, and the temperatures are plotted directly (see Fig. 20).

Heat Penetration and Processing Parameters

(See Figs. 19 and 20)

Initial Temperature (T_{ih}).   This is the temperature of the food at zero time of heating or cooling. It influences greatly the time of heating required to administer a process of any given lethality; the higher the initial temperature, the shorter is the time required.

Retort Temperature (T_r), *Come-up Time* (l), *Process Time* (B), *and Operator's Process Time* (P_t).   The retort temperature, T_r, is the controlled temperature of the heating medium, in the retort, employed in processing a product. The time required to reach retort temperature (l) after steam is turned on will vary with processing conditions. In process evaluation procedures, about 40% of this come-up time may be considered as time at retort temperature. (In continuous processing methods where containers go directly into a retort already up to temperature, $l = 0$.) Process time (B) used in process evaluation procedures is taken as the time elapsing from the instant the retort is up to temperature until the instant steam is turned off plus 0.4 of the time required to bring the retort to temperature. Since the process time (P_t) used by the retort operator to gage his process is the time elapsing from the instant the retort is up to temperature until the instant steam is turned off,

$$B = P_t + 0.4l$$

The Function I_h.   This function is defined as the difference between retort temperature and initial food temperature—that is,

$$I_h = T_r - T_{ih}$$

Pseudo-Initial Temperature—Heating (T_{pih}).   This is the temperature represented by the intersection of the extension of the straight-line portion of the semilog heating curve and a vertical line representing beginning of process (zero time). When there is no lag in heating—that is, when the semilog heating curve is straight from the beginning of process—$T_{pih} = T_{ih}$.

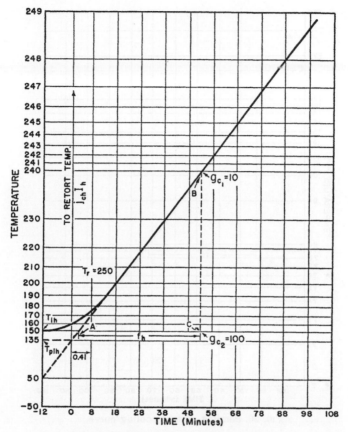

FIG. 19. Semilog plot of heating curve.

The Factor j_{ch}.   This is often referred to as the heating lag factor. It is a factor which, when multiplied by I_h, locates the intersection of the extension of the straight-line portion of the heat penetration curve and a vertical line representing beginning of process or zero time. Mathematically, it is defined as follows:

$$j_{ch} = \frac{T_r - T_{pih}}{T_r - T_{ih}}$$

The Function f_h.   This is taken as the time, in minutes, required for the straight-line portion of the heat penetration curve to traverse one log cycle (see Fig. 19). It represents the slope of the curve. Numerically, it is equal to the reciprocal of the slope.

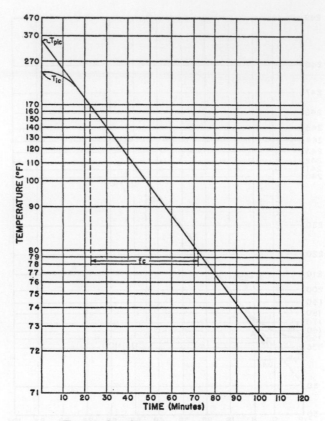

Fig. 20. Semilog plot of cooling curve.

The Function g.   This function is taken as the difference between retort temperature (T_r) and the maximum temperature reached by the food at the point of temperature measurement.

Initial Temperature (T_{ic}).   This is the temperature of the food, at the point of measurement, when cooling is started.

Cooling-Water Temperature (T_w).   This is the temperature of water used in cooling the product after the heating cycle.

The Function I_c *(Ball's m).*   This function has the same connotation with respect to cooling as I_h has with respect to heating—that is,

$$I_c = T_{ic} - T_w$$

Pseudo-Initial Temperature—Cooling (T_{pic}).   This is the temperature represented by the intersection of the extension of the straight-line portion of the semilog cooling curve and a vertical line representing the beginning of cooling (zero time). When there is no lag in cooling—that is, when the semilog cooling curve is straight from the beginning of cooling—$T_{pic} = T_{ic}$.

The Factor j_{cc}.   This is often referred to as the cooling lag factor. It is a factor which, when multiplied by I_c, locates the intersection of the extension of the straight-line portion of the cooling curve and the vertical line representing the beginning of cooling. Mathematically, it is defined as follows:

$$j_{cc} = \frac{T_w - T_{pic}}{T_w - T_{ic}}$$

The Function f_c.   This is taken as the time, in minutes, required for the straight-line portion of the cooling curve to traverse one log cycle. It represents the slope of the curve. Numerically, it is equal to the reciprocal of the slope.

SELECTED REFERENCES

Ball, C. O. (1923). Thermal process time for canned foods. *Bull. Natl. Res. Council,* **7,** Part 1, No. 37. 76 pp.

Ball, C. O. (1928). Mathematical solution of problems on thermal processing of canned foods. *Univ. Calif. Publ. Public Health,* **1,** 230 pp.

Ball, C. O., and Olson, F. C. W. (1957). "Sterilization in Food Technology," 654 pp. McGraw-Hill, New York.

Bigelow, W. D., Bohart, G. S., Richardson, A. C., and Ball, C. O. (1920). Heat penetration in processing canned foods. *Natl. Canners Assoc. Bull.,* 161.

Ecklund, O. F. (1949). Apparatus for the measurement of the rate of heat penetration in canned foods. *Food Technol.,* **3,** 231.

Fagerson, I. S., and Esselen, W. B. (1950). Heat transfer in commercial glass containers during thermal processing. *Food Technol.,* **4,** 411.

Fagerson, I. S., Esselen, W. B., and Licciardello, J. L. (1951). Heat transfer in commercial glass containers during thermal processing. II. F_0 distribution in foods heating by convection. *Food Technol.,* **5,** 261.

Gillespy, T. G. (1951). Estimation of sterilizing values of processes as applied to canned foods. I. Packs heating by conduction. *J. Sci. Food Agr.,* **2,** 107.

Hicks, E. W. (1951). On the evaluation of canning processes. *Food Technol.,* **5,** 134.

Hicks, E. W. (1952). Some implications of recent theoretical work on canning processes. *Food Technol.,* **6,** 175.

Jackson, J. M. (1940). Mechanisms of heat transfer in canned foods during thermal processing. *Proc. Food Conf. Inst. Food Technologists,* p. 39.

Jackson, J. M., and Olson, F. C. W. (1940). Thermal processing of canned foods in tin containers. IV. Studies on the mechanisms of heat transfer within the container. *Food Res.,* **5,** 409.

Stumbo, C. R. (1948). Bacteriological considerations relating to process evaluation. *Food Technol.,* **2,** 115.

Stumbo, C. R. (1949). Further considerations relating to evaluation of thermal processes for foods. *Food Technol.,* **3,** 126.

Stumbo, C. R. (1953). New procedures for evaluating thermal processes for foods in cylindrical containers. *Food Technol.,* **7,** 309.

Townsend, C. T., Somers, I. I., Lamb, F. C., and Olson, N. A. "A Laboratory Manual for the Canning Industry," 2nd ed. National Canners Association Research Laboratories, Washington, D. C.

CHAPTER 11

The General Method

In Chapter 6 we dealt with kinetics of death of bacteria subjected to moist heat and the mathematical description of these kinetics. The two parameters D and z are employed directly or indirectly in all methods of process evaluation to account for resistance of bacteria to heat. It is the purpose of this section to demonstrate how these parameters and those describing heating and cooling are employed to estimate the sterilizing capacity of heat processes.

The Original General Method

This method, described by Bigelow *et al.* (1920), is now for the most part, from the standpoint of application, of historical interest only. However, because the fundamental concepts on which it was based served as the foundation for the development of the more sophisticated procedures currently employed, a brief description seems appropriate.

The method is essentially a graphical procedure for integrating the lethal effects of various time–temperature relationships existent at some given point in a confined body of food during heat processing. Generally, the time–temperature relationships for which the lethal effects were integrated were those at the point of greatest temperature lag during heating and cooling of a product (usually a point at or near the geometrical center of a food container). From values obtained by thermocouple measurements, heating and cooling curves were constructed to represent temperatures existent during process (see Fig. 21). Each temperature represented by a point on the curves was considered to have sterilizing, or lethal, value.

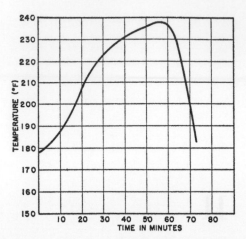

FIG. 21. Heating and cooling curves representing temperatures at geometrical center of container of product (puréed spinach in 8-ounce glass container) during process ($T_r = 240°F$).

Thermal resistance of bacteria was represented by thermal destruction (TD) curves obtained by plotting the time required to destroy a high percentage of spores in a given microbial population against temperature of heating (see Fig. 22). (At the time the method was developed, it was believed that the times represented were those required to destroy all

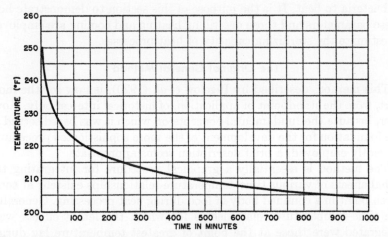

FIG. 22. Hypothetical thermal destruction curve typical in form of curves obtained for *C. sporogenes* and related organisms.

spores in all equal spore populations—in reality, the times were those required to destroy all spores in 90% or more, but not all, of such spore populations.)

Lethal Rate. From time–temperature relationships represented by a TD curve, it is possible to assign a lethal-rate value to each temperature represented by a point on the curves describing heating and cooling of a product during process. The lethal-rate value assigned to each temperature is numerically equal to the reciprocal of the number of minutes required to destroy some given percentage of spores at this temperature—the percentage of destruction being that represented by all points on the TD curve. Hence, the destruction time corresponding to any given temperature is taken from a TD curve for the organism for which a sterilizing process is being designed. For example, if the TD curve indicated that 10 minutes were required at 239°F., the lethal-rate value assigned to this temperature would be 0.1. It follows that lethality may be defined as the product of lethal rate and the time (in minutes) during which the corresponding temperature is operative. A process of unit lethality, therefore, is a process that is adequate to accomplish the same percentage of destruction, of an identical population, as represented by the TD curve.

Accordingly, it may be said that each point on the curves describing heating and cooling of a container of food during process represents a time, a temperature, and a lethal rate. In the original general method, times represented were plotted against corresponding lethal rates to obtain a lethality curve. Figure 23 shows such a curve on plain coordinate paper, lethal rate being represented in the direction of ordinates, and time in the direction of abscissae. (Because the product of lethal rate and time is equal to lethality, the area beneath the curve may be expressed directly in units of lethality.) To determine what process time must be employed to give unit lethality, the "cooling" portion of any given lethality curve was shifted to the right or left so as to give an area equal to 1. (An area equal to 1 in this case might be equal to 1, 2, or more square inches, depending on how the coordinates are labeled. For example, an area of 1 would be that area occupied by a lethal rate of 0.1 operating for a period of 10 minutes, or a lethal rate of 0.05 operating for a period of 20 minutes, etc.) When the curve was so adjusted, the time required to accomplish "sterilization" was taken as the time represented by the intersection of the "cooling curve" and the x-axis. This

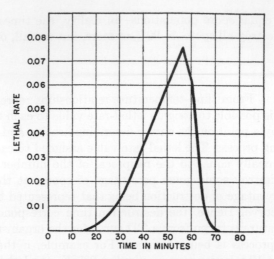

FIG. 23. Lethality curve based on values from curves in Figs. 21 and 22.

was a trial-and-error procedure, and for this reason the method is some-times referred to as the graphical trial-and-error method.

The Improved General Method

Contributions by Ball (1928) and by Schultz and Olson (1940) resulted in a much improved general method. Perhaps the chief contribution by Ball was the construction of a hypothetical TD curve passing through 1 minute at 250°F. (see Fig. 13). Lethal rates ascertained from such a curve when plotted against time, to obtain a lethality curve similar to that in Fig. 24, give a curve the area beneath which is proportional to the equivalent of the entire process in minutes at 250°F. The symbol F, hereafter given as F_c, had previously been introduced (Ball, 1923) to designate the equivalent, in minutes at 250°F., of the combined lethality of all time–temperature relationships at the point of slowest heating (about the geometrical center of a container), represented by heating and cooling curves for a product during process. It was con-sidered that this designation permitted a direct comparison of the relative sterilizing capacities of different heat processes. For example, a process equivalent to 4 minutes at 250°F. ($F_c = 4$) was considered to have twice the sterilizing capacity as one equivalent to 2 minutes at 250°F. ($F_c = 2$), and any two processes of equal F_c value would be equivalent in sterilizing capacity. It was later shown (Stumbo, 1948,

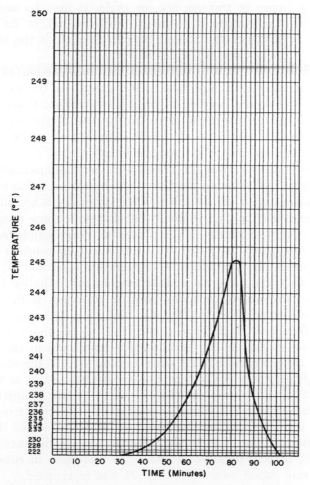

Fig. 24. Lethality curve on specially ruled lethal-rate paper.

1949, 1953) that this is not the case for conduction-heating products. For convection-heating products it is virtually true.

A second improvement contributed by Ball (1928) was the following equation for calculating lethal rate:

$$L = \log^{-1} \frac{T - 250}{z} \qquad (1)$$

in which T = any lethal temperature.

z = the number of Fahrenheit degrees required for the TD

curve to traverse one log cycle. It may be said that z
characterizes an organism with respect to its relative
resistance to different temperatures. It is the reciprocal
of the slope of the TD curve.

Schultz and Olson (1949) described the construction of special
coordinate or lethal-rate paper that added greatly to the convenience
of applying the general method of process evaluation. Resende *et al.*
(1969) presented an improved procedure for use of the paper. Both
improvements are accounted for in the description to follow. The
paper is constructed in the following manner.

First, two horizontal lines are drawn unit distance apart on a plain
sheet of paper. Unit distance may have any numerical value in inches,
but it is convenient to use 10 inches. The bottom line is considered the
zero line, and the top line is finally labeled with some temperature
above that likely to obtain in the processing range employed. However, in constructing the paper it is convenient to consider the top
line as representing the reference temperature, i.e., 250°F., 212°F.,
150°F., or whatever. In so doing it represents unit lethality (1.00) as
well as the reference temperature. Also, if 10 inches is the distance
between the bottom line and the top line, unit lethality (1.00) is
equal to 10 inches. Next, by equation 1, L is calculated (or found
in tables—see Appendix) for each lethal temperature in the processing range. Now, by multiplying each L value by 10, the distance of
horizontal lines, above the bottom line are determined so as to locate
a horizontal line to represent each L value or the temperature to
which it corresponds. For example, consider the reference temperature
as 250°F., the z value characterizing relative resistance of the organism of concern to be 18 Fahrenheit degrees and the distance between
the top and bottom lines as 10 inches, as was done in constructing
the lethal rate paper from which Fig. 24 was reproduced.

From tables (Appendix) or by equation 1, we find L values for
249°F., 248°F., and 247°F. to be 0.880, 0.774, and 0.681, respectively.
Each of these is now multiplied by 10 to obtain the equivalents in
inches, namely, 8.80, 7.74, and 6.81, respectively. Then a horizontal
line is drawn 8.80 inches above the bottom line to represent 249°F.,
or better, the lethal rate corresponding to this temperature. Another
line is drawn 7.74 inches above the bottom line to represent the lethal
rate corresponding to 248°F. Still another line is drawn 6.81 inches
above the bottom line to represent the lethal rate corresponding to
247°F. This procedure is continued for still lower temperatures until

the distance between the bottom line and the last line drawn is very small, comparatively (Fig. 24). Next, vertical lines are drawn to represent time units, as shown in Fig. 24. Obviously, a specific lethal rate paper must be constructed for each z value considered.

Choice of Units for the Vertical Scale.   Once a lethal rate paper has been constructed for a given z value, the top line may be labeled with a temperature of choice and the other lines labeled accordingly. The temperature chosen for labeling the top line should be such that, when the lethal rate curve is drawn, a convenient area for measurement results. Obviously, it has to be above the maximum temperature reached at the point of temperature measurement in the food during processing.

Use of Lethal-Rate Paper.   Points are placed on the paper corresponding to the time–temperature points of the heat-penetration data. A smooth curve is then drawn through these points (see Fig. 24), and the area beneath the curve is determined. This area is conveniently measured by a planimeter. If a planimeter is not available, the area may be cut from the paper and weighed on an analytical balance. A known area is then cut from the same paper, weighed, and the weight per square inch computed. Dividing the weight of the area under the curve by the weight per square inch gives the area under the curve in square inches. This method is quite accurate if the paper is uniform in density. If neither a planimeter nor a balance is available, the area may be estimated by counting squares.

After the area under the lethality curve has been obtained, the F value of the process may be computed from the following equation:

$$F = \frac{mA}{F_i d} \tag{2}$$

in which m = the number of inches represented by 1 inch on the time scale.

A = the area under the lethality curve, in square inches.

d = the number of inches from the bottom line to the top line.

F_i = the F_i corresponding to the temperature used to label the top line of the lethal rate paper. The value of F_i is calculated by using the following equation.

$$F_i = \log^{-1} \frac{T_x - T}{z}$$

in which,

  T_x = reference temperature.

  T = temperature used to label the top line of the lethal rate paper.

  z = as defined above.

  F = the equivalent, in minutes at the designated reference temperature, of all heat considered, with respect to its capacity to destroy spores or vegetative cells of some given organism the relative heat resistance of which is characterized by a specific z value.

It should be reiterated that the general methods are of value only for evaluating time–temperature effects at the point of temperature measurement. For evaluating lethal effects throughout containers of conduction-heating products, the mathematical procedure presented in Chapter 11 is the method of choice.

Temperature Conversion for Different Retort and Initial Temperatures

The second major contribution of Schultz and Olson (1940) was the development of equations for converting food (can) temperature, when the retort temperature is changed and the initial food temperature remains the same, or when the initial temperature is changed and the retort temperature remains the same. Suppose that an experimental heat penetration curve has been determined for a product processed at a particular retort temperature, and it is desired to find corresponding can temperatures that should be obtained if a different retort temperature were used—the initial food temperature being the same. These may be calculated from the following equation:

$$T'_c = T'_r - \frac{T'_r - T_i}{T_r - T_i}(T_r - T_c) \qquad (3)$$

in which T_r = original retort temperature.

  T'_r = new retort temperature.

  T_i = initial temperature.

  T_c = a can temperature of the original set.

  T'_c = a new can temperature corresponding to T'_r.

As was pointed out by the authors, if T'_c is plotted against T_c, the graph will be a straight line, since T'_c is a linear function of T_c. Therefore, it is necessary to use the formula for determining only two temperatures and join the points with a straight line from which other temperatures may be taken.

If the retort temperature remains the same and the initial temperature is changed, the following equation is used:

$$T'_c = T_r - \frac{T_r - T'_i}{T_r - T_i} (T_r - T_c)$$

These equations are valid for either conduction–heating or convection-heating products.

SELECTED REFERENCES

Ball, C. O. (1923). Thermal process time for canned foods. *Bull. Natl. Res. Council*, **7**, Part 1, No. 37, 76 pp.

Ball, C. O. (1928). Mathematical solution of problems on thermal processing of canned food. *Univ. Calif. Publ. Public Health*, **1**, 230 pp.

Bigelow, W. D., Bohart, G. S., Richardson, A. C., and Ball, C. O. (1920). Heat penetration in processing canned foods. *Natl. Canners Assoc. Bull.*, No. 16L.

Resende, R., Stumbo, C. R., and Francis, F. J. (1969). Calculation of thermal processes for vegetable puree in capillary tubes at temperatures up to 350°F. *Food Technol.*, **23**, 51.

Schultz, O. T., and Olson, F. C. W. (1940). Thermal processing of foods in tin containers. III. Recent improvements in the general method of thermal process calculations—a special coordinate paper and methods of converting initial and retort temperatures. *Food Res.*, **5**, 399.

Stumbo, C. R. (1948). Bacteriological considerations relating to process evaluation. *Food Technol.*, **2**, 115.

Stumbo, C. R. (1949). Further considerations relating to evaluation of thermal processes for foods. *Food Technol.*, **3**, 126.

Stumbo, C. R. (1953). New procedures for evaluating thermal processes for foods in cylindrical containers. *Food Technol.*, **7**, 309.

12

Mathematical Methods

Lethality of Heat at a Single Point

Products for Which the Semilog Heating Curve is One Straight Line, after the Initial Lag. Ball (1923) developed an equation of the straight-line semilog heating curve. Consistent with symbols and definitions given in Chapter 8, the equation may be written as follows

$$B = f_h \log \frac{j_{ch} I_h}{g_c} \tag{1}$$

in which B = process time, in minutes, when no time is required to bring the retort to processing temperature.

f_h = time, in minutes, required for the straight-line heating curve to traverse one log cycle.

j_{ch} = heating lag factor—a factor that, when multiplied by I_h, locates the intersection of the extension of the straight-line portion of the heating curve and a vertical line representing beginning of heating (zero time). The intersection is at T_{pih}. Mathematically, j_{ch} may be expressed as follows:

$$j_{ch} = \frac{T_r - T_{pih}}{T_r - T_{ih}}$$

I_h = difference, in degrees Fahrenheit, between retort temperature and initial food temperature. That is,

$$I_h = T_r - T_{ih}$$

Then

$$j_{ch}I_h = T_r - T_{pih}$$

g_c = difference, in degrees Fahrenheit, between retort temperature and maximum temperature reached by the food at the geometrical center of the container.

That the equation of the heating curve may be so written becomes obvious on study of Fig. 19. It will be noted that $1/f_h$ is the slope of the curve, process time B is the x-coordinate, and $\log j_{ch}I_h - \log g_c$ is the y-coordinate.

From triangle ABC (Fig. 19),

$$\text{Slope} = \frac{\log g_{c_2} - \log g_{c_1}}{f_h} = \frac{1}{f_h}$$

Then the general equation may be written

$$\log j_{ch}I_h - \log g_c = \frac{1}{f_h} B$$

or

$$B = f_h(\log j_{ch}I_h - \log g_c)$$

What is not obvious from inspection of equation 1 is how lethal heat conferred during cooling is accounted for by the equation of the heating curve. This is done through the relationship of g to the ratio f_h/U, U being the time required at retort temperature to accomplish the same amount of bacterial destruction as would be accomplished by a heat process of some given F value. This is best understood by constructing a hypothetical TD curve passing through 1 minute at 250°F., as was done by Ball (1928) (see Fig. 13). It is obvious from this curve that time at any other temperature is equivalent to 1 minute at 250°F. This time at any other temperature is termed F_i. Then, if U is as defined above,

$$U = FF_i$$

When the reference temperature (T_x) is taken as 250°F., mathematically F_i may be expressed by the equation of the TD curve that passes through 1 minute at 250°F. That is,

$$F_i = \log^{-1} \frac{250 - T}{z}$$

Tables of F_i for the more common processing temperature ranges are given in the Appendix. Values of F_i not given should be computed; interpolation does not yield values of sufficient accuracy.

Let us now return to $f_h/U:g$ relationships. Ball (1923) discovered that, for a single value of z and a single value of $I_c + g$, any given ratio of f_h/U had a value of g corresponding to it. It should be noted that, whereas variations in the value of z very significantly affects the $F_h/U:g$ relationship, normal variations in $I_c + g$ have only minor, really negligible, effects on the relationship. In computing values of $f_h/U:g$, Ball (1928) considered the j value of the cooling curve to be 1.41. It was later found that, though the relationship was virtually independent of process parameters such as retort temperature, container dimensions, initial food temperature, cooling water temperature, and the j value of the product heating curve, it was greatly dependent upon the j value of the product cooling curve. In recognition of this, Stumbo and Longley (1965) published limited tables (z of 12 to 22) of $f_h/U:g$ that accounted for variations in the j value of cooling curves. Relationships for these tables were arrived at manually employing the general method of integration. These same tables appeared in the first edition of this book and values from them were used to solve example problems.

Jen *et al.* (1971) presented representative tables of $f_h/U:g$, the values for which were obtained with computer aid. In obtaining values for these tables, hypothetical heating and cooling curves were generated by the computerized finite difference method of Teixeira *et al.* (1969). Values from these were then transferred into a computer program for integrating by the general method. Purohit and Stumbo (1971) refined the method of Jen *et al.* (1971). The resulting tables of $f_h/U:g$ relationships appear in the Appendix. It may be noted that these account for the magnitude of the j of the cooling curve. It should also be noted that, in computing values for these tables, the influence of temperature gradient at the end of heating (beginning of cooling) has been accounted for. Corresponding values of g and f_h/U not appearing in the tables may be obtained by interpolation between, or among, values given therein—it has been found that errors in values so obtained are negligible.

Because the general method of integration was used to obtain U values (including lethal value of heat during cooling) in the $f_h/U:g$ relationships, it follows that for any given value of g, the value of U in the $f_h/U:g$ ratio accounts for all lethal heat during both heating and cooling. Then g as used in the equation of the heating curve also accounts for all lethal heat of both heating and cooling. This can best be exemplified by a process calculation.

Suppose, for processing a particular product, the following data are given:

$$T_x = 250°F. \qquad j_{cc} = 1.40$$
$$T_r = 250°F. \qquad f_h = f_c = 25 \text{ minutes}$$
$$T_{ih} = 170°F. \qquad B = 40 \text{ minutes}$$
$$j_{ch} = 2.00 \qquad z = 14F.°$$

What is the F_c value of the process?

$$I_h = T_r - T_{ih} = 250 - 170 = 80$$
$$j_{ch}I_h = 2.00 \times 80 = \mathbf{160}$$

Because

$$B = f_h(\log j_{ch}I_h - \log g_c)$$
$$40 = 25(\log 160 - \log g_c)$$
$$g_c = \mathbf{4.02}$$

Now find in $f_h/U:g$ tables, for $z = 14$, $j_{cc} = 1.40$, and $g_c = 4.02$,

$$f_h/U_c = \mathbf{4.35}$$

Then

$$U_c = 25/4.35 = \mathbf{5.75}$$

and

$$U_c = F_cF_i$$

Because F_i at 250°F. = 1.00,

$$F_c = 5.75/1.00 = \mathbf{5.75}$$

Now suppose that, instead of process time B, we had been given $F_c = 5.75$. What would be the value of B if all other data were the same? This is just the reverse of the above calculation:

$$U_c = F_cF_i$$

Because $F_i = 1.00$,

$$U_c = \mathbf{5.75}$$

and

$$f_h/U_c = 25/5.75 = \mathbf{4.35}$$

Now find in the $f_h/U:g$ tables, for $z = 14$, $j_{cc} = 1.40$, and $f_h/U_c = 4.35$

$$g_c = \mathbf{4.02}$$

Then, since

$$B = f_h(\log j_{ch}I_h - \log g_c)$$
$$B = 25(\log 160 - \log 4.02)$$
$$B = \textbf{40 minutes}$$

It will be noted that variations in $I_c + g$ (Ball's $m + g$) are not accounted for in the $f_h/U\!:\!g$ tables. Before publication of the tables a series of calculations was made to determine the influence of variations in $I_c + g$ on calculated sterilizing values of processes. It was found that, on the average, a difference of 10 degrees Fahrenheit in $I_c + g$ causes about a 1% difference in the calculated F value. It was therefore decided to base the tables on an $I_c + g$ value of 180°F. If one wishes to correct for variations in this function, he may add 1% to F values for every 10 degrees the $I_c + g$ value is below 180°F. and subtract 1% for every 10 degrees the $I_c + g$ value is above 180°F.

It was also determined, before the $f_h/U\!:\!g$ tables were established, that variations in the j of the heating curve (j_{ch}) do not cause significant variations in f_h/U relationships. This is in contrast to the great influence that variations in the j of the cooling curve (j_{cc}) have on these relationships.

The tables were established on the assumption that f_h would always equal f_c for products that exhibit only one straight-line semilog heating curve and one straight-line semilog cooling curve, and that f_2 would always equal f_c for products that exhibit a broken-line heating curve.

It should be emphasized that equation 1 is valid only for calculating processes for products that heat wholly by conduction or wholly by convection—that is, products that exhibit only simple straight-line semilogarithmic heating and cooling curves. Processes for foods that do not exhibit these simple heating and cooling curves will be considered later.

CORRECTION FOR TIME REQUIRED TO BRING RETORT TO PROCESSING TEMPERATURE. The initial temperature (T_{ih}) is the temperature of the food at the time steam begins to enter the retort. The pseudo-initial temperature (T_{pih}) is indicated by the intersection of the extension of the straight-line portion of the heating curve and a vertical line representing zero time (see Fig. 19). When time is required to bring the retort to processing temperature, T_{pih} is determined as follows: (1) A vertical line is drawn on the heat penetration graph so that it passes through a point 0.40 of the distance from a vertical line representing the time at which retort temperature was reached, to another vertical line representing the time when steam started to enter the retort. This line represents the beginning of process—that is, zero time. (2) The straight-line portion of the heating curve is then extended to intersect the zero-time line. (3) The temperature indicated at this point of intersection is taken as T_{pih} (see Fig. 19).

Considering the vertical line representing zero time, located in this manner, as the beginning of process amounts to including 40% of the time required to bring the retort to processing temperature as processing time at retort temperature. As expressed by Ball (1923): "The time taken to bring a retort to processing temperature after steam has been turned on is time during which heat is entering the can, and therefore this period must have some time value as a part of the process. This value may be expressed in per cent of the actual length of time consumed. That is, the period of increasing the temperature of the retort, will shorten the length of time necessary to process a can of food after the retort has reached processing temperature, by a certain percentage of the period." Ball experimentally determined that 42% of the "come-up" time should be considered as process time at retort temperature. This was an average value from a number of runs. In practice, values usually range from about 30 to about 50%, indicating that 40% is a reasonable value to employ routinely.

In practice, the operator usually times his process from the time the retort reaches process temperature to the time steam is turned off. In accordance with the above, therefore, the time (B) used in formulas for calculating the sterilizing capacity of the operator's process would be operator's process time (P_t) plus 0.40 of the minimum time (l) it takes for the retort to reach process temperature after steam is turned on. That is,

$$B = P_t + 0.4l$$

Conversely, when B is calculated, the process time (P_t) given the operator should be

$$P_t = B - 0.4l$$

Integrated Lethality of Heat at All Points in the Container

Convection-Heating Products. Owing to product movement in the container, all bacteria in these products are subjected to essentially the same amount of lethal heat. In other words, product at all points in the container receives heat processes of essentially the same F value. It is considered that heating at the geometrical center of the container approximates closely the effective mean of heating throughout the container. The integral sum, expressed as its equivalent in minutes at 250°F., of lethal effects at all points in the container is designated F_s. Then, for convection-heating products, it follows that

$$F_c = F_s = D_r(\log a - \log b)$$

in which D_r = time at reference temperature required to destroy 90% of the cell (spore) population of the organism of concern.

   a = initial number of spores or vegetative cells of the organism of concern in the entire container.

   b = Number of spores or vegetative cells of the organism in the entire container after the container has received a process of designated F value.

By way of example, suppose we are given the following data for a convection-heating product.

Product contains, per gram, ten spores characterized by $z = 18$ and $D_r = 2.00$.

Container capacity = 570 gm.

$$T_x = 250°F.$$
$$T_r = 240°F.$$
$$T_{ih} = 160°F.$$
$$j_{ch} = j_{cc} = 1.00$$
$$l = 15 \text{ minutes}$$
$$f_h = f_c = 12$$

What process time (P_t) would the operator use to reduce the rate of survival to one spore per 1000 containers?

First, find F_c required:

$$a = 570 \times 10 = \mathbf{5700}$$
$$b = \mathbf{0.001}$$
$$F_c = F_s = D_r(\log a - \log b)$$
$$F_c = 2.00(\log 5700 - \log 0.001)$$
$$F_c = \mathbf{13.5}$$

Next, determine U_c desired:

$$U_c = F_c F_i$$

$$F_i = \log^{-1} \frac{250 - T_r}{z}$$

$$F_i = \log^{-1} \frac{250 - 240}{18}$$

$$F_i = \log^{-1} 0.555$$

$$F_i = \mathbf{3.59}$$

Then

$$U_c = 13.5 \times 3.59 = \mathbf{48.5}$$

and

$$f_h/U_c = 12/48.5 = \mathbf{0.248}$$

Now, from the $f_h/U:g$ tables, for $z = 18$, $j_{cc} = 1.00$, and $f_h/U_c = 0.248$,

$$g_c = 1.18 \times 10^{-3}$$

Now, determine process time B:

$$B = f_h(\log j_{ch} I_h - \log g_c)$$
$$I_h = T_r - T_{ih} = 240 - 160 = 80$$
$$j_{ch} I_h = 1.0 \times 80 = 80$$

Then

$$B - 12[\log 80 - \log(1.18 \times 10^{-3})]$$
$$B = 12(1.903 + 2.928)$$
$$B = \mathbf{58}$$

and, because

$$P_t = B - 0.4l$$
$$P_t = 58 - (0.4 \times 15)$$
$$P_t = \mathbf{52 \text{ minutes}} = \mathbf{operator's \ process \ time}$$

For processes, as the one just calculated, where the f_h/U value is below 0.5 and the g value is below 0.1, satisfactory approximations are sometimes obtained with the following equation:

$$B = U_c + f_h(\log j_{ch} I_h - 0.85)$$

Substituting values from the problem just solved, we obtain

$$B = 48.5 + 12(1.903 - 0.85)$$
$$B = \textbf{61.1}$$
$$P_t = 61.1 - 6.0 = \textbf{55.1}$$

The approximation is generally within a few percent. However, use of the approximation equation is recommended only for obtaining quick estimates.

Products That Exhibit Broken Heating Curves.   Heating and cooling of these products are often far from uniform throughout the container. There is no method available for integrating lethal effects at all points in containers of such products. Until a method is available, it is suggested that these products be considered as heating and cooling throughout as they heat and cool at the geometrical center of the container. In other words, again consider $F_c = F_s$. This will result in some over-processing, but until better methods are available it is the safest procedure. In other words, for want of a more precise method of evaluation, all such products should be treated as if they heat by convection.

Many products exhibit heating curves that may be represented by two straight lines, after the initial lag. Ball (1923) developed equations for use in evaluating processes for products for which the heating curve shows more than one break. For evaluating the latter, it is recommended that the improved general method be employed, because of uncertainties attendant with use of the mathematical method. Even for evaluating processes for products for which the heating curve shows only one break, the mathematical procedure should be used with a great deal of caution. It should be shown that the break in the curve consistently occurs at about the same place. When the heating curve is broken, attempts should not be made to convert heat penetration data for one set of conditions to another—that is, from one retort temperature, one initial temperature, or one can size to another. More often than not, the break in the curve will not occur at the same place.

When the semilog heating curve is a simple straight line, the F value for heating at a single point is computed by the following equation:

$$F = U/F_i$$

or

$$F = \frac{f_h}{(f_h/U)F_i} \tag{2}$$

When the heating curve is broken,

$$F = \frac{f_2}{(f_h/U_{h2})F_i} - \frac{r(f_2 - f_h)}{(f_h/U_{bh})F_i} \tag{3}$$

in which (consult Fig. 25) f_2 = time, in minutes, for the second straight-line portion of the heating curve to traverse one log cycle.

f_h/U_{h2} = f_h/U value corresponding to the value of g at the end of heating. To obtain this value from the f_h/U:g tables, the observed j_{cc} should be used.

f_h = time, in minutes, for the first straight-line portion of the heating curve to traverse one log cycle.

f_h/U_{bh} = f_h/U value corresponding to the value of g at the time the break in the curve occurs. To obtain this value from the f_h/U:g tables, the observed j_{cc} should also be used, because the change in the product has occurred, and, if it were cooled at this point, cooling would proceed in a manner similar to that at the end of process.

r = a proportionality factor.

F_i = as defined elsewhere.

When the semilog heating curve is a simple straight line, process time, B, is computed by the following equation:

$$B = f_h(\log j_{ch}I_h - \log g_c)$$

When the heating curve is broken,

$$B = f_h \log j_{ch}I_h + (f_2 - f_h) \log g_{bh} - f_2 \log g_{h2} \tag{4}$$

in which (consult Fig. 25) f_h = as defined above.

j_{ch} = lag factor of heating at point of temperature measurement.

$I_h = T_r - T_{ih}$.

f_2 = as defined above.

g_{bh} = g at the time the break occurs in the curve.

g_{h2} = g at the end of heating.

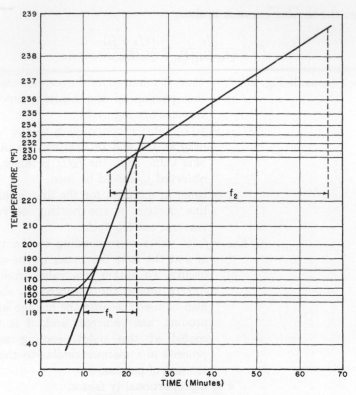

FIG. 25. A broken heating curve.

Application of these equations will now be demonstrated. Heat penetration data were obtained for a product in 404×700 cans processed at 240°F. They were plotted to obtain the curves in Figs. 25 and 26, from which it was determined that

$$
\begin{aligned}
T_r &= 240 & f_h &= 12.2 \\
T_{ih} &= 140 & f_2 &= f_c = 50 \\
j_{ch} &= 1.21 & g_{bh} &= 9.2 \\
j_{cc} &= 2.00 & I_h &= 100
\end{aligned}
$$

With a z value of 18, what would be the process time, B, for a process of $F_c = 10$?

$$
F_c = \frac{f_2}{(f_h/U_{h2})F_i} - \frac{r(f_2 - f_h)}{(f_h/U_{bh})F_i}
$$

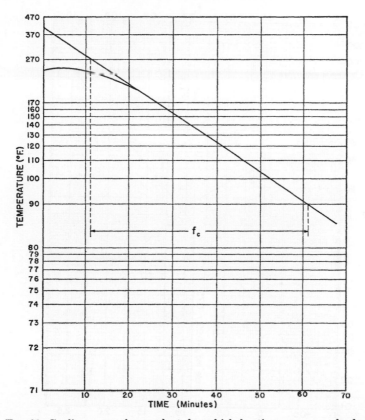

Fig. 26. Cooling curve for product for which heating curve was broken.

Then,

$$\frac{f_h}{U_{h2}} = \frac{f_2}{F_c F_i + \dfrac{r(f_2 - f_h)}{f_h/U_{bh}}}$$

For a z of 18, find $F_i = \textbf{3.594}$. Find in the $f_h/U{:}g$ tables, for $z = 18$, $j_{cc} = 2.00$, $g_{bh} = 9.2$,

$$f_h/U_{bh} = \textbf{7.17}$$

Find from Fig. 27, for $\log g_{bh} = 0.964$,

$$r = \textbf{0.74}$$

$$\frac{f_h}{U_{h2}} = \frac{50}{10 \times 3.594 + \dfrac{0.74(50 - 12.2)}{7.17}} = \textbf{1.25}$$

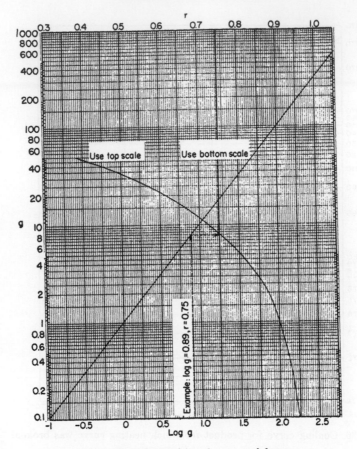

FIG. 27. Relationship of r, g, and $\log g$.

Find in the $f_h/U : g$ tables, for $z = 18$, $j_{cc} = 2.00$, and $f_h/U_{h2} = 1.25$,

$$g_{h2} = \mathbf{1.19}$$

Now compute B:

$$B = f_h \log j_{ch} I_h + (f_2 - f_h) \log g_{bh} - f_2 \log g_{h2}$$
$$B = 12.2 \log 121 + (50 - 12.2) \log 9.2 - 50 \log 1.19$$
$$B = \mathbf{58}$$

The F value of this 58 minute process was checked by the improved general method and found to be approximately 10 (readings varied some because of slight errors in measurement).

In this problem, if the process time had been given instead of the F value, the value of F would be determined by reversing the order of the above steps to obtain values for substituting in

$$F_c = \frac{f_2}{(f_h/U_{h2})F_i} - \frac{r(f_2 - f_h)}{(f_h/U_{bh})F_i}$$

Take the same data as above, except assume that F_c is unknown and $B = 58$. Rearrange equation 4 to obtain U_{h2} as follows:

$$\log g_{h2} = \frac{f_h \log j_{ch}I_h + (f_2 - f_h) \log g_{bh} - B}{f_2}$$

$$\log g_{h2} = \frac{12.2 \log 121 + (50 - 12.2) \log 9.2 - 58}{50}$$

$$\log g_{h2} = \mathbf{0.076}$$

$$g_{h2} = \mathbf{1.19}$$

Find in $f_h/U\!:\!g$ tables, for $z = 18$, $j_{cc} = 2.00$, and $g_{bh} = 9.2$,

$$f_h/U_{bh} = \mathbf{7.17}$$

Also, from Fig. 27, for $\log g_{bh} = 0.964$,

$$r = \mathbf{0.74}$$

Then

$$F_c = \frac{50}{1.25 \times 3.594} - \frac{0.74(50 - 12.2)}{7.17 \times 3.594}$$

$$F_c = \mathbf{10.0}$$

It should be noted that the value of r varies slightly as $I_c + g$ (Ball's $m + g$) varies. Considering other variables, variations in r because of variations in $I_c + g$ are negligible.

Conduction-Heating Products. Stumbo (1948a) presented certain bacteriological considerations relating to the evaluation of thermal processes for foods in cylindrical containers. He proposed that greater attention should be given to the order of death of bacteria subjected to moist heat, and to the mechanism of heat transfer within the food container during process. He then (1949a) developed a graphical procedure for estimating the effectiveness of thermal processes in terms of their capacities to reduce the number of viable bacteria in containers of food for which the heating characteristics were known. This procedure, though reasonably accurate, is tedious and time-consuming to the extent of greatly limiting its practical value.

Hicks (1951) presented a procedure for evaluating thermal processes based on essentially the same principles regarding bacterial destruction and mechanism of heat transfer as was the procedure referred to above. Gillespy (1951) also presented a procedure for evaluating processes for conduction-heating foods, based on these same principles. As was pointed out by Hicks (1952) in a comparison of the procedures, all three were designed to accomplish the same objective, though the algebra employed in their development differed. He demonstrated that the three procedures give essentially the same results in application. The procedures of Hicks and Gillespy are essentially mathematical in operation, each employing process parameters different from those used in the other, and different from those used here for evaluating the lethal effect of heat at a single point in a container.

Stumbo (1953) presented a mathematical procedure for integrating lethal effects of time–temperature combinations existent throughout a container of conduction-heating food during process. In application, this procedure employs essentially the same process parameters as used for evaluating the lethal effect of heat at a single point in a container. Because most people are familiar with these parameters, the procedure employing them will be given here rather than the procedure of Hicks or Gillespy. And because the procedure embodies somewhat broader concepts than that for evaluating the lethal effect of heat at a single point, its development will be described in some detail.

DEVELOPMENT OF PROCEDURE. During a thermal process, the severity of heating of a conduction-heating food in a container varies from the wall to the geometrical center of the container. Stumbo (1949a) described this gradient of severity by picturing a series of imaginary containers nested inside the real container. These containers were considered as cylindrical in shape and decreasing in size from wall to geometrical center of the real container progessively in accordance with a uniform decrease in length and diameter. He further considered that severity of heating was essentially the same at all points on the surface of any one imaginary cylinder. He pointed out that such a description did not represent the true picture, and that this would involve considering imaginary containers gradually changing from cylindrical in shape near the wall of the real container to ellipsoidal or spherical in shape near the geometrical center. This is how they were considered in the mathematical procedure presented later (Stumbo, 1953).

In considering the severity of heating at points on the surface of these imaginary containers, the terminology suggested by Ball et al. (1948) was adopted; that is, the surface of any imaginary container was considered an iso-F surface. (F here refers to the effectiveness of a heat treatment, in terms of its equivalent in minutes at 250°F., with respect to its capacity to reduce the number of a particular kind of bacteria.) Any surface that is iso-F is also iso-j. This is apparent from a study of the equation for calculating process time:

$$B = f_h \log \frac{j_{ch}I_h}{g}$$

For all points in a container of conduction-heating food subjected to any specified heat process, B, I_h, and f_h are invariant. The value of F depends on the value of g attained during process. Then, since g can vary only as j varies, any heating that is iso-F is necessarily iso-j. For this reason the regions of the container referred to above were termed iso-j surfaces or regions. These regions may be pictured as shells, the walls of which are infinitesimal in thickness.

This procedure, like those of Stumbo (1949a), Hicks (1951), and Gillespy (1951), is based on the hypothesis that, in order to evaluate the capacity of a heat process to reduce the number of bacteria in a container of food, consideration only of the heat treatment given any single point in the container is not sufficient; and, further, that accurate evaluation can be accomplished only if all points in the container are considered, the result of the heat process as it concerns the whole container being the sum of effects at all points throughout the container.

In view of this, it was necessary to describe carefully iso-j regions within a container of food. It was possible to describe these regions by employing the following equation developed by Olson and Jackson (1942):

$$j = j_c J_0 \left(R_1 \frac{r}{a} \right) \cos \frac{\pi y}{2b} \tag{5}$$

in which $j = j$ at any designated point in the container.

$\quad j_c = j$ at the geometrical center of the container.

$\quad J_0(x) = $ zero-order Bessel function of x.

$\quad R_1 = $ the first positive root of $J_0(x) = 0$. It is a constant, the numerical value of which is 2.4048.

$\quad r = $ distance, along the container radius, of the designated point from the vertical axis of the container.

$\quad a = $ radius of the container.

$\quad y = $ distance of the designated point above or below a horizontal plane bisecting the container midway between its two ends.

$\quad 2b = $ container height, or length.

From this equation, a number of iso-j regions from center to wall of a No. 10 (603 × 700) can of a conduction-heating food product were described. These regions are depicted graphically in Fig. 28. In the graph, a plane bisecting the container through its longitudinal axis is represented. Since the regions are symmetrical about the geometrical center of the can, regions were indicated in the figure only for the upper right-hand quadrant. Each trace in the quadrant was considered to represent an iso-j region, or shell, of infinitesimal thickness. It is obvious that, if any one of the traces near the geometrical center of the can were extended through the other three quadrants and rotated about the vertical axis, an ellipsoidal-like figure would be described. Also, if the same were done with a trace near the can wall, a figure more nearly resembling a cylinder would be described.

Next, the volume enclosed by each of the iso-j regions depicted was ascertained by the method for determining the volume of a solid of revolution. Though this is not an uncommon method, a short description of how it is carried out would seem appropriate. Figure 29 depicts the upper half of an iso-j trace. If this were rotated around the vertical axis, it would describe a conelike figure, the volume of which would be

$$v = \int_0^{y_{max}} \pi r^2 \, dy$$

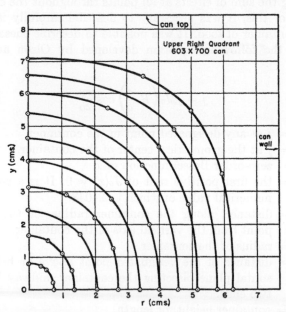

FIG. 28. Curves representing a number of iso-j regions in a 603 × 700 can of conduction-heating food.

And the volume of the ellipsoid-like figure described by rotating the entire trace would be

$$v = 2 \int_0^{y_{max}} \pi r^2 \, dy \qquad (6)$$

in which y_{max} = one-half the length of the solid of revolution.

dy = infinitesimal element of y.

r = radii of dy elements.

Accordingly, if $2y$ is plotted against πr^2, as in Fig. 30, the area under the curve will be proportional to the volume enclosed by the iso-j surface. The volume enclosed by each of a number of iso-j surfaces was determined in this manner.

The F value (F_λ) characterizing the heat treatment received during process by any iso-j region may be determined by the procedure given above for evaluating the lethal effect of heat at a single point.

Earlier, the following equation was presented to describe death of bacteria as occurring logarithmically with time of exposure to a constant temperature:

$$t = D(\log a - \log b) \qquad (7)$$

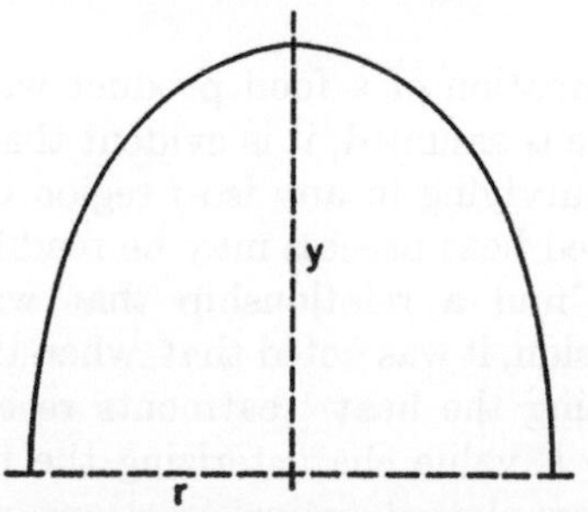

Fig. 29. Upper half of an iso-j trace.

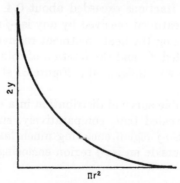

Fig. 30. Area under curve is proportional to volume enclosed by one iso-j trace.

in which t = time of heating.

$\quad\quad D$ = time to kill 90% of the cells.

$\quad\quad a$ = number of viable cells initially present in some given volume.

$\quad\quad b$ = number of viable cells present in the same volume after heating time, t.

Equation 7 was rearranged to obtain an expression equal to the fraction of cells remaining viable after heating time, t—that is:

$$\frac{b}{a} = \frac{1}{\log^{-1} t/D} \tag{8}$$

Time, t, was then expressed in terms of its equivalent in minutes at 250°F. to obtain

$$\frac{b}{a} = \frac{1}{\log^{-1} F/D_r} \tag{9}$$

The term D, when referring to destruction at 250°F., was designated D_r for convenience.

If uniform contamination of a food product with bacterial spores of known heat resistance is assumed, it is evident that, by the use of equation 9, the fraction surviving in any iso-j region of a container of food subjected to a specified heat process may be readily calculated.

In attempting to find a relationship that was subject to simple mathematical expression, it was noted that, when the differences between F values characterizing the heat treatments received by the different iso-j regions and the F value characterizing the heat treatment at the geometrical center were plotted against fractions of can volume enclosed by respective iso-j regions, virtually a straight line was obtained, at least until volume fractions exceeded about 0.4. The F value characterizing the heat treatment received by any iso-j region was designated F_λ, that characterizing the heat treatment received by the geometrical center was designated F_c, and the fraction of total can volume enclosed by any iso-j region was designated v. Figure 31 shows a plot of $F_\lambda - F_c$ against v.

A study of probable survival distribution in a container of food after heat processing revealed that, comparatively, survival would be negligible outside an iso-j region enclosing much less than 0.4 of the can volume—in fact, outside an iso-j region enclosing about 0.1 of the can volume.

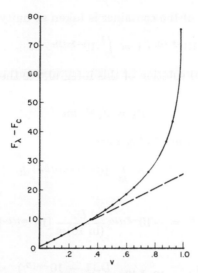

FIG. 31. A typical curve showing the relationship between $(F_\lambda - F_c)$ and v for a cylindrical container of conduction-heating food.

An equation of the curve shown in Fig. 31 may be written as follows:

$$F_\lambda - F_c = mv \tag{10}$$

in which m = slope of the curve.

Equation 9 may be rearranged as follows:

$$b/a = 10^{-F/D_r} \tag{11}$$

If the F value representing the equivalent of all heat received by the entire container with respect to its capacity to reduce the bacterial population of the container is designated as F_s, then

$$b/a = 10^{-F_s/D_r} \tag{12}$$

in which b/a = the fraction of the number of spores, initially present in the container, that should be viable at the end of the heat process.

Also,

$$b/a = 10^{-F_\lambda/D_r} \tag{13}$$

in which b/a = the fraction of the number of spores, initially present in any designated iso-j region, that should be viable at the end of the heat process. (Recall that each iso-j region is infinitesimal in thickness.)

Then, if the volume of the container is taken as unity.

$$10^{-F_s/D_r} \cdot 1 = \int_0^1 10^{-F_\lambda/D_r} \cdot dv \tag{14}$$

(For a detailed solution of this integral, see the Appendix)
But, from equation 10,

$$F_\lambda = F_c + mv \tag{15}$$

By substitution, equation 14 becomes

$$10^{-F_s/D_r} = \int_0^1 10^{-(F_c+mv)/D_r} \cdot dv \tag{16}$$

from which

$$10^{-F_s/D_r} = -10^{-F_c/D_r} \frac{D_r}{(\ln 10)m} \left[10^{-mv/D_r}\right]_{v=0}^{v=1} \tag{17}$$

or

$$10^{-F_s/D_r} = 10^{-F_c/D_r} \cdot \frac{D_r(1 - 10^{-m/D_r})}{2.303m}$$

Equation 17 was next simplified in the following manner. Equation 10 was rearranged to obtain

$$m = \frac{F_\lambda - F_c}{v} \tag{18}$$

After substituting and evaluating for a number of extreme processing conditions, it was found that $10^{-m/D_r}$ was always negligible as compared with unity. This then allowed equation 17 to be reduced to

$$10^{-F_s/D_r} = \frac{10^{-F_c/D_r} \cdot D_r v}{2.303(F_\lambda - F_c)}$$

or

$$-F_s/D_r = (-F_c/D_r) + (\log D_r + \log v) - [\log 2.303 + \log (F_\lambda - F_c)]$$

and

$$F_s = F_c - D_r\{(\log D_r + \log v) - [\log 2.303 + \log (F_\lambda - F_c)]\} \tag{19}$$

Now, since $F_\lambda - F_c$ plotted against v was a straight line for the portion of the container that must be considered, a numerical value was assigned to v, with the designation that F_λ used in the equation was always to be the F value of the heat treatment received by the iso-j region enclosing a fraction of the can volume equal to the assigned value of v. It was found convenient to employ 0.19 as the value of v for further simplification of equation 19. The reason this value was chosen will be explained later. Equation 19 was simplified as follows:

$$F_s = F_c + D_r[-\log D_r - \log 0.19 + \log 2.303 + \log (F_\lambda - F_c)]$$

or

$$F_s = F_c + D_r \left(1.084 + \log \frac{F_\lambda - F_c}{D_r}\right) \tag{20}$$

Why was 0.19 chosen as the value of v? It is apparent from the equation

$$B = f_h \log \frac{jI}{g}$$

that F varies across the container only with variations in j and g—B, f_h, and I being invariant. As was pointed out by Ball et al., (1948), for any value obtained by dividing the value of j associated with an iso-j region by the value of j associated with the center, there is always a corresponding value of v (fraction of can volume enclosed by the iso-j region).

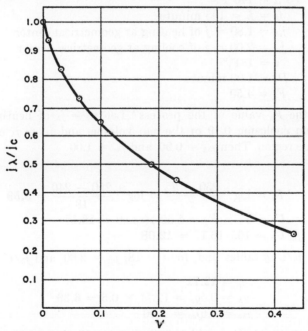

Fig. 32. A curve showing the relationship of j_λ/j_c to v for any cylindrical food container.

This relationship for cylindrical containers is independent of container size and dimensions. A curve describing the relationship appears in Fig. 32. The symbol j_λ refers to the j associated with any iso-j region, and j_c to the j associated with the geometrical center. From the curve in Fig. 32 and by calculation, it was found that, when $v = 0.19$, $j_\lambda/j_c = 0.5$ and $g_\lambda/g_c = 0.5$. Therefore, with $v = 0.19$, equation 20 could be written with very convenient use specifications—that is,

$$F_s = F_c + D_r \left(1.084 + \log \frac{F_\lambda - F_c}{D_r} \right)$$

when $g_\lambda = 0.5g_c$ and $j_\lambda = 0.5j_c$. The convenience of these specifications is apparent in the following application of the equation.

Given the following data for a conduction-heating product in a 603 × 700 can:

$$T_x = 250°\text{F.}$$
$$T_r = 240°\text{F.}$$
$$T_{ih} = 140°\text{F.}$$
$$f_h = f_c = 195 \text{ minutes}$$
$$j_{ch} = 1.80 = j \text{ of heating at geometrical center}$$
$$j_{cc} = 2.00 = j \text{ of cooling at geometrical center}$$
$$z = 18\text{F.}°$$
$$D_r = 1.00 \text{ minute}$$
$$F_c = 4.50$$

What is the F_s value of the process? Let $j_{\lambda h} = j_\lambda$ of heating of the iso-j region enclosing 0.19 of the can volume, and $j_{\lambda c} = j_\lambda$ of cooling of the same region. Then $j_{\lambda h} = 0.90$, and $j_{\lambda c} = 1.00$
Solution:

$$F_i = \log^{-1} \frac{250 - T_r}{z} = \log^{-1} \frac{250 - 240}{18} = \textbf{3.59}$$
$$U_c = F_c \times F_i = 4.50 \times 3.59 = \textbf{16.17}$$
$$f_h/U_c = 195/16.17 = \textbf{12.08}$$

From the $f_h/U:g$ tables find, for $z = 18$, $j_{cc} = 2.00$, and $f_h/U_c = 12.08$,

$$g_c = \textbf{12.71}$$
$$g_\lambda = 0.5g_c = 12.71 \times 0.5 = \textbf{6.36}$$
$$j_{\lambda c} = 0.5j_{cc} = \textbf{1.00}$$

Again, from the $f_h/U:g$ tables find, for $z = 18$, $j_{\lambda c} = 1.00$, and $g_\lambda = 6.36$,

$$f_h/U_\lambda = \textbf{6.15}$$

Then

$$U_\lambda = \frac{f_h}{6.15} = \frac{195}{6.15} = \textbf{31.7}$$

$$F_\lambda = \frac{U_\lambda}{F_i} = \frac{31.7}{3.59} = \textbf{8.84}$$

and

$$F_s = F_c + D_r\left(1.084 + \log \frac{F_\lambda - F_c}{D_r}\right)$$

$$F_s = 4.5 + 1.0\left(1.084 + \log \frac{8.84 - 4.5}{1.0}\right)$$

$$F_s = \textbf{6.22}$$

Now, suppose that there were 3000 gm. of food in the container, contaminated with ten spores per gram initially. How many spores per can should survive the above process?

Because

$$F_s = D_r(\log a - \log b)$$
$$6.22 = 1.0(\log 30{,}000 - \log b)$$
$$\log b = -1.743$$
$$b = \textbf{0.018} \text{ spores per can}$$

or 18 cans per thousand should not have been sterilized.

EQUIVALENCE OF PROCESSES FOR CONDUCTION-HEATING FOODS. Equivalent processes, from the standpoint of their allowing the same percentage of spoilage, should be those that leave the same number of spores per container. Then how should F_s values compare for equivalent processes to be applied to a given product in containers of different size? Because the number of spores per container is proportional to container volume, and because F_s is dependent on D_r, the following equation (Gillespy, 1951) expresses the relationship of one F_s value to another:

$$F_{sb} = F_{sa} + D_r \log \frac{V_b}{V_a}$$

in which V_a and V_b = volumes of two containers of different capacity. F_{sa} and F_{sb} — F_s values of equivalent processes.

In the above calculation, the F_s of the process for the 603 × 700 can of product was found to be 6.22. What would be the F_s of an equivalent process for the same product in a 211 × 304 can?

Net capacities:

$$\text{For } 603 \times 700 \text{ can, } 3000 \text{ gm.}$$
$$\text{For } 211 \times 304 \text{ can, } 240 \text{ gm.}$$

Then

$$F_{sb} = 6.22 + 1.0 \log \frac{240}{3000}$$

$$F_s \text{ for } 211 \times 304 \text{ can} = 6.22 - 1.10 = \mathbf{5.12}$$

DETERMINATION OF PROCESS TIME WHEN F_s VALUE IS KNOWN. To determine process time, the F_c value must be known. The F_c value corresponding to any F_s value may be determined by interpolation by using the following equation:

$$F_c \text{ desired} = F_{c_1} + \frac{F_{c_2} - F_{c_1}}{F_{s_2} - F_{s_1}} (F_s \text{ desired} - F_{s_1})$$

The F_c desired is usually somewhat more than one D_r value less than the F_s desired. Suppose that the F_s desired is 5.00. The F_c desired should be between 3 and 4, since $D_r = 1.00$. Now choose an F_c value somewhat greater than the expected F_c desired, and another somewhat lower than the expected F_c desired, say

$$F_{c_1} = 3.00$$
$$F_{c_2} = 5.00$$

(It should be noted that the values chosen need not necessarily be above and below the real F_c to be obtained; however, they should be reasonably close to it in value and not widely different in value themselves.)

The following process data for product in the 211×304 can are known:

$$
\begin{aligned}
T_x &= 250°\text{F.} & j_{ch} = j_{cc} &= 1.6 \\
T_r &= 250°\text{F.} & I_h = T_r - T_{ih} &= 100 \\
T_{ih} &= 150°\text{F.} & z &= 18 \\
f_h &= 35 & D_r &= 1.00
\end{aligned}
$$

Now calculate an F_s value corresponding to each F_{c_1} and F_{c_2} from the following equation:

$$F_s = F_c + D_r \left(1.084 + \log \frac{F_\lambda - F_c}{D_r} \right)$$

when $g_\lambda = 0.5 g_c$ and $j_{\lambda c} = 0.5 j_{cc}$.

It is found that,

$$\text{when } F_{c_1} = 3.0, F_{s_1} = \textbf{4.58}$$
$$\text{when } F_{c_2} = 5.0, F_{s_2} = \textbf{6.66}$$

Substituting in the interpolation equation gives

$$F_c \text{ desired} = 3.0 + \frac{5.0 - 3.0}{6.66 - 4.58} (5.0 - 4.58)$$

$$F_c \text{ desired} = \textbf{3.40}$$

Now, with F_c known, the process time may be determined. Since $T_r = 250$,

$$U_c = F_c F_i = 3.40 \times 1.00 = \textbf{3.40}$$

and

$$f_h/U_c = 35/3.40 = \textbf{10.3}$$

From the $f_h/U : g$ tables find, for $z = 18$, $j_{cc} = 1.6$, and $f_h/U_c = 10.3$

$$g_c = \textbf{10.55}$$

Then

$$B = f_h(\log j_{ch} I_h - \log g_c)$$
$$B = 35[\log (1.6 \times 100) - \log 10.55]$$
$$B = \textbf{41.3 minutes}$$

Estimation of Mass Average Temperature

A number of canned food products, if not adequately cooled subsequent to heat processing, will undergo serious discoloration ("stack burn") during storage or transport. Also, in many products, high temperatures will promote spoilage by thermophiles surviving the heat process. On the other hand, if sufficient heat is not left in the product to promote rapid drying, serious rusting of cans and jar caps often results. Rapid drying after water cooling also minimizes postcooling contamination through can leakage. For these reasons it is generally considered desirable to cool canned foods to an average temperature between 90° and 110°F.

For convection-heating products it is a relatively simple matter to withdraw cans during cooling, stir the contents, and determine average temperatures. For conduction-heating products, particularly those of heavy consistency, this can be a difficult task accompanied by a great deal of uncertainty. Consequently, it was felt that a simple and convenient method for calculating the average temperature of conduction-heating products should prove quite useful.

The general problem of calculating the temperature distribution in and mass average temperature of canned foods during heating and cooling has been considered by a number of workers (among others, Olson and Schultz, 1942; Ball and Olson, 1957; Charm, 1961). Some of the procedures developed are complicated and cumbersome in operation; however, they do not require experimental determination of time-temperature relationships at the can center, as is the case with the method to be described here. The simplest method before that presented by Stumbo (1964) is perhaps that suggested by Ball and Olson (1957), which still requires a number of computations. The method (Stumbo, 1964) described here, though somewhat complicated in derivation, is extremely simple in application.

Development of Procedure.   BASIC ASSUMPTIONS. (1) Heat transfer takes place uniformly through the entire can surface. Though deviations will occur in some processes, it is believed that, for most applications, this is a more realistic assumption than that of Charm (1961), which considers the can ends insulated during cooling. (2) Cooling will have progressed beyond the lag period—usually 5 to 10 minutes, the exact time depending on temperature distribution in the container at the time cooling is started.

GROUNDWORK. Stumbo (1953) and Ball and Olson (1957) described iso-j traces within a container of conduction-heating food during heating. The relationships described are equally valid for cooling. Each region described by a trace is not only iso-j throughout heating or cooling but, at any given instant during heating or cooling, it is isothermal. Figure 28 shows the traces as described by Stumbo (1953). In this same work, the relationship of $(j_\lambda/j_c):v$ was described (see Fig. 33). It may be noted from Fig. 33 that this curve also describes the temperature distribution throughout a can of conduction-heating product during heating or cooling. [It should be pointed out that the total increment of $(T_r - T_\lambda)$ or $(T_\lambda - T_w)$ represented in Fig. 33 could be of any magnitude—the increment of 10 was arbitrarily chosen for demonstration.] These relationships apply because, during heating,

$$\frac{j_{\lambda h}}{j_{ch}} = \frac{T_r - T_\lambda}{T_r - T_c}$$

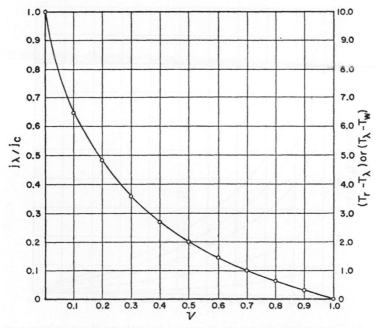

Fig. 33. Relationships of $(j_\lambda/j_c):v$ and temperature differences across a can of conduction-heating food during heating or cooling when the total gradient from geometrical center to wall covers 10°F.

where $j_{\lambda h}$ = j of semilog heating curve for any iso-j surface.

j_{ch} = j of semilog heating curve for geometrical center of the can.

T_r = temperature of heating medium.

T_λ = temperature of any iso-j surface.

T_c = temperature at the geometrical center of the can.

Also, during cooling,

$$\frac{j_{\lambda c}}{j_{cc}} = \frac{T_\lambda - T_w}{T_c - T_w}$$

where T_λ and T_c are as defined above.

$j_{\lambda c}$ = j of semilog cooling curve for any iso-j surface.

j_{cc} = j of semilog cooling curve for geometrical center of the can.

T_w = temperature of the cooling medium.

From these basic relationships it was possible to construct the curves depicted in Fig. 34, describing temperature distribution in the can at various intervals during heating or cooling. Figure 35 is a semilogarithmic

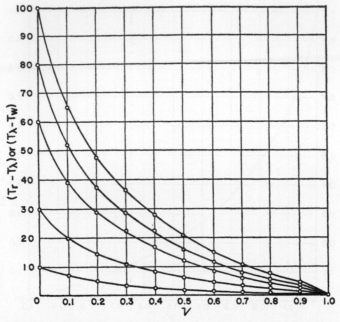

FIG. 34. Temperature differences as related to fractions of can volume enclosed by isothermal surfaces in a can of conduction-heating food during heating or cooling.

plot of three of these curves and of straight lines of best fit. It will be noted that the straight-line portions of the true curves are parallel, and that they represent temperature distribution for a large proportion of the can contents. It was reasoned, therefore that straight-line curves of best fit, with respect to temperature contribution throughout the entire can, might well be parallel. In the derivation to follow, this was assumed to be the case. The validity of the assumption is demonstrated later.

DERIVATION OF EQUATION. If the total volume of a cylindrical container is taken as unity, then

$$T_s \cdot 1 = \int_0^1 T_\lambda \, dv \tag{21}$$

(For a detailed solution of this integral, see the Appendix.)
in which T_s = mass average temperature.

T_λ = temperature at any isothermal surface

v = fraction of can volume enclosed by any isothermal surface.

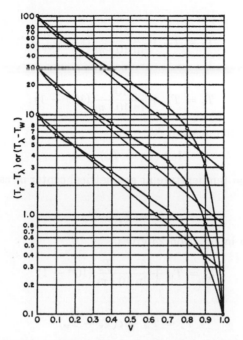

Fig. 35. Semilog plot of three of the curves shown in Fig. 34, with straight lines of best fit, with respect to temperature contribution, drawn in.

The equation of any of the straight lines depicted in Fig. 35 may be written as follows:

$$\log (T_\lambda - T_w) = mv + b \qquad (22)$$

in which T_λ = as defined for equation 21.

T_w = temperature of the cooling medium.

m = slope of the parallel straight-line curves.

b = constant.

Then

$$T_\lambda = T_w + (10^{mv} \cdot 10^b) \qquad (23)$$

Substitution in equation 21 gives

$$T_s = \int_0^1 [T_w + (10^{mv} \cdot 10^b)] \, dv$$

or

$$T_s = T_w + 10^b \int_0^1 10^{mv} \, dv$$

from which

$$T_s = T_w + \frac{10^b}{\ln 10 \cdot m} \left[10^{mv}\right]_{v=0}^{v=1}$$

or

$$T_s = T_w + \frac{10^b}{\ln 10 \cdot m} (10^m - 1) \tag{24}$$

By setting $v = 0$ in equation 22,

$$\log (T_\lambda - T_w) = b$$

where $T_\lambda = T_c =$ temperature at geometrical center of can.
Then

$$10^b = T_c - T_w$$

and substitution in equation 24 gives

$$T_s = T_w + \frac{T_c - T_w}{\ln 10 \cdot m} (10^m - 1) \tag{25}$$

or:

$$T_s = T_w + c(T_c - T_w) \tag{26}$$

in which

$$c = \text{Constant} = \frac{10^m - 1}{2.303m} \tag{27}$$

DETERMINATION OF STRAIGHT-LINE CURVES OF BEST FIT. From the relationships depicted in Figs. 34 and 35, T_λ values were computed for temperature distribution in the container when $T_c - T_w = 100°$, $30°$, and $10°$F., respectively. Curves as depicted in Fig. 36 were then constructed by plotting T_λ against v. According to the basic integral (equation 21), the area beneath each curve would be proportional to the average can temperature. By graphical integration the following values were obtained:

$$T_c - T_w = 100°\text{F.,} \qquad T_s = 97°\text{F.}$$
$$T_c - T_w = 30°\text{F.,} \qquad T_s = 78.2°\text{F.}$$
$$T_c - T_w = 10°\text{F.,} \qquad T_s = 72.7°\text{F.}$$

Substitution of these values in equation 26 gives values of 0.270, 0.274, and 0.270, respectively, for the constant c. Substitution of $c = 0.27$ in equation 27 gives -1.565 as the value of m. It may be concluded that

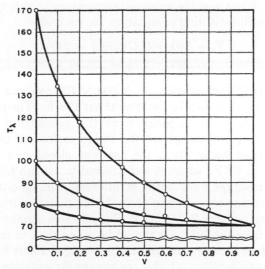

Fig. 36. Temperatures of isothermal regions as related to fractions of can volume enclosed by isothermal surfaces.

the straight-line curves of best fit with respect to temperature contribution are parallel, with a common slope of -1.565 (see Fig. 35). Also, a general equation for calculating average temperature in a can of conduction-heating food at any given time during cooling takes the simple form

$$T_s = T_w + 0.27(T_c - T_w) \tag{28}$$

in which T_w = temperature of cooling medium.
T_c = temperature at geometrical center of can.
and, for average temperature at any time during heating,

$$T_s = T_r + 0.27(T_c - T_r) \tag{29}$$

in which T_r = temperature of heating medium.
T_c = temperature at geometrical center of can.

Validity of Equations 28 and 29. Ball and Olson (1957), by a stepwise integration procedure based on the same theoretical considerations as the procedure presented here, posed and solved the following problem.

A conduction-heating product in a cylindrical can had a center temperature of 233.5°F. after heating for 69.7 minutes at a retort temperature of 240°F. What was the average temperature at this time? Answer given: 238.2°F. Substituting in equation 29, we find

$$T_s = 240 + 0.27(233.5 - 240)$$
$$= 240 - 1.75 = 238.25$$

Product Cooling (Experimental). A 603 × 700 can was filled with a suspension of 5% bentonite in water at about 198°F. (The can contained one thermocouple with its junction at the geometrical center, and another with its junction against the inside surface of the can wall.) Immediately after being filled, the can was closed and immersed in running water at 70°F., where it was allowed to remain for 58 minutes. Temperature readings were taken at intervals throughout the cooling period. The average temperature of the inside surface during the period was 79°F., varying from 76° to 83°F. The effective cooling temperature was therefore taken as 79°F. (It should be noted that, in repeated tests, the inside surface temperature generally averaged about 6 to 9 degrees above actual temperature of the cooling water entering the retort.) After 58 minutes of cooling, the center temperature was 170°F. At this time the can was removed from the water and shaken vigorously, and the center temperature was noted. The can was again vigorously shaken, and, since no lowering of the center temperature was noted, the center temperature was recorded as the average temperature. The following values were thereby obtained:

$T_c = 170°F.$ = center temperature after 58 minutes of cooling.
$T_w = 79°F.$ = effective cooling temperature (9 degrees above temperature of cooling water entering retort).
$T_s = 104.5°F.$ = observed temperature after cooling for 58 minutes and shaking.

The experiment was repeated except that the product was filled at 189°F. and cooled for 50 minutes. The following comparative values were observed:

$$T_c = 170°F.$$
$$T_w = 79°F.$$
$$T_s = 105.0°F.$$

Taking T_w as 79°F., we find, from equation 28, for both trials:

$$T_s = 79 + 0.27(170 - 79) = 103.6$$

Additional tests, with 25% tomato paste as well as 5% bentonite in 603 × 700 and 307 × 409 cans, gave similar results, the widest deviation noted being 3.8% of the observed average temperature, so long as the temperature of the inside can surface was used as T_w in the equation. It is readily noted that, if the actual cooling-water temperature (70°F.) had been used in the above calculation, the T_s value obtained would have been 97°F., about 8 degrees below the observed—a deviation of about 7%.

For most practical applications, using the actual cooling-water temperature as T_w in the equation would probably give acceptable approximations. However, as indicated above, closer approximations are obtained when the inside can surface temperature is used as T_w. Limited tests indicate that good agreement may be expected if T_w is routinely taken as cooling-water temperature plus 6 to 9 degrees. The exact factor would, of course, depend on the nature of the cooling operation—amount of water used, extent of water circulation, extent of contact of water with total can surface, etc.

Equations 28 and 29 in Heating and Cooling Operations. These equations are generally applicable for conduction-heating products in cylindrical cans, being independent of can size or dimensions. They should be similarly applicable for such products packed in cylindrical glass containers.

For products hot-filled at temperatures below 212°F., it is convenient to remove containers periodically during cooling and measure the center temperature by thermometer or thermocouple. The cooling-water temperature also is readily obtainable. With these, the average temperature is readily computed by equation 28.

For products at retort temperatures above 212°F., and for products hot-filled at temperatures above 212°F. and "still"-cooled, center temperature and temperatures of heating and cooling media are most conveniently measured with a thermocouple. However, if the heat penetration parameters are known, the necessary values for equations 28 and 29 may be readily computed, as was pointed out by Ball and Olson (1957). For products during heating by conduction, the following equations would be employed:

$$g_c = j_{ch} I_h \log^{-1} \left(-\frac{t}{f_h} \right) \tag{30}$$

in which g_c = difference between retort temperature and center food temperature at time t, or $(T_r - T_c)$.

j_{ch} = lag factor for semilog heating curve.

I_h = difference between retort temperature and food temperature at start of heating, or $(T_r - T_{ih})$.

f_h = time, in minutes, required for straight-line portion of semilog heating curve to traverse one log cycle.

Then

$$T_c = T_r - g_c \tag{31}$$

in which T_c = center temperature at time t.

T_r = retort temperature.

With T_c and T_r known, the average temperature may be readily calculated by substituting these values in equation 29.

Similarly, for products during cooling by conduction:

$$y_c = j_{cc}I_c \log^{-1}\left(-\frac{t}{f_c}\right) \tag{32}$$

in which y_c = difference between center food temperature and cooling-water temperature at cooling time t, or $(T_c - T_w)$.

j_{cc} = lag factor of semilog cooling curve.

I_c = difference between center food temperature and cooling-water temperature at start of cooling $(T_{ic} - T_w)$.

f_c = time, in minutes, required for straight-line portion of semilog cooling curve to traverse one log cycle.

Then

$$T_c = T_w + y_c \tag{33}$$

With T_c and T_w known, the average temperature may be readily calculated by substituting these values in equation 28.

SAMPLE CALCULATIONS

Given:

Retort temperature $T_r = 250°F.$

Initial temperature $T_{ih} = 170°F.$

$(T_r - T_{ih}) = I_h = 80°F.$

$j_{ch} = 2.00$

$f_h = 25$

What would be the average temperature (T_s) after 50 minutes of heating?

According to equation 30,

$$g_c = j_{ch}I_h \log^{-1}\left(-\frac{t}{f_h}\right)$$

$$= 2 \times 80 \log^{-1}\left(-\frac{50}{25}\right)$$

$$= 1.6$$

And, according to equation 31,

$$T_c = T_r - g_c$$

$$= 250 - 1.6$$

$$= 248.4°F.$$

Then, by equation 29,

$$T_s = T_r + 0.27(T_c - T_r)$$

$$= 250 + 0.27(248.4 - 250)$$

$$= 249.6°F.$$

If cooling were started after 50 minutes of heating, what would be the average temperature (T_s) after cooling for 15 minutes in water entering the retort at 70°F.? Consider the effective cooling water temperature as $70° + 9° = 79°F$. From a semilog plot of center temperatures during cooling, it is found that

$$j_{cc} = 1.50$$

$$f_c = 25$$

$$I_c = T_{ic} - T_w = 248.4 - 79 = 169.4°F.$$

According to equation 32,

$$y_c = j_{cc}I_c \log^{-1}\left(-\frac{t}{f_c}\right)$$

$$= 1.5 \times 169.4 \log^{-1}\left(-\frac{15}{25}\right)$$

$$= 63.8°F.$$

And according to equation 33,

$$T_c = T_w + y_c$$

$$= 79 + 63.8 = 142.8°F.$$

Then, by equation 28,

$$T_s = T_w + 0.27(T_c - T_w)$$
$$= 79 + 0.27(63.8)$$
$$= 96.2°F.$$

Low-Temperature Processes

For processes that are equivalent in lethality to only a few minutes or less of heating at 212°F. or some lower temperature, it is usually best to change the unit of lethality from the equivalent of 1 minute of heating at 250°F. to 1 minute of heating at a lower temperature. If the F_{250} unit is used, lethality values often may be extremely small fractions of one. All that is involved is the use of different F_i and L values. For example, if it is desired to express the lethality value of a process in minutes at 212°F., use

$$L = \log^{-1} \frac{T - 212}{z}$$

$$F_i = \log^{-1} \frac{212 - T}{z}$$

And if it is desired to express the lethality value of a process in minutes at 200°F., use

$$L = \log^{-1} \frac{T - 200}{z}$$

$$F_i = \log^{-1} \frac{200 - T}{z}$$

By so doing, the F values computed will be F_{212} and F_{200} values, respectively. Procedures employed in process evaluation are the same; all other equations are unchanged.

SELECTED REFERENCES

Ball, C. O. (1923). Thermal process time for canned foods. *Bull. Natl. Res. Council*, **7**, Part 1, No. 37, 76 pp.

Ball, C. O. (1927). Theory and practice in processing. *Canner*, **64**, 27.

Ball, C. O. (1928). Mathematical solution of problems on thermal processing of canned food. *Univ. Calif. (Berkeley) Publ. Public Health*, **1**, 245.

Ball, C. O. (1943). Short time pasturization of milk. *Ind. Eng. Chem.*, **35**, 71.

Ball, C. O., and Olson, F. C. W. (1957). "Sterilization in Food Technology," 654 pp. McGraw-Hill, New York, Toronto, London.

Ball, C. O., Olson, F. C. W., and Boyd, G. D. (1948). Evaluation of the effect of heat processing upon the organoleptic quality of canned foods. Unpublished paper presented before the Eighth Annual Institute of Food Technologists Food Conference, Philadelphia, Pennsylvania, June 8, 1948.

Charm, Stanley (1961). A method for calculating the temperature distribution and mass average temperature in conduction-heated canned foods during water cooling. *Food Technol.*, **15**, 248.

Gillespy, T. G. (1951). Estimation of sterilizing values of processes as applied to canned foods. I. Packs heating by conduction. *J. Sci. Food Agr.*, **2**, 107.

Hicks, E. W. (1951). On the evaluation of canning processes. *Food Technol.*, **5**, 134.

Hicks, E. W. (1952). Some implications of recent theoretical work on canning processes. *Food Technol.*, **6**, 175.

Jen, Y., Manson, J. E., Stumbo, C. R., and Zahradnik, J. W. (1971). A procedure for estimating sterilization of and quality factor degradation in thermally processed foods. *J. Food Sci.*, **36**, 692.

Olson, F. C. W., and Jackson, J. M. (1942). Heating-curves—theory and practical application. *Ind. Eng. Chem.*, **34**, 337.

Olson, F. C. W., and Schultz, O. T. (1942). Temperatures in solids during heating or cooling. *Ind. Eng. Chem.*, **34**, 874.

Purohit, K. S., and Stumbo, C. R. (1972). Computer calculated parameters for thermal process evaluations. (Unpublished data.)

Secrist, J. L., and Stumbo, C. R. (1956). Application of spore resistance in the newer methods of process evaluation. *Food Technol.*, **10**, 543.

Stumbo, C. R. (1948a). Bacteriological considerations relating to process evaluation. *Food Technol.*, **2**, 115.

Stumbo, C. R. (1948b). A technique for studying resistance of bacterial spores to temperatures in the higher range. *Food Technol.*, **2**, 228.

Stumbo, C. R. (1949a). Further considerations relating to evaluation of thermal processes for foods. *Food Technol.*, **3**, 126.

Stumbo, C. R. (1949b). Thermobacteriology as applied to food processing. *Advan. Food Res.*, **2**, 47.

Stumbo, C. R. (1953). New procedures for evaluating thermal processes for foods in cylindrical containers. *Food Technol.*, **7**, 309.

Stumbo, C. R. (1964). Estimation of the mass average temperature during heating and cooling of conduction-heating products in cylindrical cans. *Food Technol.*, **18**, 126.

Stumbo, C. R., and Longley, Ruth E. (1966). New parameters for process calculation. *Food Technol.*, **20**, 109.

Teixeira, A. A., Dixon, J. R., Zahradnik, J. W., and Zinsmeister, G. E. (1969). Computer optimization of nutrient retention in the thermal processing of conduction-heated foods. *Food Technol.*, **23**, 845.

Townsend, C. T., Somers, I. I., Lamb, F. C., and Olson, N. A. (1965). "A Laboratory Manual for the Canning Industry," 2nd ed. National Canners Association Research Laboratories, Washington, D. C.

CHAPTER 13

Conversion of Heat Penetration Data

When some processing conditions are changed for products for which the semilog heating curve is one straight line, it is possible to make data conversions to comply satisfactorily with the new conditions. This can be done much more conveniently and rapidly than can additional heat penetration data be collected.

Change of Retort Temperature when Initial Food Temperature Remains the Same

First, we must consider whether moderate changes in retort temperature greatly affect the j of the cooling curve (j_{cc}). Generally speaking, the greater the value of g_c, the greater will be the value of j_{cc}, because of the steeper descending temperature gradient from wall to geometrical center of the container at the beginning of cooling. But how much is the value of j_{cc} influenced by across-the-container temperature gradients? An equation presented by Olson and Jackson (1942) has been found useful in this analysis. Their equation, modified to apply to the cooling situation, may be written as follows:

$$j_{cc} = 1.27 + 0.77(T_w - T_r)/(T_w - T_{ic})$$

in which $j_{cc} = j$ of cooling curve for geometrical center of container.

T_w = temperature of cooling water.

T_r = retort temperature, representing temperature at can wall just before start of cooling.

$T_{ic} = T_r - g_c$ = temperature at the geometrical center of the can just before start of cooling.

Taking $T_r = 250$ and $T_w = 70$, we may determine how j_{cc} shall be expected to vary with g_c.

When $g_c = 0$, $T_{ic} = 250$ and

$$j_{cc} = 2.04$$

When $g_c = 5$, $T_{ic} = 245$ and

$$j_{cc} = 2.06$$

When $g_c = 10$, $T_{ic} = 240$ and

$$j_{cc} = 2.09$$

Accordingly, a change in retort temperature sufficient to cause a change in g_c of 10 degrees should not change the j_{cc} more than about 0.1 unit. Generally a change of about 20 degrees in T_r is required to cause a change of 10 degrees in g_c. Theoretically, when temperatures are uniform throughout at the beginning of heating and cooling, j_{ch} and j_{cc} should be about equal, if the retort come-up time is very short and if the product doesn't "boil" at the start of cool, owing to rapid condensation of head-space gases. Boiling is not too common. For conduction-heating products, when the retort come-up time is appreciable, j_{cc} will generally be considerably greater than j_{ch}, if the j_{ch} is computed after correcting for retort come-up time. The j_{cc} is generally about the same as the j_{ch} computed before correcting for come-up time. The j_{ch} computed in this manner will be greater than one computed after correcting for come-up because of different values of T_{pih} used in the computations.

The heat penetration parameters j_{ch} and f_h are independent of retort temperature. Therefore, when the initial food temperature (T_{ih}) is the same, heat penetration data obtained with one retort temperature may be employed for calculating a process when another retort temperature is used if the retort come-up time is about the same. The only values substituted in the data would be T_r, and l if it changes. The following comparison of processes for a conduction-heating product in a 307 × 409 can illustrates the conversion.

Case 1	Case 2
$T_r = 250°F.$	$T_r = 240°F.$
$T_{ih} = 150°F.$	$T_{ih} = 150°F.$
$I_h = T_r - T_{ih} = 100°F.$	$I_h = 90°F.$
$j_{ch} = j_{cc} = 2.00$	$j_{ch} = j_{cc} = 2.00$

$$j_{ch}I_h = 200 \qquad\qquad j_{ch}I_h = 180$$
$$f_h = 60 \text{ minutes} \qquad\qquad f_h = 60 \text{ minutes}$$
$$F_i = 1.00 \qquad\qquad\qquad F_i = 3.594$$
$$l = 15 \text{ minutes} \qquad\qquad l = 12 \text{ minutes}$$
$$z = 18\text{F.}° \qquad\qquad\qquad z = 18\text{F}°.$$

What would be the operator's process time (P_t) in each case for a process having an F_c value of 5.00?

Case 1

$$U_c = F_c F_i = 5.00 \times 1.00 = \mathbf{5.00}$$
$$f_h/U_c = 60/5 = \mathbf{12}$$

From the $f_h/U:g$ tables find, for $z = 18$, $j_{cc} = 2.00$, and $f_h/U_c = 12$,

$$g_c = \mathbf{12.66}$$

Then

$$B = f_h(\log j_{ch}I_h - \log g_c)$$
$$B = 60(\log 200 - \log 12.66) = \mathbf{71.8 \ minutes}$$

and

$$P_t = B - 0.4l$$
$$P_t = 71.8 - (0.4 \times 15) = \mathbf{65.8 \ minutes}$$

Case 2

$$U_c = F_c F_i = 5.00 \times 3.594 = \mathbf{17.97}$$
$$f_h/U_c = 60/17.97 = \mathbf{3.34}$$

From the $f_h/U:g$ tables find, for $z = 18$, $j_{cc} = 2.00$, and $f_h/U_c = 3.34$,

$$g_c = \mathbf{4.80}$$

Then

$$B = f_h(\log j_{ch}I_h - \log g_c)$$
$$B = 60(\log 180 - \log 4.80) = \mathbf{94.5 \ minutes}$$

and

$$P_t = B - 0.4l$$
$$P_t = 94.5 - (0.4 \times 12) = \mathbf{89.7 \ minutes}$$

Change of Initial Food Temperature when the Retort Temperature Remains the Same

The heat penetration parameters j_{ch} and f_h are independent of initial temperature. The j_{cc} may also be considered invariant, because differences in its value, caused by differences in temperature gradient at the start of cool as a result of differences in T_{ih}, are quite small. Therefore, conversion of heat penetration data in this case is similar to that done when T_r is changed and T_{ih} remains the same. The new T_{ih} is simply substituted for the original in the equations.

Change of Container Size—Convection-Heating Products

The parameter j is influenced only slightly by a change in container size. However, f_h is influenced greatly because its value depends on container dimensions as well as on the nature of the product. It is the only parameter of concern in considering a change in container size.

Schultz and Olson (1938) developed the following equation for conversion of f_h for convection-heating products in cylindrical cans. They demonstrated that conversion values obtained by the equation agreed well with experimentally determined values.

$$\frac{f_h}{f'_h} = \left(\frac{ry}{r+y}\right) \cdot \left(\frac{r'+y'}{r'y'}\right)$$

in which $f_h = f_h$ for one can size.

$f'_h = f_h$ for another can size.

r and $r' =$ corresponding can radii.

y and $y' =$ corresponding can lengths.

The following example illustrates how the equation is applied in practice.

Experimentally, the f_h of the heating curve for green beans in a 307 × 409 can was found to be 5.4 minutes. What would be the f_h of the same product in a 603 × 700 can?

Approximate inside dimensions of the 307 × 409 can are

$$r = 1.65 \text{ inches}$$
$$y = 4.19 \text{ inches}$$

Approximate inside dimensions of the 603 × 700 can are:

$$r = 3.03 \text{ inches}$$
$$y = 6.64 \text{ inches}$$

Taking f_h as that for the 603 × 700 can and f'_h as that for the 307 × 409 can, we have

$$f_h = 5.4 \left[\left(\frac{3.03 \times 6.64}{3.03 + 6.64}\right) \times \left(\frac{1.65 + 4.19}{1.65 \times 4.19}\right) \right]$$

$f_h = $ **9.5 minutes for the 603 × 700 can.**

This agrees exactly with the experimental value obtained for the product in the 603 × 700 can. Agreement is seldom exact, but usually very close.

Change of Container Size—Conduction-Heating Products

For these products, the parameter f_h varies with container dimensions and with thermal diffusivity, k, of the food. Thermal diffusivity refers to temperature flow similarly as thermal conductivity refers to heat flow. Numerically, its value may be expressed as follows (Olson and Jackson, 1942):

$$k = \frac{\text{thermal conductivity}}{\text{specific heat} \times \text{density}}$$

Following the method of Ball (1928), Olson and Jackson (1942) developed equations for determining thermal diffusivity from experimental heating curves obtained for variously shaped objects. Of these, two are of primary concern in converting f_h from one container size to another—that is, (1) an equation for a brick (rectangular food container) and (2) an equation for a finite cylinder (cylindrical food container).

Object	Equation
Brick	$k = \dfrac{0.933}{[(1/a^2) + (1/b^2) + (1/c^2)]f_h}$
Finite cylinder	$k = \dfrac{0.398}{[(1/a^2) + (0.427/b^2)]f_h}$

in which k = thermal diffusivity.

a = radius of finite cylinder or one-half the width of brick.

b = one-half the length of finite cylinder or one-half the length of brick.

c = one-half the thickness of brick.

The value of k may be expressed either as square centimeters per second or as square inches per minute. Since can dimensions are generally given in inches and f_h in minutes, it is more convenient in process calculation procedures to express k in square inches per minute. The equations of k given above may be used as f_h conversion equations as illustrated in the following example:

Suppose that heat penetration data are available for cream-style corn in a 307 × 409 can but not for a 603 × 700 can. To calculate a process for the product in the 603 × 700 can it is necessary to know the f_h value of the heating curve. The f_h value of the heating curve for the product in the 307 × 409 can is 60 minutes. What is the f_h value of the heating curve for the product in the 603 × 700 can?

First, from f_h of the 307 × 409 can, determine the value of k by

$$k = \frac{0.398}{[(1/a^2) + (0.427/b^2)]f_h}$$

The approximate inside dimensions of the 307 × 409 can are

$$a = 1.65 \text{ inches}$$
$$b = 2.095 \text{ inches}$$

Then

$$k = \frac{0.398}{[(1/1.65^2) + (0.427/2.095^2)]60}$$

$$k = 0.398/27.9 = \textbf{0.0143 square inch/minute}$$

Since thermal diffusivity is independent of container size or dimensions, its value may be substituted back in the same equation to determine f_h for the 603 × 700 can. That is,

$$f'_h = \frac{0.398}{[(1/a^2) + (0.427/b^2)]0.0143}$$

Approximate inside dimensions of the 603 × 700 can are

$$a = 3.03 \text{ inches}$$
$$b = 3.32 \text{ inches}$$

Then

$$f'_h = \frac{0.398}{[(1/3.03^2) + (0.427/3.32^2)]0.0143}$$

$$f'_h = 0.398/0.00211 = \textbf{189 minutes}$$

From these two f_h values, an f_h conversion factor may be computed as follows:

$$\frac{f_h \text{ of } 603 \times 700}{f_h \text{ of } 307 \times 409} = \frac{189}{60} = \textbf{3.15}$$

Since k is a product characteristic, the f_h conversion may be made without actually calculating k. Taking k, f_h, a, and b representing one container and k', f'_h, a', and b' representing another container,

$$k = k' = \frac{0.398}{[(1/a^2) + (0.427/b^2)]f_h} = \frac{0.398}{[(1/a'^2) + (0.427/b'^2]f'_h}$$

Then

$$[(1/a^2) + (0.427/b^2)]f_h = [(1/a'^2) + (0.427/b'^2)]f'_h$$

and

$$f'_h = \frac{[(1/a^2) + (0.427/b^2)]f_h}{(1/a'^2) + (0.427/b'^2)}$$

Since thermal diffusivity for straight conduction-heating products varies only slightly, f_h values for product heating in a number of different can sizes may be computed, and conversion factors tabulated. A table of conversion factors is convenient for use in routine process calculation work.

If j is considered invariable for both the heating and cooling curves, conversion of f_h is all that is needed for calculating the F_c values of processes for conduction-heating products in containers of different sizes, through the following:

$$B = f_h(\log j_{ch}I_h - \log g_c)$$
$$f_h/U{:}g \text{ relationship}$$
$$F_c = U/F_i$$

Change in Value of j with Change in Container Dimensions

Olson and Jackson (1942) discussed the reasons that, in practical application, j for conduction-heating products varies with the ratio of can length to diameter, even though mathematical methods indicate that it is constant. Equations of heating curves are for asymptotes tangent to heating curves at infinity. If the slope of the asymptote were to be determined experimentally, it would be necessary to continue the heating curve to infinity. In practice, straight lines are drawn as heat penetration curves tangent to the heating curve at one or two log cycles from the origin, and slope values from these are used in the equation of the heating curve. The use of such tangents makes j dependent on the ratio of can length to diameter, the extent of the dependence depending on the point of tangency used in drawing the straight line. Fortunately, in practice, if two tangents of different slopes are drawn to one heating curve, the j value will be larger for the one having the smaller f_h value, and the difference in f_h value will tend to compensate for the difference in j in the equation of the heating curve:

$$B = f_h(\log jI - \log g)$$

As was pointed out by Olson and Jackson (1942), theoretically the j at the center of an infinite slab is 1.27, and that at the center of an infinite cylinder is 1.60. (Because the solution for a finite cylinder is the product of the solution for the infinite slab and the solution for the infinite cylinder, j at the center of a finite cylinder ought to approximate the product of these two values (1.27 and 1.60)—that is, 2.03.) They pointed out, further, that one would expect j at the center of a finite cylinder to vary somewhat between these limits as the ratio of length to diameter varies from zero to infinity. They demonstrated, with j values obtained from ninety-eight heat penetration tests, that j did vary in the manner expected.

How much does j vary for the normal range of can sizes? If the tangent is drawn to a point two log cycles from the origin of the heating curve, the variation is not serious—maximum of about 0.2 of a j unit (according to the analysis of Olson and Jackson, in which they related a change in value of j to the ratio b^2/a^2). One-half the can length was taken as b, and the can radius as a. The analysis shows that b^2/a^2 may vary from 0.2 to 2.0 without appreciable variation in the value of j. The ratio b^2/a^2 equal to 0.2 represents a ratio of can length to can diameter of about 0.5, and the ratio equal to 2.0 represents a ratio of can length to can diameter of about 1.4. The experimental data presented by Olson and Jackson indicate that variation in j over this range of ratios of can length to can diameter should be no greater than about 0.2 of a j unit.

Therefore, for conversion of heat penetration data from one can size to another, over the range of common can sizes, considering j to be invariant with can size should not cause serious errors in process calculations. The following calculation demonstrates the point.

Case 1	Case 2
$j_{ch} = j_{cc} = 1.8$	$j_{ch} = j_{cc} = 2.00$

$$T_x = 250$$
$$f_h = f_c = 60$$
$$I_h = 100$$
$$T_r = 250$$
$$F_c = 6.00$$
$$z = 18.00$$

Case 1

$$U_c = F_c F_i = 6.00$$
$$f_h/U_c = 60/6 = \mathbf{10}$$

From the $f_h/U{:}g$ tables find, for $j_{cc} = 1.8$, $z = 18.00$, and $f_h/U_c = 10$,

$$g_c = \mathbf{10.93}$$

Then

$$B = f_h(\log j_{ch} I_h - \log g_c)$$
$$B = 60(\log 180 - \log \mathbf{10.93})$$
$$B = \mathbf{72.9\ minutes}$$

Case 2

$$U_c = F_c F_i = 6.00$$
$$f_h/U_c = \mathbf{10.00}$$

From the $f_h/U{:}g$ tables find, for $j_{cc} = 2.00$, $z = 18.00$, and $f_h/U_c = 10$,

$$g_c = \mathbf{11.47}$$

Then

$$B = 60(\log 200 - \log 11.47)$$
$$B = \mathbf{74.5\ minutes}$$

SELECTED REFERENCES

Ball, C. O. (1928). Mathematical solution of problems on thermal processing of canned food. *Univ. Calif. (Berkeley) Publ. Public Health,* **1,** 245 pp.

Olson, F. C. W., and Jackson, J. M. (1942). Heating curves—theory and practical application. *Ind. Eng. Chem.,* **34,** 337.

Schultz, O. T., and Olson, F. C. W. (1938). Thermal processing of canned foods in tin containers. I. Variation in heating rate with can size for products heating by convection. *Food Res.,* **3,** 647.

CHAPTER 14

Typical Process Determination
Problems—Canned Foods

Problem I

With regard to this problem and the problems to follow, any resemblance to actual cases is not purely coincidental. However, some of the basic resistance and process parameters were chosen arbitrarily.

A food packer has developed a low-acid food product (pH 6.2) which he wishes to pack in 603 × 700 cans for institutional trade and in 307 × 409 cans for retail trade. Because a part of the pack is to be exported to tropical areas, it is agreed that the entire pack should be processed so as to reduce the rate of thermophilic incubator spoilage to 0.1% or less. By bacteriologic analysis it has been established that the product, on the average, would be expected to have, before processing, about three heat-resistant thermophilic spores per gram. The spores that are most heat resistant have been isolated, and the organism producing them has been tentatively identified as *Bacillus stearothermophilus*. Heat resistance determinations have established that the spores are characterized by a z of 14 and a D_r of 3.2.

Heat penetration determinations for the product in the 603 × 700 can have been made and the following parameters determined:

$$f_h = f_c = 200$$
$$T_{ih} = 160$$
$$T_{pih} = 70$$
$$T_x = 250$$

The product is to be given a "still" cook at 250°F., and the minimum retort come-up time (l) is to be 15 minutes. What process times should be recommended for the product in the two can sizes?

Step 1. Determine the F_s desired for the 603 × 700 can.

$$\text{Can capacity} = 3000 \text{ gm.}$$
$$a = \text{Spores per can} = 3 \times 3000 = 9000$$
$$b = 0.001$$
$$F_s = D_r(\log a - \log b)$$
$$F_s = 3.2(\log 9000 - \log 0.001)$$
$$F_s = \mathbf{22.2}$$

Step 2. Determine the F_c corresponding to the F_s of 22.2. This is done by interpolation. First choose

$$F_{c_1} = 15$$
$$F_{c_2} = 20$$

Determine F_{s_1} corresponding to F_{c_1}:

$$F_i = \log^{-1}\frac{250 - T_r}{z} = \mathbf{1.00}$$

$$U_{c_1} = F_{c_1}F_i = 15 \times 1 = \mathbf{15}$$
$$f_h/U_{c_1} = 200/15 = \mathbf{13.3}$$

Calculate the value of j_{ch} as follows:

$$j_{ch} = \frac{T_r - T_{pih}}{T_r - T_{ih}} = \frac{250 - 70}{250 - 160} = \frac{180}{90} = \mathbf{2.00}$$

Take

$j_{cc} = j_{ch}$ (heat penetration data show this to be a safe procedure)
$j_{\lambda c} = 0.5 j_{cc} = 1.00$

From the $f_h/U{:}g$ tables find, for $z = 14$, $j_{cc} = 2.00$, and $f_h/U_c = 13.3$,

$$g_{c_1} = \mathbf{10.22}$$
$$g_{\lambda_1} = 0.5g_c = \mathbf{5.11}$$

From the $f_h/U{:}g$ tables find, for $z = 14$, $j_{\lambda c} = 1.00$, and $g_{\lambda_1} = 5.11$

$$f_h/U_{\lambda_1} = \mathbf{6.89}$$
$$U_{\lambda_1} = 200/6.89 = \mathbf{29.0}$$
$$F_{\lambda_1} = U_{\lambda_1}/F_i = \mathbf{29.0}$$
$$F_{s_1} = F_{c_1} + D_r\left(1.084 + \log\frac{F_{\lambda_1} - F_{c_1}}{D_r}\right)$$

$$F_{s_1} = 15 + 3.2\left(1.084 + \log\frac{29.0 - 15}{3.2}\right) = 20.5$$

Determine F_{s_2} corresponding to $F_{c_2} = 20$:

$$U_{c_2} = F_{c_2}F_i = 20 \times 1 = 20$$
$$f_h/U_{c_2} = 200/20 = 10.0$$

From the $f_h/U{:}g$ tables find, for $z = 14$, $j_{cc} = 2.00$, and $f_h/U_{c_2} = 10.0$,

$$g_{c_2} = 8.77$$
$$g_{\lambda_2} = 0.5g_c = 4.39$$

From the $f_h/U{:}g$ tables find, for $z = 14$, $j_{\lambda c} = 1.00$, and $g_{\lambda_2} = 4.39$

$$f_h/U_{\lambda_2} = 5.61$$
$$U_{\lambda_2} = 200/5.61 = 35.7$$
$$F_{\lambda_2} = U_{\lambda_2}/F_i = 35.7$$
$$F_{s_2} = F_{c_2} + D_r\left(1.084 + \log\frac{F_{\lambda_2} - F_{c_2}}{D_r}\right)$$
$$F_{s_2} = 20 + 3.2\left(1.084 + \log\frac{35.7 - 20.0}{3.2}\right) = 25.7$$

Determine F_c desired—that corresponding to $F_s = 22.2$

$$F_c \text{ desired} = F_{c_1} + \frac{F_{c_2} - F_{c_1}}{F_{s_2} - F_{s_1}}(F_s \text{ desired} - F_{s_1})$$
$$F_c \text{ desired} = 15 + \frac{20 - 15}{25.7 - 20.5}(22.2 - 20.5)$$
$$= 16.6$$

Step 3. Determine the process time at 250°F. to obtain F_c of 16.6 in the product packed in the 603 × 700 can.

$$B = f_h(\log j_{ch}I_h - \log g_c)$$
$$I_h = T_r - T_{ih} = 250 - 160 = 90$$
$$j_{ch} = 2.00$$
$$U_c = F_cF_i = 16.6$$
$$f_h/U_c = 200/16.6 = 12.0$$

From the $f_h/U{:}g$ tables find, for $z = 14$, $j_{cc} = 2.00$, and $f_h/U_c = 12.0$

$$g_c = 9.65$$

Then

$$B = 200[\log(2.00 \times 90) - \log 9.65] = \textbf{254 minutes}$$

and

$$P_t = B - 0.4(l) = 254 - 0.4(15) = \textbf{248 minutes}$$

Step 4. Determine F_s of an equivalent process for the product packed in the 307 × 409 can.

Capacity of 603 × 700 can = 3000 gm.
Capacity of 307 × 409 can = 570 gm.

Take F_s for 603 × 700 can as F_{s_b}, and F_s for 307 × 409 can as F_{s_a}.

$$F_{s_b} = F_{s_a} + D_r \log \frac{V_b}{V_a}$$

$$22.2 = F_{s_a} + 3.2 \log \frac{3000}{570}$$

$$F_{s_a} = \textbf{19.9} = F_s \textbf{ desired}$$

Step 5. Determine the f_h value (f'_h) for the product in the 307 × 409 can.

$$f'_h = \frac{[(1/a^2) + (0.427/b^2)]f_h}{(1/a'^2) + (0.427/b'^2)}$$

in which a = can radius.
$b = \frac{1}{2}$ can length.
Using approximate inside dimensions,

For 603 × 700 can, $a = 3.03$
$b = 3.32$

For 307 × 409 can, $a = 1.65$
$b = 2.10$

Then since f_h for the product in the 603 × 700 can is 200,

$$f'_h = \frac{[(1/3.03^2) + (0.427/3.32^2)]200}{(1/1.65^2) + (0.427/2.10^2)}$$

$$f'_h = \frac{[(1/9.18) + (0.427/11.0)]200}{(1/2.72) + (0.427/4.41)}$$

$$f'_h = 29.6/0.465 = \textbf{63.6}$$

Step 6. Determine F_c corresponding to the F_s desired of 19.9. Choose

$$F_{c_1} = 13$$
$$F_{c_2} = 18$$

Determine F_{s_1} corresponding to $F_{c_1} = 13$:

$$U_{c_1} = F_{c_1}F_i = 13 \times 1 = 13$$
$$f_h/U_{c_1} = 63.6/13 = \mathbf{4.89}$$

Taking j_{cc} the same as for the 603×700 can, from the $f_h/U{:}g$ tables find, for $z = 14$, $j_{cc} = 2.00$, and $f_h/U_{c_1} = 4.89$,

$$g_{c_1} = \mathbf{5.23}$$

then

$$g_{\lambda_1} = \mathbf{2.62}$$

From $f_h/U{:}g$ tables find, for $z = 14$, $j_{\lambda c} = 1.00$ and $g_{\lambda_1} = \mathbf{2.62}$,

$$f_h/U_{\lambda_1} = \mathbf{3.11}$$
$$U_{\lambda_1} = 63.6/3.11 = \mathbf{20.4}$$
$$F_{\lambda_1} = U_{\lambda_1}/F_i = \mathbf{20.4}$$

$$F_{s_1} = F_{c_1} + D_r\left(1.084 + \log\frac{F_{\lambda_1} - F_{c_1}}{D_r}\right)$$

$$F_{s_1} = 13 + 3.2\left(1.084 + \log\frac{20.4 - 13}{3.2}\right) = \mathbf{17.6}$$

Determine F_{s_2} corresponding to $F_{c_2} - 18$

$$U_{c_2} = F_{c_2}F_i = 18 \times 1 = \mathbf{18}$$
$$f_h/U_{c_2} = 63.6/18 = \mathbf{3.53}$$

From the $f_h/U{:}g$ tables find, for $z = 14$, $j_{cc} = 2.00$, and $f_h/U_{c_2} = 3.53$,

$$g_{c_2} = 3.83$$

then

$$g_{\lambda_2} = \mathbf{1.92}$$

From the $f_h/U{:}g$ tables find, for $z = 14$, $j_{\lambda c} = 1.00$, and $g_{\lambda_2} = 1.92$,

$$f_h/U_{\lambda_2} = \mathbf{2.37}$$
$$U_{\lambda_2} = 63.6/2.37 = \mathbf{26.8}$$
$$F_{\lambda_2} = \frac{U_{\lambda_2}}{F_i} = \mathbf{26.8}$$

$$F_{s_2} = 18 + 3.2\left(1.084 + \log\frac{26.8 - 18}{3.2}\right) = \mathbf{22.9}$$

Determine F_c desired—that corresponding to $F_s = 19.9$:

$$F_c \text{ desired} = F_{c_1} + \frac{F_{c_2} - F_{c_1}}{F_{s_2} - F_{s_1}} (F_s \text{ desired} - F_{s_1})$$

$$F_c \text{ desired} = 13 + \frac{18 - 13}{22.9 - 17.6} (19.9 - 17.6) = \mathbf{15.2}$$

Step 7. Determine process time at 250°F. to obtain F_c of 15.2.

$$B = f_h(\log j_{ch} I_h - \log g_c)$$
$$I_h = T_r - T_{ih} = 250 - 160 = \mathbf{90}$$
$$j_{ch} = \mathbf{2.00}$$
$$U_c = F_c F_i = \mathbf{15.2}$$
$$f_h/U_c = 63.6/15.2 = \mathbf{4.18}$$

From the $f_h/U{:}g$ tables find, for $z = 14$, $j_{cc} = 2.00$, and $f_h/U_c = 4.18$,

$$g_c = \mathbf{4.54}$$

Then

$$B = 63.6[\log (2.00 \times 90) - \log 4.54] = \mathbf{102 \ minutes}$$

and

$$P_t = B - 0.4(l) = 102 - 0.4(15) = \mathbf{96 \ minutes}$$

Cooling Time.   It is desired to water-cool these products so that at the end of cooling the mass average can temperature will be approximately 100°F. The average temperatures of the cooling water to be used is about 70°F. What time would be required to cool the product in each size container?

COOLING PRODUCT IN 603×700 CAN. Assume that the effective cooling-water temperature is 79°F. (see Chapter 11). Then

$$T_{sc} = T_w + 0.27(T_c - T_w).$$

in which T_{sc} = mass average temperature at the end of water cooling.
   T_w = effective cooling-water temperature.
   T_c = temperature at geometrical center of can at the end of water cooling.

Then

$$100 = 79 + 0.27(T_c - 79)$$
$$T_c = \mathbf{157°F.}$$

Find y_c from

$$y_c = T_c - T_w = 157 - 79 = \mathbf{78}$$

But, also

$$y_c = j_{cc}I_c \log^{-1}(-t/f_c)$$

in which $j_{cc} = j$ of cooling curve.

I_c = difference between center food temperature and cooling water temperature at start of cooling, or $(T_{ic} - T_w)$.

t = time, in minutes, of cooling.

$f_c = f$ of cooling curve.

Since

$$T_{ic} = T_r - g_c$$
$$T_{ic} = 250 - 9.65 = \mathbf{240.35}$$

and

$$I_c = 240.35 - 79 = \mathbf{161.35}$$

Since

$$y_c = j_{cc}I_c \log^{-1}(-t/f_c)$$
$$\log y_c = \log j_{cc} + \log I_c - t/f_c$$
$$t = (\log j_{cc} + \log I_c - \log y_c)f_c$$
$$j_{cc} = \mathbf{2.00}$$
$$f_c = \mathbf{200}$$

Then

$$t = (\log 2.00 + \log 161.35 - \log 78)200$$

$t = \mathbf{123}$ **minutes = cooling time to obtain mass average temperature of 100°F.**

Not including time to load and unload the retort, the complete processing cycle would require 386 minutes $(15 + 248 + 123)$. This is a severe process, and considerable product damage might be expected.

COOLING PRODUCT IN 307 × 409 CAN

$$T_{sc} = T_w + 0.27(T_c - T_w)$$
$$100 = 79 + 0.27(T_c - 79)$$
$$T_c = \mathbf{157}$$
$$y_c = T_c - T_w = 157 - 79 = \mathbf{78}$$
$$y_c = j_{cc}I_c \log^{-1}(-t/f_c)$$

$$t = (\log j_{cc} + \log I_c - \log y_c)f_c$$
$$f_c = 63.6$$
$$j_{cc} = 2.00$$
$$I_c = T_{ic} - T_w$$
$$T_{ic} = T_r - g_c = 250 - 4.54 = \mathbf{245.46}$$
$$I_c = 245.46 - 79 = \mathbf{166.46}$$
$$t = (\log 2.00 + \log 166.46 - \log 78)63.6$$
$$t = \mathbf{40\ minutes = cooling\ time\ to\ obtain\ mass}$$
$$\mathbf{average\ temperature\ of\ 100°F.}$$

Not including time to load and unload the retort, the complete processing cycle would require 151 minutes (15 + 96 + 40).

These processes were tried, and it was found that severe degradation in product quality occurred even with the product packed in the 307 × 409 can. That packed in the 603 × 700 can was considered unsalable. Facilities were available for agitating cans of both sizes during process, by axial rotation of the 307 × 409 can through a continuous cooker, and by end-over-end rotation of the 603 × 700 can in a "batch-type" cooker. Heat penetration determinations were made in pilot-scale models of the cookers. The product was found to agitate well, as indicated by the following data:

307 × 409 can	603 × 700 can
$f_h = f_c = 8.00$	$f_h = f_c = 17.00$
$j_{ch} = j_{cc} = 1.00$	$j_{ch} = j_{cc} = 1.2$

What process times should be used for products in these types of cookers? It was decided that a retort temperature of 260°F. would be used for processing the 307 × 409 can in the continuous cooker, and a retort temperature of 255°F. for processing the 603 × 700 can in the batch cooker.

PROCESS FOR PRODUCT IN 307 × 409 CAN. Because of virtually uniform heating under agitation, take

$$F_c = F_s$$

Then

$$F_c \text{ desired} = \mathbf{19.9}$$

and

$$B = f_h(\log j_{ch}I_h - \log g_c)$$

First, determine F_i:

$$F_i = \log^{-1} \frac{250 - T_r}{z}$$

$$F_i = \log^{-1} \frac{250 \quad 260}{14} = \log^{-1}(-0.715)$$

$$F_i = \mathbf{0.193}$$

Next, determine g_c:

$$U_c = F_c F_i$$
$$U_c = 19.9 \times 0.193 = \mathbf{3.84}$$
$$f_h/U_c = 8.00/3.84 = \mathbf{2.08}$$

From the $f_h/U{:}g$ tables find, for $z = 14$, $j_{cc} = 1.00$, and $f_h/U_c = 2.08$,

$$g_c = \mathbf{1.64}$$

Now determine I_h:

$$I_h = T_r - T_{ih} = 260 - 160 = \mathbf{100}$$

and

$$B = 8[\log 1.00 \times 100) - \log 1.64]$$
$$B = \mathbf{14.3 \text{ minutes}}$$

Because $l = 0$ for continuous cook,

$$P_t = B = \mathbf{14.3 \text{ minutes}}$$

PROCESS FOR PRODUCT IN 603×700 CAN

Again,

$$F_c = F_s$$

and in this case,

F_c desired $= \mathbf{22.2}$

$$F_i = \log^{-1} \frac{250 - 255}{14} = \log^{-1}(-0.357)$$

$$F_i = \mathbf{0.439}$$

$$U_c = F_c F_i = 22.2 \times 0.439 = \mathbf{9.75}$$

$$f_h/U_c = 17/9.75 = \mathbf{1.74}$$

From the $f_h/U:g$ tables find, for $z = 14$, $j_{cc} = 1.2$, and $f_h/U_c = 1.74$,

$$g_c = \mathbf{1.33}$$
$$I_h = 255 - 160 = \mathbf{95}$$

Then

$$B = 17 \log(1.2 \times 95) - \log 1.33 = \mathbf{32.8 \ minutes}$$

Minimum retort come-up time was found to be 15 minutes. Therefore,

$$P_t = 32.8 - 0.4(15) = \mathbf{26.8 \ minutes}$$

The product given these processes was of very satisfactory quality, though the quality of the product in the 307 × 409 can was superior.

Problem II

A food packer had been for many years packing a low-acid formulated food product successfully, receiving almost no consumer complaints. Suddenly, consumer complaints began arriving by the dozens. Though variously worded, all complaints were very similar. The product had a sour off-flavor. Shortly after consumer complaints started arriving, the quality control department reported that random samples of the product incubated at 122°F. for one week were showing a high rate of flat-sour spoilage, on the average about 18%. Prior incubator spoilage had been less than 1%.

This was one of the company's best selling products—a major item. Something had to be done quickly. It was recognized that a thorough study of the situation was necessary to determine the cause of the sudden outbreak of spoilage. It was also recognized that this would probably take valuable time and that, if possible, something must be done immediately. The entire problem was placed in the hands of research and quality control personnel with instructions to drop everything else until it was solved.

What immediate steps were taken to alleviate the situation while a thorough investigation could be conducted? First, cans of spoiled food were examined microscopically for evidence of can leakage. Rather slender rod-shaped bacteria were observed, many of which were granulated. This was an indication that spoilage was not due to can leakage subsequent to the heat process. Samples of the spoiled food were cultured and placed under incubation (see Chapter 3). Next, the cans that contained spoiled product were carefully examined for dents, internal and external corrosion, etc. Can seams were stripped and carefully examined.

No imperfections could be found that would account for serious can leakage. Records showed that a residual chlorine concentration of 3 p.p.m. was being maintained in the cooling water. Determinations of pH showed that, on the average, pH of the spoiled product was about one unit lower than that of the unspoiled product. On the basis of this evidence it was tentatively concluded that spoilage was due to underprocessing. But with a process that had been satisfactory for more than ten years?

For some reason the thermophilic spore load must suddenly have increased. Examination of records of routine spore counts on ingredients and finished unprocessed product denied this. Then, in some way a thermophile must have been introduced that had a higher resistance than the one for which the thermal process was established. Had the formula of the product been changed, had some other ingredient or ingredients been introduced? The foreman of the canning department said no. Then there must have been a failure in retort temperature and/or time control. Examination of automatic recorder charts denied this.

On the basis of this preliminary evidence, and in spite of it, it was reasoned that a thermophile had been introduced that had a higher resistance than the one for which the thermal process was established. An increase in process was indicated, at least temporarily. But how much?

The product, packed in 307 × 409 cans (570 gm. of product per can), had been receiving a process based on an average maximum initial spore load of five spores per gram. The process was designed to reduce this load to one spore per 100 cans. Based on a number of heat resistance determinations over the years, maximum heat resistance of the spores was considered to be characterized by $D_r = 3$ and $z = 18$. The product was being given a still-cook process. Heat penetration determinations had established the following parameters:

$$f_h = f_c = 60$$
$$j_{ch} = j_{cc} = 2.00$$
$$T_{ih} = 160$$
$$T_x = 250$$

The product was being processed at 250°F. The minimum retort come-up time was 15 minutes. The process being given had been arrived at as follows:

$$F_s = D_r(\log a - \log b)$$
$$F_s = 3[\log (5 \times 570) - \log 0.01] = \mathbf{16.3} = F_s \text{ desired}$$

The F_c corresponding to this F_s was determined next:

$$F_{c_1} = 10 \text{ was chosen.}$$
$$F_{c_2} = 15 \text{ was chosen.}$$
$$U_{c_1} = F_{c_1} F_i = \mathbf{10}$$
$$f_h/U_{c_1} = 60/10 = \mathbf{6}$$
$$U_{c_2} = F_{c_2} F_i = \mathbf{15}$$
$$f_h/U_{c_2} = 60/15 = \mathbf{4}$$

From the $f_h/U : g$ tables find, for $z = 18$ and $j_{cc} = 2.00$,

$$g_{c_1} = \mathbf{8.07}$$
$$g_{c_2} = \mathbf{5.75}$$

Then

$$g_{\lambda_1} = 0.5 \times 8.07 = \mathbf{4.04}$$
$$g_{\lambda_2} = 0.5 \times 5.75 = \mathbf{2.88}$$

From the $f_h/U : g$ tables find, for $z = 18$ and $j_{\lambda_c} = 1.00$,

$$f_h/U_{\lambda_1} = \mathbf{3.68}$$
$$f_h/U_{\lambda_2} = \mathbf{2.71}$$

Then

$$U_{\lambda_1} = 60/3.68 = \mathbf{16.3} = F_{\lambda_1}$$
$$U_{\lambda_2} = 60/2.71 = \mathbf{22.1} = F_{\lambda_2}$$

$$F_s = F_c + D_r \left(1.084 + \log \frac{F_\lambda - F_c}{D_r} \right)$$

$$F_{s_1} = 10 + 3 \left(1.084 + \log \frac{16.3 - 10}{3} \right) = \mathbf{14.2}$$

$$F_{s_2} = 15 + 3 \left(1.084 + \log \frac{22.1 - 15}{3} \right) = \mathbf{19.4}$$

$$F_c \text{ desired} = F_{c_1} + \frac{F_{c_2} - F_{c_1}}{F_{s_2} - F_{s_1}} (F_s \text{ desired} - F_{s_1})$$

$$F_c \text{ desired} = 10 + \frac{15 - 10}{19.4 - 14.2} (16.3 - 14.2)$$

$$F_c \text{ desired} = \mathbf{12.0}$$

$$B = f_h (\log j_{ch} I_h - \log g_c)$$

$$I_h = T_r - T_{ih} = 250 - 160 = \mathbf{90}$$
$$f_h/U_c = 60/12.0 = \mathbf{5.00}$$

From the $f_h/U:g$ tables find, for $z = 18$, $j_{cc} = 2.00$, and $f_h/U_c = \mathbf{5.00}$,

$$y_c = \mathbf{6.99}$$

Then

$$B = 60[\log (2.00 \times 90) - \log 6.99]$$
$$B = \mathbf{84.7 \ minutes}$$
$$P_t = B - 0.4l$$
$$P_t = 84.7 - (0.4 \times 15) = \mathbf{78.7 \ minutes}$$

A process time of 79 minutes was used by the operator. This should have kept the average rate of incubator spoilage below one can per 100 if there were no adverse change in the contamination picture. What adverse change or changes in the contamination picture could account for the sudden outbreak of spoilage? As was pointed out above, there was no increase in the initial spore load and therefore a different organism must have been introduced. How must this organism differ in resistance in order to have survived the process in greater numbers. Naturally it had to be more resistant, but how could this greater resistance be characterized? One of three things had to be. The z value for the organism could be the same and the D_r value greater; or the D_r value could be the same and the z value smaller; or the D_r value could be greater and the z value smaller. How much greater would the D_r value have to be to account for 18% spoilage compared to 1% spoilage, if the z value were assumed to be the same?

As was shown above, for a spore load of five spores per gram, characterized by $D_r = 3.00$, in order to reduce the load to one spore per 100 cans,

$$F_s = 3[\log (5 \times 570) - \log 0.01] = \mathbf{16.3}$$

What was the most probable number of spores surviving to cause spoilage of 18 cans per 100?

$$\bar{x} = 2.303 \log \frac{n}{q} \times 100$$

$$\bar{x} = 2.303 \log \frac{100}{82} \times 100 = \mathbf{20}$$

Then, since

$$D_r = \frac{F_s}{\log a - \log b}$$

$$D_r = \frac{16.3}{\log 2850 - \log 0.20} = \mathbf{3.93}$$

How much would the process have to be increased if spores of the new organism encountered had a D_r value of 4.00? In this case,

$$F_s = 4(\log 2850 - \log 0.01) = \mathbf{21.8}$$

A new process time was computed similarly as was the original process time. The F_c corresponding to an F_s of 21.8 was first determined.

$$F_{c_1} = 10 \text{ was chosen.}$$
$$F_{c_2} = 15 \text{ was chosen.}$$

Since these are the same as those used in the original calculation, determine corresponding F_{s_1} and F_{s_2} values for $D_r = 4.00$ and find,

$$F_c \text{ desired} = 10 + \frac{15 - 10}{20.4 - 15.1}(21.8 - 15.1) = \mathbf{16.3}$$
$$f_h/U_c = 60/16.4 = \mathbf{3.68}$$

From the $f_h/U : g$ tables find, for $z = 18$, $j_{cc} = 2.00$, and $f_h/U_c = 3.68$,

$$g_c = \mathbf{5.29}$$

Then

$$B = 60(\log 180 - \log 5.29)$$
$$B = \mathbf{92 \text{ minutes}}$$
$$P_t = 92 - 6 = \mathbf{86 \text{ minutes}}$$

Before this process was recommended as, at least, a temporary expedient, it was decided to determine what the effect on process time would be if the D_r of the new organism were 3.00 but the z value were 12 instead of 18. Seldom have z values of less than 12 been observed for flat-sour spores. This required another complete calculation. The F_s desired, or required, in this case would be the same as for the original process—that is, 16.3.

$$F_{c_1} = 10 \text{ was chosen.}$$
$$F_{c_2} = 15 \text{ was chosen.}$$

Then

$$U_{c_1} = F_{c_1}F_i = \mathbf{10}$$
$$f_h/U_{c_1} = 60/10 = \mathbf{6}$$
$$U_{c_2} = F_{c_2}F_i = \mathbf{15}$$
$$f_h/U_{c_2} = 60/15 = \mathbf{4}$$

From the $f_h/U:g$ tables find, for $z = 12$ and $j_{cc} = 2.00$,

$$g_{c_1} = \mathbf{5.05}$$
$$g_{c_2} = \mathbf{3.59}$$

Then

$$g_{\lambda_1} = \mathbf{2.53}$$
$$g_{\lambda_2} = \mathbf{1.80}$$

From the $f_h/U:g$ tables find, for $z = 12$ and $j_{\lambda_1} = 1.00$,

$$f_h/U_{\lambda_1} = \mathbf{3.47}$$
$$f_h/U_{\lambda_2} = \mathbf{2.51}$$
$$U_{\lambda_1} = 60/3.47 = \mathbf{17.3} = F_{\lambda_1}$$
$$U_{\lambda_2} = 60/2.51 = \mathbf{23.9} = F_{\lambda_2}$$
$$F_s = F_c + D_r\left(1.084 + \log\frac{F_{\lambda_1} - F_c}{D_r}\right)$$
$$F_{s_1} = 10 + 3\left(1.084 + \log\frac{17.3 - 10}{3}\right) = \mathbf{14.4}$$
$$F_{s_2} = 15 + 3\left(1.84 + \log\frac{23.9 - 15}{3}\right) = \mathbf{19.7}$$
$$F_c \text{ desired} = F_{c_1} + \frac{F_{c_2} - F_{c_1}}{F_{s_2} - F_{s_1}}(F_s \text{ desired} - F_{s_1})$$
$$F_c \text{ desired} = 10 + \frac{15 - 10}{19.7 - 14.4}(16.3 - 14.4) = \mathbf{11.8}$$
$$B = f_h(\log j_{ch}I_h - \log g_c)$$
$$U_c = F_cF_i = \mathbf{11.8}$$
$$f_h/U_c = 60/11.8 = 5.09$$

From the $f_h/U:g$ tables find, for $z = 12$, $j_{cc} = 2.00$, and $f_h/U_c = \mathbf{5.09}$,

$$g_c = \mathbf{4.42}$$

Then

$$B = 60(\log 180 - \log 4.42)$$
$$B = \mathbf{96.6 \text{ minutes}}$$
$$P_t = 96.6 - 6 = \mathbf{90.6 \text{ minutes}}$$

On the basis of this, as a temporary expedient, it was recommended to the canning foreman that he increase his process time from the 79 minutes he had been using to 91 minutes. The process time was so increased, and thermophilic incubator spoilage decreased to less than 1%. However, the increase in process very noticeably reduced the organoleptic quality of the product; both color and flavor were impaired. It was important to return to the original process as soon as possible.

Investigation of the problem was continued. Cultures of the spoiled product yielded a spore-forming thermophilic flat-sour organism in pure culture. A spore suspension was prepared, and a thermal resistance determination made. The spores were suspended in the product during heating. Resistance values obtained were $D_r = 3.97$ and $z = 17.9$. In view of this and the above analysis, the process of 91 minutes was reduced to 85 minutes. This resulted in some improvement in product quality. Incubator spoilage remained below 1%.

In reviewing ingredient records, one ingredient change was discovered that coincided with the spoilage incident. Just prior to the beginning of spoilage troubles, imported sugar was substituted for domestic sugar. This was done with the sanction of the quality control department on the basis that the imported sugar met NCA standards for thermophilic spores, The company was interested in using the imported sugar because it was approximately one dollar per hundred cheaper than domestic. In further work, a flat-sour thermophile was isolated from the imported sugar which had almost the same heat resistance parameters as the one that had caused the spoilage.

The administrative decision was to use up the inventory of imported sugar, with the process of 85 minutes at 250°F., then change back to a domestic supply and the 79-minute process.

Problem III

Consider that you recently were hired in the research department of a major food packing company. Your first assignment is to review the sterilization processes given various canned food items and to recommend corrections where indicated. In your review, one of the first things that comes to your attention is that one glass-packed low-acid food item is being given a process of 160 minutes at 240°F., a process having an

$F_c{}^{18}$ value of about 26. Other similar glass-packed items are being given processes with $F_c{}^{18}$ values between 21 and 25. These processes appear, on the surface at least, to be overly severe. You decide that an investigation is indicated.

On inquiring as to why such severe processes are being given, you are told that sporadic but important thermophilic flat-sour spoilage was encountered with less severe processes. You still feel that such severe processes ought not to be necessary, but before recommending any reduction certain studies are necessary. What should they be?

First, information regarding the normal product spore load is needed. The quality control department has thermophilic spore load data for a number of product ingredients but not for the finished product prior to heat processing. You proceed to obtain these over a period of two months on product samples taken three times daily. Time permits sampling only one product, so you choose the one being processed at $F_c{}^{18} = 26$. The average maximum thermophilic spore load determined was three spores per gram, the range being from one to five spores per gram. (The average maximum spore count was taken as the average of counts for the half of the samples showing the highest counts.) During the sampling period, you obtain, by selective heating, ten isolates of reasonably high heat resistance.

You now proceed to prepare spore suspensions of these ten isolates and determine their thermal resistance at 250°F. You find D_{250} values ranging from 1.00 to 3.00 minutes. You select an isolate having a D_{250} value of 3.00 for determination of resistance at 240°, 230°, and 220°F. From the data obtained, plotted as a thermal destruction curve, you find that the organism is characterized by a z value of 16. According to these findings, what process should be adequate to reduce the number of thermophilic spores to less than one spore per 100 containers of this product?

First you determine the F_s value that should be required of the process. The product is being packed in 303 × 411 glass jars. The average net weight of the product per jar is 425 gm.

$$425 \times 3 = 1275 = \text{average maximum spore load per jar.}$$

Then

$$F_s = D_r(\log a - \log b)$$
$$F_s = 3(\log 1275 - \log 0.01) = \mathbf{15.3}$$

Now you determine the F_c corresponding to this F_s. The following data for the product are available:

$$T_x = 250$$
$$T_r = 240$$
$$T_{ih} = 160$$
$$f_h = f_c = 55$$
$$j_{ch} = j_{cc} = 2.00$$
$$l = 22$$

You compute the F_c value as follows:

$$F_s \text{ desired} = \mathbf{15.3}$$

$$F_i = \log^{-1} \frac{250 - 240}{16} = \mathbf{4.22}$$

You choose

$$F_{c_1} = 10 \qquad U_{c_1} = 42.2$$
$$F_{c_2} = 13 \qquad U_{c_2} = 54.9$$

Then

$$f_h/U_{c_1} = 55/42.2 = \mathbf{1.30} \qquad\qquad f_h/U_{c_2} = 55/54.9 = \mathbf{1.00}$$
$$g_{c_1} = \mathbf{1.10} \qquad\qquad g_{c_2} = \mathbf{0.609}$$
$$g_{\lambda_1} = \mathbf{0.550} \qquad\qquad g_{\lambda_2} = \mathbf{0.305}$$

$$j_{\lambda c} = \mathbf{1.00}$$

$$f_h/U_{\lambda_1} = \mathbf{1.04} \qquad\qquad f_h/U_{\lambda_2} = \mathbf{0.832}$$
$$U_{\lambda_1} = 55/1.04 = \mathbf{52.9} \qquad\qquad U_{\lambda_2} = 55/0.832 = \mathbf{66.2}$$
$$F_{\lambda_1} = 52.9/4.22 = \mathbf{12.5} \qquad\qquad F_{\lambda_2} = 66.2/4.22 = \mathbf{15.7}$$

$$F_{s_1} = 10 + 3\left(1.084 + \log \frac{12.5 - 10}{3}\right) = \mathbf{13.0}$$

$$F_{s_2} = 13 + 3\left(1.084 + \log \frac{15.7 - 13}{3}\right) = \mathbf{16.1}$$

$$F_c \text{ desired} = 10 + \frac{13 - 10}{16.1 - 13.0}(15.3 - 13.0) = \mathbf{12.2}$$

You compute the process time to obtain this F_c vaue as follows:

$$B = f_h(\log j_{ch}I_h - \log g_c)$$
$$I_h = 240 - 160 = \mathbf{80}$$
$$U_c = 12.2 \times 4.22 = \mathbf{51.5}$$

$$f_h/U_c = 55/51.5 = \mathbf{1.07}$$
$$g_c = \mathbf{0.723}$$
$$B = 55(\log 160 - \log 0.723) = \mathbf{129\ minutes}$$
$$P_t = B - 0.4l = 129 - 9 = \mathbf{120\ minutes}$$

This value is so much lower than the 160 minutes currently being used that you immediately suspect that something may be wrong with the process control. You examine the automatic recorder charts for the past year and find them in good order. Next you check the retort thermometers against standardized thermometers. They check within $\pm0.5°F$. Could the problem be one of nonuniform temperature distribution? A study of the retort hookup might give some clue. The retorts are horizontal and 22 feet long. In the bottom of each retort there are two steam–air perforated spreader pipes. There is only one main air–steam line feeding these. (Steam enters the air line just outside the retort.) This main line enters near the front end of the retort. Both the thermometer and the controller element are positioned on one side of the retort above the steam and air connections. To get satisfactory temperature distribution in a "water" retort, the total area of the holes in a spreader pipe should be the same or slightly less than the cross-sectional area of the pipe. This is particularly important for a long horizontal retort. However, you find that in these retorts the ratio of hole area to pipe area is about 1.5. This is a satisfactory ratio for "steam" retorts, but not for "water" retorts—it impedes continuous purging of water from the spreader pipes. Continuous air flow is depended on to agitate and circulate the water to maintain uniform temperature distribution.

In view of this retort hookup, you suspect poor retort temperature distribution. You obtain permission to run temperature distribution studies. Results of these studies confirm your suspicion. Maximum temperatures reached toward the back end of some of the retorts loaded with product were as much as 10 degrees below the designated 240°F. retort temperature. Toward the front, in the area of the thermometer and controller element, temperatures generally were no more than 1 degree low.

What F value was product in the coldest spots getting? You compute this, using, as T_r, 230°F. instead of 240°F. That is,

$$T_r = 230$$
$$T_{ih} = 160$$

$$f_h = f_c = 55$$
$$j_{ch} = j_{cc} = 2.00$$
$$l = 22 \text{ minutes}$$
$$B = P_t + 0.4l = 160 + 9 = 169$$
$$z = 16$$

Then

$$169 = 55[\log (70 \times 2) - \log g_c]$$
$$\log g_c = -0.934$$
$$g_c = 0.116$$

From the $f_h/U{:}g$ tables find, for $z = 16$, $j_{cc} = 2.00$, and $g_c = 0.116$,

$$f_h/U_c = \textbf{0.584}$$
$$U_c = 55/0.584 = \textbf{94.1}$$
$$F_c = U_c/F_i$$
$$F_i = \log^{-1} \frac{250 - 230}{16} = \textbf{17.8}$$
$$F_c = 94.1/17.8 = \textbf{5.29}$$

This is in contrast to the $F_c{}^{18}$ of 28 received by product in areas where the T_r was 240°F. It is also in contrast to the $F_c{}^{16}$ of 12.2 that you have shown should be required for sterilizing the product. The $F_s{}^{16}$ value of the process in the lowest temperature areas would be about 6.5. This is in contrast to the 15.3 you have determined as necessary. What rate of thermophilic incubator spoilage might be expected in product from these coldest areas?

$$F_s = D_r(\log a - \log b)$$
$$6.5 = 3(\log 1275 - \log b)$$
$$b = \textbf{8.7} = \text{Average number of spores per container}$$

This tells you that almost 100% spoilage might be expected (cf. $\bar{x}$ equation, Chapter 8). What rate of spoilage might be expected in areas where $T_r = 240$°F.? The $F_c{}^{16}$ value of the process where $T_r = 240$°F. is 21.5. The $F_s{}^{16}$ value corresponding to this is 24.6.
Then

$$24.6 = 3(\log 1275 - \log b)$$
$$b = \textbf{0.81} \times \textbf{10}^{-5}$$

This tells you that no more than 1 or 2 containers out of every 100,000 should spoil.

Spoilage expected of product from other areas of the retorts would fall between these wide limits. The problem is now well-defined. How do you propose to solve it? There would seem to be only one solution—that is, alteration of the retort hookup. You decide to recommend alterations to make the retorts conform to basic requirements for horizontal retorts used in processing glass-packed products, including a recirculation pump as recommended by the National Canners Association. This will cost money and must be justified to management. You estimate the cost of the proposed alterations, and ask that one retort only be altered until it can be shown that the alterations will solve the problem with respect to this retort. In presentation of your proposal you emphasize the following points:

1. Much of the product processed in the retorts is being severely overprocessed, with attendant loss of quality.
2. There is variation in product quality due to variation in process.
3. Obtaining uniform temperature distribution would correct these situations and would allow, on the average, a reduction of process times by at least 40 minutes.
4. Such a reduction in process times would allow three processing cycles per day instead of the two now possible; throughput could be increased by 50%.

Approval for altering the hookup of one retort is obtained. Subsequent to the alterations, temperature distribution shows variations no greater than ±0.5°F. from the control point. In consequence, hookups of the other retorts are likewise corrected. Processes are reduced to 129 min. at 240°F. as indicated by the above analysis. As predicted, thermophilic incubator spoilage of product so processed is no greater than 1%.

Not often do we find retorts as far out of line as the ones described in the above example. However, it is not uncommon, even with steam retorts, to find cold spots as much as 5 degrees low because of improper retort hookup or improper operation. Let us consider how much the spoilage chances would be increased in product processed at 235°F. rather than at 240°F. In the above example, it was determined, for the one product considered, that an F_s of 15.3 would be required if the rate of thermophilic incubator spoilage were to be kept to 1% or less. Then in a cold retort area at 235°F., the following values would obtain:

$$T_r = 235$$
$$T_{ih} = 160$$
$$f_h = f_c = 55$$
$$j_{ch} = j_{cc} = 2.00$$
$$B = 129$$
$$l = 22$$
$$D_r = 3.00$$
$$z = 16.00$$

Then, because

$$B = f_h(\log j_{ch}I_h - \log g_c)$$
$$129 = 55(\log 150 - \log g_c)$$
$$g_c = \mathbf{0.670}$$
$$f_h/U_c = \mathbf{1.04}$$
$$U_c = 55/1.04 = \mathbf{52.9}$$
$$F_i = \log^{-1}\frac{250 - 235}{16} = \mathbf{8.66}$$
$$F_c = U_c/F_i = 52.9/8.66 = \mathbf{6.11}$$
$$j_{\lambda_o} = \mathbf{1.00}$$
$$g_\lambda = \mathbf{0.335}$$
$$f_h/U_\lambda = \mathbf{0.862}$$
$$U_\lambda = 55/0.862 = \mathbf{63.8}$$
$$F_\lambda = 63.8/8.66 = \mathbf{7.37}$$
$$F_s = 6.11 + 3\left(1.084 + \log\frac{7.37 - 6.11}{3}\right) = \mathbf{8.23}$$
$$F_s = D_r(\log a - \log b)$$
$$8.23 = 3(\log 1275 - \log b)$$
$$\log b = \mathbf{0.363}$$
$$b = \mathbf{2.3} = \text{averager per container}$$

This tells us that, in the cold spot, about two spores per container should survive, and about 90% thermophilic incubator spoilage might be expected of the product from this area of the retort. (cf. $\bar{x}$ equation, Chapter 8). These examples amply demonstrate the great importance of accurate retort temperature control.

Problem IV

Not infrequently, because of breakdown or inadequate retort capacity, foods filled at a relatively high temperature (140°F. to 180°F.) cool to relatively low temperatures before being placed in the retort. How may corrections be made in process time to account for a drop in T_{ih}? Suppose that the process for a given product is based on a minimum initial temperature of 160°F. An electricity failure causes a cessation of operations for 2 hours; and, at the time of the failure, a retort load of product had just left the closing machine. During the 2 hour delay, the temperature of the product, on the average, dropped to 100°F. How should process time be adjusted? Take, for example, the product and process considered in problem III—that is,

$$T_r = 240 \qquad\qquad D_r = 3.00$$
$$T_{ih} = 160 \qquad\qquad z = 16$$
$$f_h = f_c = 55 \qquad\qquad F_s = 15.3$$
$$j_{ch} = j_{cc} = 2.00 \qquad\qquad F_c = 12.2$$
$$l = 22 \qquad\qquad F_i = 4.22$$
$$B = 129 \qquad\qquad \text{Spores per container} = 1275$$
$$P_t = 120$$

Now take $T_{ih} = 100$,

$$U_c = 12.2 \times 4.22 = 51.5$$
$$f_h/U_c = 55/51.5 = 1.07$$
$$g_c = 0.723$$
$$B = f_h(\log j_{ch}I_h - \log g_c)$$
$$B = 55 \log(2 \times 140) - \log 0.723 = \mathbf{142}$$
$$P_t = 142 - 9 = \mathbf{133}$$

This is an appreciable increase in process time. What would the F_s value be if the process time were not increased, and what would be the expected rate of thermophilic incubator spoilage?

$$B = 129 = 55(\log 280 - \log g_c)$$
$$\log g_c = 0.097$$
$$g_c = 1.25$$
$$f_h/U_c = 1.39$$
$$U_c = 55/1.39 = \mathbf{39.6}$$
$$F_c = 39.6/4.22 = \mathbf{9.38}$$
$$j_\lambda = \mathbf{1.00}$$
$$g_\lambda = \mathbf{0.625}$$
$$f_h/U_\lambda = \mathbf{1.10}$$
$$U_\lambda = 55/1.10 = \mathbf{50.0}$$
$$F_\lambda = 50.0/4.22 = \mathbf{11.9}$$

$$F_s = F_c + D_r \left(1.084 + \log \frac{F_\lambda - F_c}{D_r} \right)$$

$$F_s = 9.38 + 3 \left(1.084 + \log \frac{11.9 - 9.38}{3} \right) = \mathbf{12.40}$$

$$F_s = D_r(\log a - \log b)$$

$$12.40 = 3(\log 1275 - \log b)$$

$$\log b = \mathbf{-1.027}$$

$$b = \mathbf{0.094}$$

If the rate of thermophilic incubator spoilage were 1% under normal conditions ($T_{ih} = 160$), it should be about 9% under these conditions ($T_{ih} = 100$).

This example demonstrates the importance of process time adjustments for drops in T_{ih}. In practice it is a good idea to supply the retort operator with a chart showing how much time should be added for each 5-degree drop in T_{ih}. Because the correction varies with products and processes, a chart should be supplied with the process instruction for each product in each container size.

15

Evaluation and Equivalency of
Pasteurization Processes

As discussed in Chapter 4, the term *pasteurization* has become, through common usage, the term generally applied to mild heat treatments designed to sterilize foods with respect to organisms that are low in thermal resistance compared to those for which sterilization processes are designed to eliminate. Basically, as mentioned briefly in Chapter 12, graphical and mathematical procedures for evaluating pasteurization processes and for establishing equivalency of time–temperature relationships, to accomplish some given degree of sterilization with respect to a particular organism, are the same as employed when more resistant organisms are of chief concern.

Units of Lethality

In comparing the sterilizing capacity of different pasteurization processes, it is convenient for the unit of lethality to be such that the total sterilizing value of any given process can be expressed in minutes rather than as small fractions of 1 minute. For example, if the unit of lethality were taken as 1 minute of heating at 250°F. when considering pasteurization processes designed to free foods of pathogenic microorganisms, the total sterilizing (F) value of any given process would be a very small, and cumbersome, fraction of one. On the other hand, if the unit of lethality were taken as 1 minute of heating at 150°F., F values of the pasteurization processes would generally be in the range of 1 to 10 minutes—values convenient for comparison. Therefore, the unit of lethality chosen for evaluation computations should be chosen with consideration of the thermal resistance of the organism or organisms of concern. The unit may be the equivalent of heating for 1 minute at any designated temperature. For example one may choose 250°F., 180°F., or 150°F., or whatever, as the reference temperature. In stating the F value of a process the reference temperature is commonly designated in subscript; e.g., F_{250}, F_{212}, F_{180}, or F_{150}. Also, as noted in Chapter 10, the z value characterizing relative resistance of the organism, or organisms, of concern are commonly designated in superscript; e.g., F^8, F^{10}, F^{12}, F^{18}, etc. The definitive designation of the F value of a process, unless clearly understood from usage, would therefore carry both subscript and superscript, e.g., F^8_{150} when the z value is 8F.° and the reference temperature is 150°F.

Evaluation of Pasteurization Processes by the General Method

Many pasteurization processes, as sterilization processes, are such that a very significant amount of lethal heat is active during the heating of the product to some given temperature and during the cooling of the product from this temperature. Often the heating and cooling curves are not subject to exact mathematical description, thereby indicating the general method of process evaluation as the method of choice.

The same basic procedure, discussed in Chapter 11, is followed. In constructing lethal rate paper, L values are calculated by the equation

$$L = \log^{-1} \frac{T - T_z}{z} \tag{1}$$

in which T_z = designated reference temperature.

$\quad\quad T$ = any temperature used to label a principal horizontal line (see Fig. 24).

$\quad\quad L$ = lethal rate at any temperature, T. (Refer to Table A.1 in the Appendix for computed values of L for use in evaluating pasteurization and sterilization processes. Values of L corresponding to values of z not appearing in the table should be computed—sufficient accuracy is not obtained by interpolation.) For example, if 150°F. were employed as the reference temperature, L at any temperature T would be calculated by

$$L = \log^{-1} \frac{T - 150}{z} \tag{2}$$

Lethal rate paper is constructed exactly as described in Chapter 11. It shall be noted that lethal rate paper constructed for one given value of z and designed for use for a given processing temperature range may be used for any other processing temperature range so long as the z value remains the same. For example, suppose a lethal rate paper has been constructed for evaluating the lethality of a sterilization process wherein the maximum temperature reached by the food was about 247°F. and the z value characterizing relative resistance of the organism of concern was 10F.° In constructing the paper the top line was considered to represent unit lethality—in this case 1 minute of heating at 250°F. Then, 10 inches on the vertical scale would represent 1 unit of lethality and any principal horizontal line between the top and bottom lines would be placed a distance equal to the lethal rate, corresponding to any lower temperature, multiplied by 10. This is as explained in Chapter 11.

Now suppose one desires to evaluate a pasteurization process in terms of its equivalent (F value) in minutes at 150°F. instead of in minutes at 250°F. as for the sterilization process. Here, the reference temperature is taken as 150°F. and 1 minute of heating at 150°F. is considered 1 unit of lethality. The same lethal rate paper, constructed for evaluating the sterilization process, may be used as long as z is the same—in this case 10F.° This is true because

When 1 minute at 250°F. =   *When 1 minute at 150°F. =*
1 unit of lethality: *1 unit of lethality:*

$$L_{250} = \log^{-1} \frac{250 - 250}{10} = 1 = L_{150} = \log^{-1} \frac{150 - 150}{10}$$

and

$$L_{249} = \log^{-1} \frac{249 - 250}{10} = 0.794 = L_{149} = \log^{-1} \frac{149 - 150}{10}$$

(and so on for other relative temperatures)

Once a lethal rate paper is constructed for a given z value, the top line of the paper may be labeled with any temperature above the maximum temperature reached during heating of a product, and other principal horizontal lines, below the top line, may be labeled accordingly as explained in Chapter 11. Then, when time–temperature data from heating and cooling curves are transferred to the lethal rate paper to obtain a lethality curve, the area under the curve is proportional to the lethality of the process in terms of its equivalent in minutes at any designated reference temperature, as long as the following equation is used to compute lethality.

$$F = \frac{mA}{F_i d} \tag{3}$$

in which F = total lethality of all heat, represented by time–temperature relationships existent during heating and cooling of the product, expressed in terms of minutes at some designated reference temperature.

m = number of minutes per inch on the x axis.

A = area in square inches under the lethality curve.

F_i = the F_i corresponding to the temperature used to label the top line of the lethal rate paper.

d = number of inches from the bottom line to the top line of the lethal rate paper.

As an example, suppose we wish to determine the F value of a pasteurization process in terms of its equivalent in minutes at 150°F., with respect to its capacity to kill organisms the relative resistance of which is characterized by $z = 10$F.° (Here, the unit of lethality is taken as the equivalent of 1 minute heating at 150°F., the reference temperature.) Then, F would be designated as F_{150}^{10}. Further suppose that, in order to obtain an area under the lethality curve that is convenient to measure, the top line of the appropriately constructed lethal rate paper is labeled 146 and the principal horizontal lines below this labeled accordingly. Then equation 3 becomes

$$F_{150}^{10} = \frac{mA}{F_{i_{146}} d}$$

Now, consider the following:

The lethal rate paper has been constructed such that 1 inch on the x axis represents 10 minutes (m) and the distance between the top line and the bottom line is 10 inches (d).

Time–temperature data from heating and cooling curves have been transferred to the lethal rate paper and the area under the resulting lethality curve has been found equal to 6.5 square inches.

$$F_{i_{146}} = \log^{-1} \frac{T_x - T}{z} = \log^{-1} \frac{150 - 146}{10} = \log^{-1} 0.4 = 2.51$$

Then,

$$F_{150}^{10} = \frac{10 \times 6.5}{2.51 \times 10} = \mathbf{2.59} = \text{the equivalent, in minutes at } 150°\text{F.,}$$

of all heat at the point of temperature measurement.

Evaluation of Pasteurization Processes by Mathematical Methods

The same basic mathematical procedures are followed for evaluating pasteurization processes that are followed for evaluating sterilization processes. Any pasteurization process, for which the heating and cooling may be described by straight semilogarithmic curves, may be evaluated mathematically. The only variation from the evaluation of a sterilization process results from the use of a different unit of lethality in evaluating the pasteurization process. This variation is accounted for through the following equation,

$$F_i = \log^{-1} \frac{T_x - T}{z} = \log^{-1} \frac{T_x - T_r}{z}$$

Wherein

F_i, T_x, and z are as defined elsewhere.

$T = T_r$ = retort temperature or temperature of the external heating medium (as in a tubular heat exchanger).

(Refer to Table A.2 in the Appendix for some computed values of F_i. Values of F_i corresponding to z values not appearing in the table should be computed—sufficient accuracy is not obtained by interpolation.)

Consider a process for milk to be pasteurized in a tubular heat exchanger. Heating and cooling data have been determined by thermocouple measurements. When these are plotted on semilogarithmic coordinates the curves are straight lines and may be described by the following parameters.

$$f_h = f_c = 3.20$$
$$j_{ch} = j_{cc} = 1.00$$

Other processing data are as follows:

$$T_x = 150°\text{F.}$$
$$T_r = 160°\text{F.} = \text{temperature of external heating medium.}$$
$$T_{ih} = 40°\text{F.}$$
$$F_{150}^8 \text{ desired} = 9.00$$
$$l = 0$$

Parameters defining resistance of the organism of concern are:

$$D_{150} = 0.60$$
$$z = 8F.°$$

What would be the required process time (B)?
Since,

$$B = f_h(\log j_{ch}I_h - \log g_c)$$

And

$$I_h = T_r - T_{ih} = 160 - 40 = 120$$
$$B = 3.2(\log 120 - \log g_c)$$

Also

$$U_c = F_cF_i = F_{150}^8 F_{i_{160}}$$

in which

$$F_{i_{160}} = \log^{-1} \frac{T_x - T_r}{z} = \log^{-1} \frac{150 - 160}{8} = 0.0562$$

then

$$U_c = 9.00 \times 0.0562 = \textbf{0.506}$$

and

$$f_h/U = \frac{3.2}{0.506} = 6.32$$

Now, from $f_h/U:g$ table find for $f_h/U = 6.32$, $j_{cc} = 1$, and $z = 8$,

$$g_c = 2.93$$

Then

$$B = 3.2(\log 120 - \log 2.93)$$
$$B = 3.2 \times 1.612 = \textbf{5.16 minutes} = \text{process time}$$

This process would be adequate to accomplish a 15 log cycle reduction in the number of viable cells of the organism of concern. (It should be noted that, because this is a convection heating product under the conditions of process, F_c is considered equal to F_s.)

In this example, if the process time (B) had been given, instead of the F_{150}^8 desired, the F_{150}^8 value of the process could have been determined by simply reversing the order of the calculations, as demonstrated in sterilization process evaluations discussed in Chapters 12 and 14.

Not infrequently in pasteurization, the product is rapidly heated to some given temperature, held at this temperature for some prescribed period, and then rapidly cooled. Very often, when such processes are employed, lethal heat active during heating and cooling is taken as a "safety factor" and the F value of the process is determined for heat of the holding period only. In this case the F value of the process is determined simply by dividing the holding time t by the F_i corresponding to the holding temperature. Suppose, for example, a product is rapidly heated to 145°F. and held at this temperature for 30 minutes, then rapidly cooled. What would be the F_{150}^8 value of the process?

$$F_{i_{145}} = \log^{-1} \frac{T_x - T}{z} = \log^{-1} \frac{150 - 145}{8}$$

$$F_{i_{145}} = \textbf{4.22}$$

and

$$F_{150}^8 = 30/4.22 = \textbf{7.11}$$

It is often interesting to determine how much is being considered as safety factor in a process of this nature. For example take the process discussed above and assume that a heating medium temperature of 160°F. is used to heat the product to the holding temperature of 145°F., after which the heating medium temperature is rapidly lowered to 145°F. and kept there during the holding period. Now assume that the heat transfer parameters are:

$$f_h = f_c = 3.2 \quad \text{and} \quad j_{ch} = j_{cc} = 1.00$$

and that the product is immediately cooled after reaching 145°F. Then, the F_{150}^8 value of the heating up and cooling down portions of the process may be determined as follows:

$$g_c = 160 - 145 = \textbf{15°F.}$$

From $f_h/U:g$ table for $z = 8$, $j_{cc} = 1.00$ and $g_c = 15$, find

$$f_h/U_c = \textbf{500}$$

Then

$$U_c = 3.2/500 = \textbf{0.0064}$$

and

$$F_{150}^8 = \frac{U_c}{F_{i_{160}}} = \frac{0.0064}{0.0562} = \textbf{0.114}$$

In such a process this could very well be considered as no more than a moderate safety factor.

Equivalency of Pasteurization Processes

With regard to the destruction of a particular organism in equal product volumes, processes involving different time–temperature relationships are equivalent if they have the same F_s value. For convection-heating products, because heating is virtually uniform throughout the product mass, F_s is considered to equal F_c. In the case of conduction-heating products in cylindrical containers, F_s values of pasteurization processes may be determined in the same manner as described in Chapter 12 for evaluating sterilization processes. For products that heat by conduction, and for which F_s values of processes cannot be ascertained, considering F_s to equal F_c results in some overprocessing; however, if the evaluation is done judiciously, the applied processes is always on the safe side, because F_s values are always greater than corresponding F_c values. For the evaluation of most pasteurization processes, because of the inability to determine F_s values, F_s is considered equal to F_c and the lethality of processes is simply designated as F, appropriately defined with respect to z and T_x; i.e., F_{150}^8, F_{180}^{10}, F_{250}^{18}, etc. Then in general terms it may be said that

$$F = D(\log a - \log b)$$

Here, the expression $\log a - \log b$ is often considered in terms of some designated log cycle reduction, such as $8D$, $10D$, $12D$, or $15D$, in the number of viable cells of the organism of chief concern. For example, when pathogens are of chief concern, a reduction in the range of $12D$ to $15D$ is generally considered appropriate; when the organisms of concern are neither pathogenic nor toxigenic, i.e., when prevention of economically important spoilage is the only consideration, a reduction in the range of $5D$ to $10D$ may be judged adequate.

In view of the above considerations, the lethality value of a reference base pasteurization process may be expressed as,

$$F = D \log 10^n \tag{4}$$

in which

n = number of log cycles specifying the total reduction considered.

With respect to a given organism, the reference temperature (T_x) and the z value characterizing relative resistance of the organism to different temperatures would be designated in subscript and superscript. Equation 4 would then become

$$F_{T_x}^z = D_{T_x} \log 10^n \qquad (5)$$

For example, assume that *Coxiella burnetti* is the organism of concern and its resistance is characterized by $D_{150} = 0.60$ and $z = 9$. What F_{150} value is required of a process in order to accomplish a 15 log cycle reduction in the number of viable cells of this organism in a product? Substituting in equation 5, we obtain

$$F_{150}^9 = D_{150} \log 10^{15}$$

Then

$$F_{150}^9 = 0.60 \times 15 = \textbf{9.0 minutes}$$

With regard to pasteurization processes for which the lethal heat active during heating of a product to some given temperature and during cooling of the product from this temperature is considered only as a safety factor, i.e., processes for which only lethality of some given holding time at a given temperature is considered in evaluation, equivalency of processes (holding times) may be readily computed by use of the following equation:

$$U = t = FF_i \qquad (6)$$

For example, the holding time at 145°F. equivalent to $F_{150} = 9.0$ would be computed as follows:

$$t_{145}^9 = F_{150}^9 F_{i_{145}}^9$$

$$F_{i_{145}}^9 = \log^{-1} \frac{150 - 145}{9} = \textbf{3.60 minutes}$$

Then

$$t_{145}^9 = 9 \times 3.60 = \textbf{32.4 minutes}$$

It may be recalled from Chapter 4 that present milk pasteurization standards specify the equivalent of 30 minutes heating at 145°F. Small differences in the value of z used in computations would cause some variations in the value of t_{145}: e.g., if the z value were taken as 8, the t_{145} would be 38.0; and, if the z value were taken as 10, the t_{145} would be 28.5. Small differences in D_r would cause similar variations.

Equivalent pasteurization holding times (assuming instantaneous heating and cooling of product to and from holding temperature) to

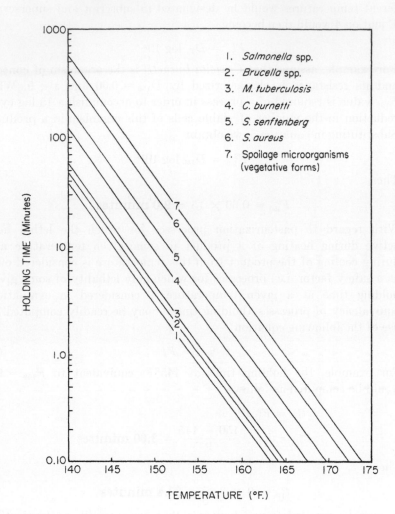

FIG. 37. Relative holding times at various temperatures to accomplish a 15 log cycle reduction in the number of different microorganisms of greatest importance in food pasteurization—based on approximate maximum resistance and $z = 9F.°$

accomplish a 15 log cycle reduction in number of viable cells of non-spore-forming organisms most important in pasteurization, are graphically depicted in Fig. 37. To obtain time–temperature relationship represented in Fig. 37, a z value of 9F.° was assumed, to account for

the relative resistance of all the organisms to the different temperatures. The maximum D_{150} values, as given in Table 6, Chapter 8, were
used as reference resistance values.

It should be emphasized that, with some products that are pasteurized by processes during which the maximum food temperature
reached is above about 150°F., the active lethal heat during heating
and cooling may be very substantial and logically should be considered in determining the total lethality of the processes. For example,
assume the following data for evaluating the lethality conferred during heating and cooling of a convection-heating product.

$$T_x = 150°F.$$
$$T_r = 170°F. = \text{temperature of external heating medium.}$$
$$g_c = 10°F. = T_r \text{ minus maximum temperature (160°F.) reached}$$
during heating.
$$f_h = f_c = 3.00 \text{ minutes}$$
$$j_{ch} = j_{cc} = 1.00$$
$$z = 9F.°$$
$$D_{150} = 0.60$$

Now, further assume that the product is immediately cooled upon
reaching 160°F. What would be the F^9_{150} of the process, i.e., of the
active lethal heat during heating and cooling?

$$U^9_{170} = F^9_{i_{170}} \times F^9_{150}$$

Then

$$F^9_{150} = U^9_{170}/F^9_{i_{170}}$$

in which

$$U^9_{170} \text{ is obtained by use of } F_h/U:g \text{ tables.}$$

Because $g_c = 10$, find from f_h/U tables for $z = 9$, $j_{cc} = 1.00$, and
$g_c = 10$

$$f_h/U^9_{170} = 80.2$$
$$U^9_{170} = 3.00/80.2 = \mathbf{0.0374}$$

Hence

$$F^9_{150} = 0.0374/0.006 = \mathbf{6.23}$$

This tells us that, under the designated conditions, simply heating the
product to 160°F. and cooling it therefrom would result in an F^9_{150}
process lethality value of 6.23; and further, that this alone would be

sufficient to accomplish slightly more than a 10 log cycle reduction (6.23/0.60) in the number of viable cells of the organism of concern. It is suggested that, for pasteurizing products with similar heat transfer characteristics where product temperatures above 150°F. are attained, heating and cooling temperature curves ought to be established by thermocouple measurement and the total lethality of the process evaluated by either the general method or by mathematical methods.

16

Quality Factor Degradation

The foregoing chapters have dealt entirely with heat inactivation of bacteria, in foods, that might if not destroyed prove hazardous to the health of food consumers or cause economically important spoilage of foods during storage and distribution. In many cases the heat applied for bacterial inactivation degrades desirable organoleptic and nutritive properties, such as color, texture, flavor, heat labile vitamins, etc. A number of these factors have been found to degrade exponentially with time of heating at a constant temperature; i.e., their concentration decreases similarly as the number of viable cells of a given bacterial species decreases. Unfortunately, there is a grave paucity of information relating to the degradation kinetics of many of them; yet, there is enough known about some, e.g., thiamine, to warrant consideration as a counterpart of thermobacteriology.

Because there is probably more known about the heat degradation kinetics of thiamine than about any other organoleptic or nutritive factor in foods, discussions here are confined for the most part to this important nutrient. However, it must be emphasized that the mathematical procedures presented for evaluating the effect of heat on thiamine are equally applicable for evaluating the effect of heat on any food property that degrades exponentially with time of heating at a constant temperature. Also, it must be emphasized that although the original F_s equation of Stumbo (1953) has been retained for sterility evaluations as treated in the foregoing chapters, the new F_s equation to be presented here is equally applicable for sterility evaluations. The reverse is not true—the original F_s equation is not applicable for evaluating heat effects on organoleptic and nutritive properties the heat stability of which is characterized by much higher D and z values than characterize the heat resistance of bacteria. It was retained for sterility evaluations because of prior familiarity with it and because it is slightly more convenient in application. In using the original F_s equation, lethal effects are integrated accurately from the geometrical center outward, through all iso-j regions where comparatively there can be significant bacterial survival. In conduction-heating canned foods the volume of concern in sterilization lies within an iso-j region that encloses somewhat less than one tenth of the total can volume. This is not true when considering heat-vulnerable quality factors; very significant amounts of these may be retained in the outermost regions of the container. Therefore, in determining the effects of heat on these factors, no portion of the container contents may be ignored, i.e., accurate integration must encompass all regions of the container.

A number of methods (Ball and Olson, 1957; Teixeira et al., 1969; Hayakawa, 1969; and Jen et al., 1971) have been presented for estimating degradation, or retention, during thermal processing, of heat-labile food constituents that are inactivated or degraded exponentially with time of heating at a constant temperature. The method of Ball and Olson requires major computational effort. Also, because of the finite layer procedure used in integrating destructive or lethal effects and because of inadequate accounting for cooling curve lags, highest accuracy in estimations would not be expected. The finite difference method of Teixeira et al., although it should give good accuracy in estimations, requires computerization in application—it is too complicated for manual manipulation. Beyond this, although good accuracy may be expected in calculating the degree of sterility or the extent of quality factor degradation when the processing parameters (such as time and temperature of processing) are known, it is a most difficult and expensive task, even with a computer, to calculate the processing parameters that when applied result in a designated degree of sterility, some given retention of nutrient factors, or some given degree of enzyme degradation. The dimensionless group method of Hayakawa, although it may yield acceptably accurate estimations, requires major computational effort for calculating and interpolating values for five dimensionless groups.

The method of Jen et al. obviates the major disadvantages of the other three methods and therefore it is presented here and its application is demonstrated. The method is reasonably simple and versatile in application, and although it may be readily computer programed, all calculations involved in its use are easily carried out manually.

The Jen *et al.* method is an extension, and in certain respects a refinement, of the method of Stumbo discussed in Chapter 12 in connection with sterility evaluations. It is an extension in the sense that it may be applied to estimating nutrient and organoleptic factor degradation as well as to microbial inactivation. It is a refinement in the sense that the f_h/U:g relationships essential in its application are developed from heating and cooling curves for which the nature of the temperature lag periods are described by the method of Teixeira *et al.* (1969) rather than being estimated by inspection. Tables of f_h/U:g were further expanded by Purohit and Stumbo (1972) as discussed in Chapter 12. Also, all values in these latter tables were derived so as to account for temperature gradients across the container at any designated time during heating, as well as to account for variations in the magnitude of the j value of the cooling curve. These tables as they now appear in the Appendix cover a range of z values from 8F.° to 200F.° and a range of cooling j values from 0.40 to 2.00. They ought to prove adequate for evaluating any heat process with respect to its lethal or degradation effects on any food component (bacterium or quality factor) that is destroyed exponentially with time of heating at a constant temperature. Relationships not appearing in the tables may be satisfactorily obtained by interpolation or extrapolation.

Basic Considerations

With respect to conduction-heating products in cylindrical containers, the method is based on the iso-j, and therefore iso-F, concept of distribution in foods during heating and cooling. (For details regarding this concept see Chapter 12.) It is further based on the concept that, to integrate lethal or degradative effects throughout a container of food, all points in the food mass must be considered.

Reduction in concentration of any heat vulnerable factor, that reduces exponentially with time of heating at a constant temperature, is described by the equation of the survivor curve (equation 4, Chapter 7), that is

$$t = D(\log a - \log b) \tag{1}$$

in which

t = time of heating at a constant temperature.

D = time of heating, at the designated temperature, required for a 90% reduction in concentration.

a = initial concentration expressed in appropriate units.

b = concentration after heating time, t, expressed in the same units.

Therefore, if the lethal or degradative effectiveness of all heat considered is expressed in terms of its equivalent in minutes at some given reference temperature, and the equivalent represented by the symbol F, equation 1 becomes:

$$F = D(\log a - \log b) \tag{2}$$

For sterility evaluation (see Chapter 12), the integral expression used to represent lethal effects throughout a container of a conduction-heating food was as follows:

$$10^{-F_s/D_r} \cdot 1 = \int_0^1 10^{-F_\lambda/D_r} \cdot dv \tag{3}$$

in which F_s = F value of heat received by the entire container during process.

F_λ = F value of heat received by any single point in the container.

and in which $10^{-F_s/D_r}$ = concentration fraction of any given heat-vulnerable factor remaining in the entire container at the end of process.

$10^{-F_\lambda/D_r}$ = concentration fraction of the same factor remaining at any single point in the container after process.

This integral simply specifies that the fraction remaining in the entire container must equal the sum of the fractions remaining at all points in the container.

To obtain a usable equation for sterility calculations, Stumbo (1953) solved the above integral by using the equation of the early straight-line portion of a curve obtained by plotting $(F_\lambda - F_c)$ versus v—see Fig. 31, Chapter 12. This, as pointed out above, was permissible because it was found that microbial survival beyond iso-F regions in the container wherein the relationships were not represented by a straight line was always negligible compared with survival in the container regions where the relationships were represented by a straight line. It was recognized at the time that the use equation obtained (equation 20, Chapter 12) was not applicable for determining quality factor degradation, or retention, because, unlike in the case of microbial survival, there was no portion of the container contents in which retention of these factors could be considered negligible compared with their retention in any other portion. It was necessary, therefore, to solve the basic integral so as to obtain a use equation that would accurately integrate heat effects at all points in the container.

Development of Equation

It was discovered that for conduction-heating foods in cylindrical containers, when the logarithm of $(1 - v)$ was plotted against $(F_\lambda - F_c)$ all points representing corresponding values of F_λ and v, from center to wall of container, could be well represented by a straight line (Fig. 38). The equation of this straight line may be written as follows:

$$\ln(1 - v) = m(F_\lambda - F_c) \qquad (4)$$

in which 1 = volume of container.

v = volume enclosed by any iso-j region, or shell.

F_λ = F value of heat received by any iso-j region, or shell.

F_c = F value of heat received by the geometrical center of the container.

Starting with the basic integral

$$10^{-F_s/D_r} \cdot 1 = \int_0^1 10^{-F_\lambda/D_r} \cdot dv$$

and taking $v = 0.19$ as a point of reference on the straight line, as in Stumbo's original solution, the integral was solved by substituting equivalent values from equation 4. The following use equation was obtained:

$$F_s = F_c + D_r \log \frac{D_r + 10.93(F_\lambda - F_c)}{D_r} \qquad (5)$$

(The detailed derivation of this equation is given in the Appendix.) As for the original F_s equation (equation 20, Chapter 12), equation 5 is valid when, and only when, values of j_λ and g_λ are taken as equal to one-half the values of j_c and g_c, respectively. These relationships are set by taking the integral solution reference point on the straight line as $v = 0.19$. The reason for this restriction is purely for convenience in problem solving as explained in Chapter 12.

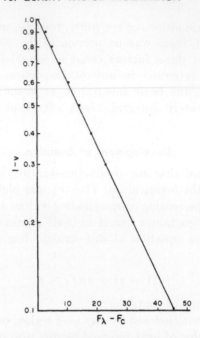

FIG. 38. A typical curve showing the relationship between the logarithm of $(1 - v)$ and $(F_\lambda - F_c)$ for a cylindrical container of conduction-heating food.

Sample Calculations and Comparisons

For demonstration and comparative purposes it is interesting to consider processes discussed in problem I, Chapter 14. In this problem sterility considerations were given to heat processes for a low-acid conduction-heating product packed in two different can sizes; 603 × 700 and 307 × 409.

Product Packed in 603 × 700 Can.   Processing data for the product packed in this can size and still cooked were taken as:

$$f_h = f_c = 200$$
$$T_{ih} = 160°F.$$
$$T_x = 250°F.$$
$$T_r = 250°F.$$
$$j_{ch} = j_{cc} = 2.00$$
$$l = 15 \text{ minutes}$$

From assumed spore load and spore resistance data, it was determined that an F_s^{14} value of 22.2 would be required to accomplish the desired degree of sterilization. It was further determined that a process time (B) at 250°F. of 254 minutes would be required to obtain the desired $F_s^{14} = 22.2$.

What thiamine degradation, or retention, is expected in the product so processed? Let us assume that thiamine stability in this product is characterized by $D_r = 154$ and $z = 46$ (cf. Feliciotti and Esselen, 1957).

First determine the F_s^{46} value of the process.

$$F_i = \log^{-1} \frac{250 - T_r}{46} = \log^{-1} \frac{250 - 250}{46} = \mathbf{1.00}$$
$$I_h = 250 - 160 = \mathbf{90}$$

Then, since

$$B = f_h(\log j_{ch}I_h - \log g_c)$$
$$254 = 200[\log (2.00 \times 90) - \log g_c]$$
$$\log g_c = 2.255 - 1.270 = \mathbf{0.985}$$

and

$$g_c = \mathbf{9.66}$$

From $f_h/U{:}g$ table for $z = 46$ (Table A.22) find for $j_{cc} = 2.00$ and $g_c = 9.66$

$$f_h/U_c = \mathbf{2.34}$$

Then, because $F_i = 1.00$

$$F_c = U_c = 200/2.34 = \mathbf{85.5}$$

Since

$$g_\lambda = 0.5g_c \quad \text{and} \quad j_{\lambda_c} = 0.5j_{cc}$$
$$g_\lambda = \mathbf{4.83} \quad \text{and} \quad j_{\lambda_c} = \mathbf{1.00}$$

Now, from $f_h/U{:}g$ Table A.22, find for $j_{\lambda_c} = 1.00$ and $g_{\lambda_c} = 4.83$,

$$f_h/U_\lambda = \mathbf{1.88}$$

Then

$$F_\lambda = U_\lambda = 200/1.88 = \mathbf{106.3}$$

Now, because

$$F_s = F_c + D_r \log \frac{D_r + 10.93(F_\lambda - F_c)}{D_r}$$
$$F_s^{46} = 85.5 + 154 \log \frac{154 + 10.93(106.3 - 85.5)}{154}$$
$$F_s^{46} = \mathbf{146.1}$$

Also

$$F_s^{46} = D_r(\log a - \log b)$$

taking $a = 100\%$, then

$$146.1 = 154(\log 100 - \log b)$$
$$\log b = 1.051$$
$$b = \mathbf{11.25} = \textbf{percent thiamine retained at end of process}$$

How much improvement in thiamine retention might be realized if this product were agitated during processing at a retort temperature of 255°F.? In problem I, Chapter 14, the equivalent sterilizing process for this product in the 603 × 700 can was determined considering that the product was agitated in a batch type cooker during process and that its heating was characterized by

$$f_h = f_c = 17.00$$
$$j_{ch} = j_{cc} = 1.20$$

Under these conditions the equivalent sterilizing process was found to be **32.8** (B) minutes at the retort temperature of 255°F. Because the product was considered to heat by convection during agitation, F_s was considered to equal F_c. First determine the F_s^{46} value of this process.

$$F_i = \log^{-1} \frac{250 - 255}{46} = 0.778$$
$$I_h = 255 - 160 = 95$$

Then, since

$$B = f_h(\log j_{ch}I_h - \log g_c)$$
$$32.8 = 17[\log (1.20 \times 95) - \log g_c]$$
$$\log g_c = 2.057 - 1.930 = \mathbf{0.127}$$
$$g_c = \mathbf{1.34}$$

From $f_h/U{:}g$ Table A.22, find for $z = 46$, $j_{cc} = 1.20$ and $g_c = 1.34$,

$$f_h/U_c = 0.934$$

Then

$$U_c = 17/0.934 = 18.2$$

and since

$$U_c = F_c F_i$$
$$F_c = 18.2/0.778 = \mathbf{23.4} = F_s^{46}$$

Again

$$F_s^{46} = D_r(\log a - \log b)$$

then

$$23.4 = 154(\log 100 - \log b)$$
$$\log b = \mathbf{1.848}$$
$$b - \mathbf{70.5} = \textbf{percent thiamine retained at end of process}$$

Note that this retention is great compared to the 11.25% retained in still cook. It may be expected that retention of other heat vulnerable quality factors will be improved accordingly, although to different extents.

Same Product Packed in 307 × 409 Cans. Processing data (problem I, Chapter 14) for the product packed in this can size and still cooked were taken as:

$$f_h = f_c = 63.6$$
$$T_{ih} = 160°\text{F.}$$
$$T_x = 250°\text{F.}$$
$$T_r = 250°\text{F.}$$
$$j_{ch} = j_{cc} = 2.00$$
$$l = 15 \text{ minutes}$$

It was determined (Chapter 14) that a process for this product in the 307 × 409 can would have an F_s^{14} value of 19.9 and that the process time (B) at 250°F. to obtain this would be 102 minutes. What thiamine retention would be expected in the product so processed?

First determine the F_s^{46} value of the process.

$$F_t = \log^{-1} \frac{250 - 250}{46} = 1.00$$
$$I_h = 250 - 160 = 90$$

Then, since

$$B = f_h(\log j_{ch}I_h - \log g_c)$$
$$102 = 63.6[\log (2.00 \times 90) - \log g_c]$$
$$\log g_c = \mathbf{0.651}$$

and

$$g_c = \mathbf{4.48}$$

From $f_h/U{:}g$ Table A.22, find for $z = 46$, $j_{cc} = 2.00$, and $g_c = 4.48$

$$f_h/U_c = 1.41$$
$$F_c = U_c = 63.6/1.41 = \mathbf{45.1}$$

Since

$$g_\lambda = 0.5g_c \quad \text{and} \quad j_{\lambda_c} = 0.5j_{cc}$$
$$g_\lambda = \mathbf{2.24} \quad \text{and} \quad j_{\lambda_c} = \mathbf{1.00}$$

From $f_h/U:g$ Table A.22, find for $z = 46$, $j_{\lambda_c} = 1.00$, and $g_\lambda = 2.24$

$$f_h/U_\lambda = \mathbf{1.21}$$

Then

$$F_\lambda = U_\lambda = 63.6/1.21 = \mathbf{52.6}$$

Therefore

$$F_s^{46} = 45.1 + 154 \log \frac{154 + 10.93(52.6 - 45.1)}{154}$$

$$F_s^{46} = \mathbf{73.6}$$

Also

$$F_s^{46} = D_r(\log a - \log b)$$

then

$$73.6 = 154(\log 100 - \log b)$$
$$\log b = \mathbf{1.522}$$
$$b = \mathbf{33.2} = \textbf{percent thiamine retained at end of process}$$

This is not good but considerably better than the 11.25% estimated as retained in the product packed in the 603 × 700 can and given an equivalent still cook sterilizing process.

How much improvement in thiamine retention might be realized if the product were agitated during processing at a retort temperature of 260°F. instead of 250°F? In problem I, Chapter 14, the equivalent sterilizing process for this product in the 307 × 409 can was determined considering that the product was agitated in a continuous cooker during process and that its heating was characterized by

$$f_h = f_c = 8.00$$
$$j_{ch} = j_{cc} = 1.00$$

Under these conditions the equivalent sterilizing process was found to be 14.3 (B) minutes at the retort temperature of 260°F. First determine the F_s^{46} value of the process.

$$F_i = \log^{-1} \frac{250 - 260}{46} = 0.606$$

$$I_h = 260 - 160 = 100$$

Then, since

$$B = f_h(\log j_{ch}I_h - \log g_c)$$
$$14.3 = 8[(\log 1 \times 100) - \log g_c]$$
$$\log g_c = 2.00 - 1.789 = \mathbf{0.211}$$
$$g_c = \mathbf{1.63}$$

From $f_h/U:g$ Table A.22, find for $z = 46$, $j_{cc} = 1.00$, and $g_c = 1.63$

$$f_h/U_c = \mathbf{1.05}$$

Then

$$U_c = 8/1.05 = \mathbf{7.62}$$

and since

$$U_c = F_cF_i$$
$$F_c = 7.62/0.606 = \mathbf{12.6} = F_s^{46}$$

Again

$$F_s^{46} = D_r(\log a - \log b)$$

then

$$12.6 = 154(\log 100 - \log b)$$
$$\log b = 1.918$$
$$b = 82.8 = \textbf{percent thiamine retained at end of process}$$

Here we see not only the influence of more uniform heating, attained by agitation, but the influence of container size on thiamine retention.

Influence of Processing Temperature on Quality Retention

Heat stability of most heat-labile food quality factors is characterized by much higher z values than characterize the relative heat resistance of bacteria. Because of this fact, high temperatures applied for short times, which result in processes equivalent in sterilizing capacities to processes employing lower temperatures for longer times, are relatively less deleterious to food quality. Therefore, in systems, such as turbulent flow heat exchangers, wherein virtually uniform temperature distribution can be maintained during process, much is often gained in quality by using high-temperature short-time (HTST) processes. This may be illustrated by a simple process calculation for which it is assumed that a food product is instantaneously heated to some given lethal temperature, held at this temperature for some designated time, and then cooled instantaneously to a nonlethal temperature.

Suppose that for reference the times at two different temperatures (say 240°F. and 280°F.) to accomplish a 6 log cycle reduction in the number of P.A.3679 spores $(D_{250} = 1.5,\ z = 18)$ in a low-acid food product are considered.

$$F_{250} = D_{250}(\log a - \log b)$$
$$F_{250} = 1.5(\log 10^6)$$
$$F_{250} = \mathbf{9.00}$$

Then, the time required at 240°F. would be computed as follows:

$$t_{240} = F_{250}F_{i_{240}}$$
$$F_{i_{240}} = \log^{-1}\frac{250 - 240}{18} = 3.59$$
$$t_{240} = 9.00 \times 3.59 = \mathbf{32.3\ minutes}$$

The equivalent time at 280°F. would be computed as follows:

$$t_{280} = F_{250}F_{i_{280}}$$
$$F_{i_{280}} = \log^{-1}\frac{250 - 280}{18} = 0.0215$$
$$t_{280} = 9.00 \times 0.0215 = \mathbf{0.194\ minutes}$$

What would be the comparative effect of these two processes on a heat vulnerable quality factor? Again, take thiamine as an example and assume that its stability in the product is characterized by $D_{250} = 154$ and $z = 46$. First, determine the corresponding D_{240} and D_{280} values by use of the equation of the thermal destruction curve, that is

$$\log D_2 - \log D_1 = \frac{1}{z}\,(T_1 - T_2)$$

To determine D_{240}, let $D_2 = D_{240}$ and $D_1 = D_{250}$.
Then

$$\log D_{240} - \log 154 = \frac{1}{46}\,(250 - 240)$$
$$\log D_{240} = \mathbf{2.405}$$
$$D_{240} = \mathbf{254}$$

To determine D_{280}, let $D_2 = D_{280}$ and $D_1 = D_{250}$.
Then

$$\log D_{280} - \log 154 = \frac{1}{46}\,(250 - 280)$$
$$\log D_{280} = 1.535$$
$$D_{280} = \mathbf{34.3}$$

Next determine the percent thiamine that should be retained after each process. After 32.3 minutes heating at 240°F.,

$$t = D(\log a - \log b)$$
$$32.3 = 254(\log 100 - \log b)$$
$$\log b = 1.873$$
$$b = 74.7 = \textbf{percent thiamine retained}$$

After 0.194 minutes at 280°F.,

$$0.194 = 34.3(\log 100 - \log b)$$
$$\log b = \textbf{1.994}$$
$$b = 98.6 = \textbf{percent thiamine retained}$$

Using thiamine degradation, or retention, as an example the foregoing consideration has demonstrated the influence of such factors as size of container employed to pack conduction-heating products, mechanism of heat transfer (whether conduction or convection), and temperature of processing (when heating is virtually uniform during processing) on heat-vulnerable quality factors in foods receiving processes that are equivalent in sterilizing value. As mentioned, the stability of most quality, or quality influencing factors, is characterized by z values that are much greater than those characterizing bacterial resistance; therefore, their retention is favored by HTST processing. For example, limited evidence indicates z values for chlorophyll a of the order of 90F.°, for chlorophyll b of the order of 180F.°, and for the inactivation of peroxidase in low-acid foods in the range of 45F.° to 90F.° (Yamamoto et al., 1962; Joffe and Ball, 1962; Gupte et al., 1963; and Resende et al., 1969).

It is hoped that the procedures presented here, and the implementing tables of $f_h/U:g$, will permit more meaningful estimations of the effects of heat processing on heat-vulnerable quality factors. They should be useful to the optimization of heat processes for maximum quality retention.

SELECTED REFERENCES

Ball. C. D., and Olson, F. C. W. (1957). "Sterilization in Food Technology," 654 pp. McGraw-Hill, New York, Toronto, London.

Feliciotti, E., and Esselen, W. B. (1957). Thermal destruction rates of thiamine in pureed meats and vegetables. *Food Technol.*, **11**, 77.

Gupte, S. M., El-Bisi, H. M., and Francis, F. J. (1963). Kinetics of thermal degradation of chlorophyll in spinach puree. *J. Food Sci.*, **29**, 379.

Hayakawa, Kan-Ichi. (1969). New parameters for calculating mass average sterilizing values to estimate nutrients in thermally conductive food. *Can. Ist. Food Technol. J.*, **2**, 167.

Jen, Y., Manson, J. E., Stumbo, C. R., and Zahradnik, J. W. (1971). A procedure for estimating sterilization of and quality factor degradation in thermally processed foods. *J. Food Sci.*, 36, 692.

Joffe, F. M., and Ball, C. O. (1962). Kinetics and energetics of thermal inactivation and the regeneration rates of a peroxidase system. *J. Food Sci.*, 27, 587.

Purohit, K. S., and Stumbo, C. R. (1972). Computer calculated parameters for thermal process evaluations. (Unpublished data.)

Resende, R., Francis, F. J., and Stumbo, C. R. (1969). Thermal destruction and regeneration of enzymes in green bean and spinach puree. *Food Technol.*, 23, 63.

Stumbo, C. R. (1953). New Procedures for evaluating thermal processes for foods in cylindrical containers. *Food Technol.*, 7, 309.

Stumbo, C. R. (1966). Fundamental considerations in high-temperature short-time processing of foods. *Proc. 2nd Intern. Cong. Food Sci. and Technol.*, pp. 171-176.

Teixeira, A. A., Dixon, J. R., Zahradnik, J. W., and Zinsmeister, G. E. (1969). Computer optimization of nutrient retention in the thermal processing of conduction-heated foods. *Food Technol.*, 23, 845.

Yamamoto, H. Y., Steinberg, M. P., and Nelson, A. I. (1962). Kinetic studies on the heat inactivation of peroxidase in sweet corn. *J. Food Sci.*, 27, 113.

APPENDIX

TABLE A.1

$$L = \log^{-1} \frac{T - T_z}{z}$$

				z value (F.°)				
$T - T_z$	8	10	12	14	16	18	20	22
−30	0.0002	0.0010	0.0032	0.0072	0.0133	0.0215	0.0316	0.0433
−29	0.0002	0.0013	0.0038	0.0085	0.0154	0.0245	0.0355	0.0481
−28	0.0003	0.0016	0.0046	0.0100	0.0178	0.0278	0.0398	0.0534
−27	0.0004	0.0020	0.0056	0.0118	0.0205	0.0316	0.0447	0.0593
−26	0.0006	0.0025	0.0068	0.0139	0.0237	0.0359	0.0501	0.0658
−25	0.0008	0.0032	0.0083	0.0164	0.0274	0.0408	0.0562	0.0731
−24	0.0010	0.0040	0.0100	0.0193	0.0316	0.0464	0.0631	0.0811
−23	0.0013	0.0050	0.0121	0.0228	0.0365	0.0528	0.0708	0.0901
−22	0.0018	0.0063	0.0147	0.0268	0.0422	0.0599	0.0794	0.1000
−21	0.0024	0.0079	0.0178	0.0316	0.0487	0.0681	0.0891	0.1110
−20	0.0032	0.0100	0.0215	0.0373	0.0562	0.0774	0.1000	0.1233
−19	0.0042	0.0126	0.0261	0.0439	0.0649	0.0880	0.1122	0.1369
−18	0.0056	0.0158	0.0316	0.0518	0.0750	0.1000	0.1259	0.1520
−17	0.0075	0.0200	0.0383	0.0611	0.0866	0.1136	0.1413	0.1688
−16	0.0100	0.0251	0.0464	0.0720	0.1000	0.1292	0.1585	0.1874
−15	0.0133	0.0316	0.0562	0.0848	0.1155	0.1468	0.1778	0.2081
−14	0.0178	0.0398	0.0681	0.1000	0.1334	0.1668	0.1995	0.2310
−13	0.0237	0.0501	0.0825	0.1179	0.1540	0.1896	0.2239	0.2565
−12	0.0316	0.0631	0.1000	0.1390	0.1778	0.2154	0.2512	0.2848
−11	0.0422	0.0794	0.1212	0.1638	0.2054	0.2448	0.2818	0.3162
−10	0.0562	0.1000	0.1468	0.1931	0.2371	0.2783	0.3162	0.3511
−9	0.0750	0.1259	0.1778	0.2276	0.2738	0.3162	0.3548	0.3899
−8	0.1000	0.1585	0.2154	0.2683	0.3162	0.3594	0.3981	0.4329
−7	0.1334	0.1995	0.2610	0.3162	0.3652	0.4084	0.4467	0.4806
−6	0.1778	0.2512	0.3162	0.3728	0.4217	0.4642	0.5012	0.5337
−5	0.2371	0.3162	0.3831	0.4394	0.4870	0.5275	0.5623	0.5926
−4	0.3162	0.3981	0.4642	0.5179	0.5623	0.5995	0.6310	0.6579
−3	0.4217	0.5012	0.5623	0.6105	0.6494	0.6813	0.7079	0.7305
−2	0.5623	0.6310	0.6813	0.7197	0.7499	0.7743	0.7943	0.8111
−1	0.7499	0.7943	0.8254	0.8483	0.8660	0.8799	0.8913	0.9006
0	1.0000	1.0000	1.0000	1.0000	1.0000	1.0000	1.0000	1.0000
1	1.334	1.259	1.212	1.179	1.155	1.136	1.122	1.110
2	1.778	1.585	1.468	1.389	1.334	1.292	1.259	1.233
3	2.371	1.995	1.778	1.638	1.540	1.468	1.413	1.369
4	3.162	2.512	2.154	1.931	1.778	1.668	1.585	1.520
5	4.217	3.162	2.610	2.276	2.054	1.896	1.778	1.688
6	5.623	3.981	3.162	2.683	2.371	2.154	1.995	1.874
7	7.499	5.012	3.831	3.162	2.738	2.448	2.239	2.081
8	10.00	6.310	4.642	3.728	3.162	2.783	2.512	2.310
9	13.34	7.943	5.623	4.394	3.652	3.162	2.818	2.565

TABLE A.1 (Continued)

$T - T_x$	8	10	12	14	16	18	20	22
				z value (F.°)				
10	17.78	10.00	6.813	5.179	4.217	3.594	3.162	2.848
11	23.71	12.59	8.254	6.105	4.870	4.084	3.548	3.162
12	31.62	15.85	10.00	7.197	5.623	4.642	3.981	3.511
13	42.17	19.95	12.12	8.483	6.494	5.275	4.467	3.899
14	56.23	25.12	14.68	10.00	7.499	5.995	5.012	4.329
15	74.99	31.62	17.78	11.79	8.660	6.813	5.623	4.806
16	100.0	39.81	21.54	13.89	10.00	7.743	6.310	5.337
17	133.4	50.12	26.10	16.38	11.55	8.799	7.079	5.926
18	177.8	63.10	31.62	19.31	13.34	10.00	7.943	6.579
19	237.1	79.43	38.31	22.76	15.40	11.36	8.913	7.305
20	316.2	100.0	46.42	26.83	17.78	12.92	10.00	8.111
21	421.7	125.9	56.23	31.62	20.54	14.68	11.22	9.006
22	562.3	158.5	68.13	37.28	23.71	16.68	12.59	10.00
23	749.9	199.5	82.54	43.94	27.38	18.96	14.13	11.10
24	1000	251.2	100.0	51.79	31.62	21.54	15.85	12.33
25	1334	316.2	121.2	61.05	36.52	24.48	17.78	13.69
26	1778	398.1	146.8	71.97	42.17	27.83	19.95	15.20
27	2371	501.2	177.8	84.83	48.70	31.62	22.39	16.88
28	3162	631.0	215.4	100.0	56.23	35.94	25.12	18.74
29	4217	794.3	261.0	117.9	64.94	40.84	28.18	20.81
30	5623	1000	316.2	138.9	74.99	46.42	31.62	23.10

TABLE A.2

$$F_i = \log^{-1}\frac{T_x - T}{z}$$

				z value (F.°)				
$T_x - T$	8	10	12	14	16	18	20	22
30	5623	1000	316.2	138.9	74.99	46.42	31.62	23.10
29	4217	794.3	261.0	117.9	64.94	40.84	28.18	20.81
28	3162	631.0	215.4	100.0	56.23	35.94	25.12	18.74
27	2371	501.2	177.8	84.83	48.70	31.62	22.39	16.88
26	1778	398.1	146.8	71.97	42.17	27.83	19.95	15.20
25	1334	316.2	121.2	61.05	36.52	24.48	17.78	13.69
24	1000	251.2	100.0	51.79	31.62	21.54	15.85	12.33
23	749.9	199.5	82.54	43.94	27.38	18.96	14.13	11.10
22	562.3	158.5	68.13	37.28	23.71	16.68	12.59	10.00
21	421.7	125.9	56.23	31.62	20.54	14.68	11.22	9.006
20	316.2	100.0	46.42	26.83	17.78	12.92	10.00	8.111
19	237.1	79.43	38.31	22.76	15.40	11.36	8.913	7.305
18	177.8	63.10	31.62	19.31	13.34	10.00	7.943	6.579
17	133.4	50.12	26.10	16.38	11.55	8.799	7.079	5.926
16	100.0	39.81	21.54	13.89	10.00	7.743	6.310	5.337
15	74.99	31.62	17.78	11.79	8.660	6.813	5.623	4.806
14	56.23	25.12	14.68	10.00	7.499	5.995	5.012	4.329
13	42.17	19.95	12.12	8.483	6.494	5.275	4.467	3.899
12	31.62	15.85	10.00	7.197	5.623	4.642	3.981	3.511
11	23.71	12.59	8.254	6.105	4.870	4.084	3.548	3.162
10	17.78	10.00	6.813	5.179	4.217	3.594	3.162	2.848
9	13.34	7.943	5.623	4.394	3.652	3.162	2.818	2.565
8	10.00	6.310	4.642	3.728	3.162	2.783	2.512	2.310
7	7.499	5.012	3.831	3.162	2.738	2.448	2.239	2.081
6	5.623	3.981	3.162	2.683	2.371	2.154	1.995	1.874
5	4.217	3.162	2.610	2.276	2.051	1.806	1.778	1.688
4	3.162	2.512	2.154	1.931	1.778	1.668	1.585	1.520
3	2.371	1.995	1.778	1.638	1.540	1.468	1.413	1.369
2	1.778	1.585	1.468	1.389	1.334	1.292	1.259	1.233
1	1.334	1.259	1.212	1.179	1.155	1.136	1.122	1.110
0	1.000	1.000	1.000	1.000	1.000	1.000	1.000	1.000
−1	0.7499	0.7943	0.8254	0.8483	0.8660	0.8799	0.8913	0.9006
−2	0.5623	0.6310	0.6813	0.7197	0.7499	0.7743	0.7943	0.8111
−3	0.4217	0.5012	0.5623	0.6105	0.6494	0.6813	0.7079	0.7305
−4	0.3162	0.3981	0.4642	0.5179	0.5623	0.5995	0.6310	0.6579
−5	0.2371	0.3162	0.3831	0.4394	0.4870	0.5275	0.5623	0.5926
−6	0.1778	0.2512	0.3162	0.3728	0.4217	0.4642	0.5012	0.5337
−7	0.1334	0.1995	0.2610	0.3162	0.3652	0.4084	0.4467	0.4806
−8	0.1000	0.1585	0.2154	0.2683	0.3162	0.3594	0.3981	0.4329
−9	0.0750	0.1259	0.1778	0.2276	0.2738	0.3162	0.3548	0.3899

TABLE A.2 (Continued)

$T_x - T$	z value (F.°)							
	8	10	12	14	16	18	20	22
−10	0.0562	0.1000	0.1468	0.1931	0.2371	0.2783	0.3162	0.3511
−11	0.0422	0.0794	0.1212	0.1638	0.2054	0.2448	0.2818	0.3162
−12	0.0316	0.0631	0.1000	0.1390	0.1778	0.2154	0.2512	0.2848
−13	0.0237	0.0501	0.0825	0.1179	0.1540	0.1896	0.2239	0.2565
−14	0.0178	0.0398	0.0681	0.1000	0.1334	0.1668	0.1995	0.2310
−15	0.0133	0.0316	0.0562	0.0848	0.1155	0.1468	0.1778	0.2081
−16	0.0100	0.0251	0.0464	0.0720	0.1000	0.1292	0.1585	0.1874
−17	0.0075	0.0200	0.0383	0.0611	0.0866	0.1136	0.1413	0.1688
−18	0.0056	0.0158	0.0316	0.0518	0.0750	0.1000	0.1259	0.1520
−19	0.0042	0.0126	0.0261	0.0439	0.0649	0.0880	0.1122	0.1369
−20	0.0032	0.0100	0.0215	0.0373	0.0562	0.0774	0.1000	0.1233
−21	0.0024	0.0079	0.0178	0.0316	0.0487	0.0681	0.0891	0.1110
−22	0.0018	0.0063	0.0147	0.0268	0.0422	0.0599	0.0794	0.1000
−23	0.0013	0.0050	0.0121	0.0228	0.0365	0.0528	0.0708	0.0901
−24	0.0010	0.0040	0.0100	0.0193	0.0316	0.0464	0.0631	0.0811
−25	0.0008	0.0032	0.0083	0.0164	0.0274	0.0408	0.0562	0.0731
−26	0.0006	0.0025	0.0068	0.0139	0.0237	0.0359	0.0501	0.0658
−27	0.0004	0.0020	0.0056	0.0118	0.0205	0.0316	0.0447	0.0593
−28	0.0003	0.0016	0.0046	0.0100	0.0178	0.0278	0.0398	0.0534
−29	0.0002	0.0013	0.0038	0.0085	0.0154	0.0245	0.0355	0.0481
−30	0.0002	0.0010	0.0032	0.0072	0.0133	0.0215	0.0316	0.0433

TABLE A.3[a]

$f_h/U : g$ *Relationships when* $z = 8$

Values of g when j of cooling curve is:

f_h/U	0.40	0.60	0.80	1.00	1.20	1.40	1.60	1.80	2.00
0.20	2.22-05	2.31-05	2.40-05	2.49-05	2.59-05	2.68-05	2.77-05	2.86-05	2.95-05
0.30	7.34-04	7.77-04	8.21-04	8.65-04	9.09-04	9.53-04	9.97-04	1.04-03	1.08-03
0.40	4.92-03	5.27-03	5.61-03	5.96-03	6.31-03	6.66-03	7.01-03	7.35-03	7.70-03
0.50	1.62-02	1.74-02	1.87-02	2.00-02	2.12-02	2.25-02	2.38-02	2.50-02	2.63-02
0.60	3.63-02	3.95-02	4.27-02	4.58-02	4.90-02	5.22-02	5.54-02	5.86-02	6.18-02
0.70	6.47-02	7.13-02	7.78-02	8.43-02	9.09-02	9.74-02	1.04-01	1.10-01	1.17-01
0.80	9.98-02	1.11-01	1.23-01	1.35-01	1.46-01	1.58-01	1.69-01	1.81-01	1.93-01
0.90	0.140	0.158	0.177	0.195	0.214	0.232	0.251	0.269	0.288
1.00	0.183	0.210	0.237	0.264	0.291	0.318	0.345	0.372	0.399
2.00	0.733	0.841	0.949	1.058	1.166	1.274	1.382	1.490	1.599
3.00	1.38	1.48	1.59	1.70	1.80	1.91	2.01	2.12	2.22
4.00	1.93	2.01	2.09	2.17	2.25	2.33	2.42	2.50	2.58
5.00	2.35	2.41	2.48	2.55	2.61	2.68	2.74	2.81	2.87
6.00	2.68	2.74	2.79	2.85	2.91	2.97	3.03	3.09	3.15
7.00	2.94	3.00	3.06	3.12	3.18	3.23	3.29	3.35	3.41
8.00	3.16	3.22	3.29	3.35	3.41	3.47	3.54	3.60	3.66
9.00	3.35	3.42	3.49	3.56	3.63	3.70	3.77	3.84	3.91
10.00	3.51	3.59	3.67	3.75	3.82	3.90	3.98	4.06	4.14
15.00	4.11	4.25	4.38	4.52	4.66	4.79	4.93	5.07	5.21
20.00	4.54	4.74	4.94	5.13	5.33	5.53	5.72	5.92	6.12
25.00	4.89	5.15	5.40	5.65	5.91	6.16	6.42	6.67	6.93
30.00	5.19	5.50	5.80	6.11	6.42	6.73	7.04	7.35	7.66
35.00	5.44	5.80	6.16	6.53	6.89	7.25	7.61	7.97	8.33
40.00	5.66	6.08	6.49	6.90	7.31	7.73	8.14	8.55	8.96
45.00	5.86	6.32	6.78	7.25	7.71	8.17	8.63	9.10	9.56
50.00	6.03	6.54	7.05	7.56	8.08	8.59	9.10	9.61	10.12
60.00	6.32	6.92	7.53	8.14	8.74	9.35	9.96	10.57	11.17
70.00	6.55	7.24	7.94	8.64	9.34	10.03	10.73	11.43	12.13
80.00	6.74	7.52	8.30	9.08	9.86	10.64	11.43	12.21	12.99
90.00	6.90	7.76	8.62	9.48	10.33	11.19	12.05	12.91	13.77
100.00	7.05	7.98	8.90	9.83	10.75	11.68	12.60	13.53	14.46
150.00	7.82	8.94	10.06	11.18	12.30	13.42	14.54	15.66	16.78
200.00	8.72	9.86	10.99	12.13	13.26	14.40	15.53	16.67	17.80
250.00	9.66	10.72	11.79	12.86	13.92	14.99	16.06	17.12	18.19
300.00	10.5	11.5	12.5	13.4	14.4	15.4	16.4	17.3	18.3
350.00	11.3	12.2	13.0	13.9	14.8	15.7	16.6	17.5	18.4
400.00	11.9	12.7	13.5	14.3	15.2	16.0	16.8	17.6	18.4
450.00	12.5	13.2	14.0	14.7	15.4	16.2	16.9	17.7	18.4
500.00	12.9	13.6	14.3	15.0	15.7	16.4	17.1	17.8	18.5
600.00	13.7	14.3	15.0	15.6	16.2	16.8	17.4	18.0	18.6
700.00	14.4	14.9	15.5	16.0	16.6	17.2	17.7	18.3	18.8
800.00	14.9	15.4	15.9	16.4	17.0	17.5	18.0	18.5	19.1
900.00	15.3	15.8	16.3	16.8	17.3	17.8	18.3	18.8	19.3
999.99	15.7	16.1	16.6	17.1	17.6	18.1	18.6	19.1	19.6

[a] Tables A.3 to A.59 have been reproduced directly from computer readout and consequently the lower values of g are in computer readout exponent form. To convert to conventional exponent form simply read -01, -02, -03, -04, and -05 as 10^{-1}, 10^{-2}, 10^{-3}, 10^{-4}, and 10^{-5}, respectively.

TABLE A.4

$f_h/U : g$ *Relationships when* $z = 10$

f_h/U	Values of g when j of cooling curve is:								
	0.40	0.60	0.80	1.00	1.20	1.40	1.60	1.80	2.00
0.20	2.68-05	2.78-05	2.88-05	2.98-05	3.07-05	3.17-05	3.27-05	3.36-05	3.46-05
0.30	8.40-04	9.39-04	1.04-03	1.14-03	1.24-03	1.34-03	1.43-03	1.53-03	1.63-03
0.40	5.84-03	6.51-03	7.18-03	7.85-03	8.53-03	9.20-03	9.87-03	1.05-02	1.12-02
0.50	2.01-02	2.21-02	2.40-02	2.60-02	2.79-02	2.99-02	3.18-02	3.38-02	3.57-02
0.60	4.73-02	5.11-02	5.49-02	5.87-02	6.25-02	6.63-02	7.01-02	7.39-02	7.77-02
0.70	8.85-02	9.44-02	1.00-01	1.06-01	1.12-01	1.18-01	1.24-01	1.30-01	1.36-01
0.80	0.143	0.151	0.159	0.167	0.175	0.183	0.191	0.199	0.207
0.90	0.208	0.218	0.228	0.238	0.248	0.258	0.268	0.278	0.288
1.00	0.282	0.294	0.305	0.317	0.329	0.340	0.352	0.364	0.376
2.00	1.14	1.17	1.19	1.21	1.24	1.26	1.29	1.31	1.33
3.00	1.83	1.88	1.92	1.97	2.01	2.05	2.10	2.14	2.19
4.00	2.33	2.41	2.48	2.55	2.63	2.70	2.77	2.85	2.92
5.00	2.71	2.81	2.92	3.03	3.14	3.24	3.35	3.46	3.57
6.00	3.01	3.15	3.29	3.43	3.57	3.72	3.86	4.00	4.14
7.00	3.25	3.43	3.61	3.78	3.96	4.13	4.31	4.49	4.66
8.00	3.47	3.68	3.89	4.10	4.30	4.51	4.72	4.93	5.14
9.00	3.67	3.90	4.14	4.38	4.62	4.85	5.09	5.33	5.57
10.00	3.84	4.11	4.38	4.64	4.91	5.17	5.44	5.70	5.97
15.00	4.60	4.97	5.35	5.72	6.09	6.47	6.84	7.21	7.59
20.00	5.22	5.67	6.12	6.57	7.01	7.46	7.91	8.35	8.80
25.00	5.78	6.27	6.77	7.27	7.77	8.27	8.76	9.26	9.76
30.00	6.27	6.81	7.34	7.88	8.41	8.95	9.48	10.02	10.55
35.00	6.72	7.29	7.85	8.41	8.98	9.54	10.10	10.67	11.23
40.00	7.14	7.72	8.31	8.89	9.48	10.06	10.65	11.23	11.82
45.00	7.52	8.12	8.72	9.33	9.93	10.53	11.13	11.73	12.33
50.00	7.87	8.49	9.10	9.72	10.34	10.95	11.57	12.18	12.80
60.00	8.51	9.15	9.78	10.42	11.06	11.69	12.33	12.97	13.60
70.00	9.07	9.72	10.37	11.02	11.68	12.33	12.98	13.63	14.28
80.00	9.56	10.23	10.89	11.55	12.22	12.88	13.55	14.21	14.88
90.00	10.0	10.7	11.4	12.0	12.7	13.4	14.1	14.7	15.4
100.00	10.4	11.1	11.8	12.5	13.1	13.8	14.5	15.2	15.9
150.00	11.9	12.6	13.4	14.1	14.8	15.5	16.3	17.0	17.7
200.00	13.0	13.7	14.5	15.2	16.0	16.8	17.5	18.3	19.0
250.00	13.7	14.5	15.3	16.1	16.9	17.7	18.5	19.3	20.1
300.00	14.3	15.2	16.0	16.8	17.7	18.5	19.3	20.1	21.0
350.00	14.8	15.7	16.5	17.4	18.3	19.1	20.0	20.9	21.7
400.00	15.2	16.1	17.0	17.9	18.8	19.7	20.6	21.5	22.4
450.00	15.5	16.5	17.4	18.3	19.3	20.2	21.1	22.1	23.0
500.00	15.8	16.8	17.8	18.7	19.7	20.6	21.6	22.6	23.5
600.00	16.3	17.4	18.4	19.4	20.4	21.4	22.4	23.4	24.5
700.00	16.8	17.8	18.9	19.9	21.0	22.1	23.1	24.2	25.3
800.00	17.1	18.2	19.3	20.4	21.5	22.6	23.7	24.8	25.9
900.00	17.4	18.5	19.7	20.8	22.0	23.1	24.3	25.4	26.6
999.99	17.7	18.8	20.0	21.2	22.4	23.6	24.7	25.9	27.1

TABLE A.5
$f_h/U:g$ Relationships when $z = 12$

	Values of g when j of cooling curve is:								
f_h/U	0.40	0.60	0.80	1.00	1.20	1.40	1.60	1.80	2.00
0.20	2.86-05	3.03-05	3.21-05	3.39-05	3.57-05	3.74-05	3.92-05	4.10-05	4.27-05
0.30	1.16-03	1.26-03	1.35-03	1.45-03	1.54-03	1.64-03	1.73-03	1.83-03	1.92-03
0.40	8.18-03	8.82-03	9.47-03	1.01-02	1.08-02	1.14-02	1.21-02	1.27-02	1.33-02
0.50	2.72-02	2.92-02	3.12-02	3.32-02	3.52-02	3.72-02	3.92-02	4.12-02	4.32-02
0.60	6.13-02	6.56-02	6.98-02	7.41-02	7.83-02	8.26-02	8.68-02	9.10-02	9.53-02
0.70	0.111	0.118	0.125	0.132	0.139	0.147	0.154	0.161	0.168
0.80	0.173	0.184	0.194	0.205	0.216	0.226	0.237	0.248	0.258
0.90	0.246	0.260	0.274	0.289	0.303	0.318	0.332	0.347	0.361
1.00	0.326	0.344	0.363	0.381	0.399	0.418	0.436	0.455	0.473
2.00	1.20	1.26	1.32	1.38	1.44	1.49	1.55	1.61	1.67
3.00	1.91	2.01	2.11	2.21	2.31	2.41	2.51	2.61	2.71
4.00	2.47	2.61	2.75	2.89	3.03	3.17	3.31	3.45	3.59
5.00	2.92	3.10	3.28	3.46	3.64	3.82	4.00	4.18	4.36
6.00	3.31	3.52	3.74	3.96	4.18	4.40	4.62	4.83	5.05
7.00	3.65	3.90	4.15	4.41	4.66	4.92	5.17	5.42	5.68
8.00	3.96	4.24	4.53	4.81	5.10	5.39	5.67	5.96	6.25
9.00	4.24	4.56	4.87	5.19	5.51	5.82	6.14	6.46	6.77
10.00	4.50	4.85	5.19	5.54	5.88	6.22	6.57	6.91	7.26
15.00	5.61	6.06	6.52	6.98	7.44	7.89	8.35	8.81	9.27
20.00	6.49	7.03	7.57	8.10	8.64	9.18	9.72	10.26	10.79
25.00	7.24	7.83	8.43	9.03	9.62	10.22	10.81	11.41	12.01
30.00	7.89	8.53	9.17	9.81	10.44	11.08	11.72	12.36	13.00
35.00	8.47	9.14	9.81	10.48	11.15	11.82	12.49	13.16	13.83
40.00	8.99	9.68	10.38	11.07	11.76	12.46	13.15	13.85	14.54
45.00	9.46	10.17	10.88	11.59	12.31	13.02	13.73	14.44	15.15
50.00	9.89	10.61	11.34	12.07	12.79	13.52	14.24	14.97	15.69
60.00	10.6	11.4	12.1	12.9	13.6	14.4	15.1	15.9	16.6
70.00	11.3	12.1	12.8	13.6	14.3	15.1	15.9	16.6	17.4
80.00	11.9	12.6	13.4	14.2	14.9	15.7	16.5	17.3	18.0
90.00	12.3	13.1	13.9	14.7	15.5	16.3	17.0	17.8	18.6
100.00	12.8	13.6	14.4	15.2	15.9	16.7	17.5	18.3	19.1
150.00	14.3	15.2	16.0	16.9	17.8	18.6	19.5	20.4	21.2
200.00	15.3	16.2	17.2	18.1	19.1	20.0	21.0	21.9	22.8
250.00	16.0	17.0	18.0	19.1	20.1	21.1	22.2	23.2	24.2
300.00	16.5	17.6	18.7	19.8	20.9	22.1	23.2	24.3	25.4
350.00	16.9	18.1	19.3	20.5	21.7	22.9	24.0	25.2	26.4
400.00	17.3	18.5	19.8	21.0	22.3	23.6	24.8	26.1	27.3
450.00	17.6	18.9	20.2	21.5	22.9	24.2	25.5	26.8	28.2
500.00	17.8	19.2	20.6	22.0	23.4	24.8	26.1	27.5	28.9
600.00	18.3	19.8	21.3	22.8	24.2	25.7	27.2	28.7	30.2
700.00	18.7	20.3	21.8	23.4	25.0	26.6	28.1	29.7	31.3
800.00	19.1	20.7	22.4	24.0	25.6	27.3	28.9	30.5	32.1
900.00	19.5	21.2	22.8	24.5	26.2	27.9	29.5	31.2	32.9
999.99	19.8	21.6	23.3	25.0	26.7	28.4	30.1	31.8	33.5

TABLE A.6

$f_h/U : g$ Relationships when $z = 14$

	Values of g when j of cooling curve is:								
f_h/U	0.40	0.60	0.80	1.00	1.20	1.40	1.60	1.80	2.00
0.20	3.21-05	3.44-05	3.68-05	3.91-05	4.15-05	4.38-05	4.62-05	4.85-05	5.09-05
0.30	1.40-03	1.51-03	1.63-03	1.75-03	1.87-03	1.99-03	2.10-03	2.22-03	2.34-03
0.40	9.93-03	1.07-02	1.14-02	1.22-02	1.30-02	1.37-02	1.45-02	1.52-02	1.60-02
0.50	3.30-02	3.52-02	3.74-02	3.96-02	4.19-02	4.41-02	4.63-02	4.85-02	5.07-02
0.60	7.42-02	7.87-02	8.31-02	8.76-02	9.20-02	9.64-02	1.01-01	1.05-01	1.10-01
0.70	0.133	0.140	0.148	0.155	0.162	0.169	0.177	0.184	0.191
0.80	0.207	0.217	0.228	0.238	0.249	0.259	0.270	0.280	0.291
0.90	0.291	0.306	0.320	0.334	0.348	0.362	0.376	0.390	0.404
1.00	0.384	0.402	0.420	0.438	0.456	0.474	0.492	0.510	0.528
2.00	1.35	1.42	1.49	1.56	1.64	1.71	1.78	1.85	1.93
3.00	2.11	2.25	2.39	2.53	2.67	2.81	2.95	3.09	3.23
4.00	2.71	2.92	3.12	3.33	3.54	3.74	3.95	4.16	4.36
5.00	3.23	3.49	3.76	4.02	4.28	4.55	4.81	5.07	5.34
6.00	3.69	4.00	4.32	4.63	4.94	5.25	5.57	5.88	6.19
7.00	4.11	4.47	4.82	5.17	5.53	5.88	6.24	6.59	6.94
8.00	4.50	4.89	5.28	5.67	6.06	6.45	6.84	7.23	7.62
9.00	4.87	5.29	5.71	6.13	6.55	6.96	7.38	7.80	8.22
10.00	5.21	5.66	6.10	6.55	6.99	7.44	7.88	8.33	8.77
15.00	6.68	7.22	7.75	8.29	8.82	9.36	9.89	10.43	10.97
20.00	7.85	8.44	9.03	9.63	10.22	10.81	11.40	12.00	12.59
25.00	8.80	9.43	10.07	10.71	11.34	11.98	12.62	13.25	13.89
30.00	9.59	10.26	10.94	11.61	12.29	12.96	13.63	14.31	14.98
35.00	10.3	11.0	11.7	12.4	13.1	13.8	14.5	15.2	15.9
40.00	10.8	11.6	12.3	13.1	13.8	14.5	15.3	16.0	16.8
45.00	11.3	12.1	12.9	13.7	14.4	15.2	16.0	16.7	17.5
50.00	11.8	12.6	13.4	14.2	15.0	15.8	16.6	17.4	18.2
60.00	12.5	13.4	14.3	15.1	16.0	16.8	17.7	18.5	19.4
70.00	13.1	14.1	15.0	15.9	16.8	17.7	18.6	19.5	20.4
80.00	13.6	14.6	15.6	16.5	17.5	18.4	19.4	20.4	21.3
90.00	14.1	15.1	16.1	17.1	18.1	19.1	20.1	21.1	22.1
100.00	14.5	15.5	16.6	17.6	18.7	19.7	20.7	21.8	22.8
150.00	15.9	17.1	18.3	19.5	20.7	21.9	23.1	24.4	25.6
200.00	16.8	18.2	19.5	20.8	22.2	23.5	24.8	26.2	27.5
250.00	17.6	19.0	20.5	21.9	23.3	24.7	26.1	27.6	29.0
300.00	18.3	19.8	21.3	22.8	24.2	25.7	27.2	28.7	30.2
350.00	18.8	20.4	22.0	23.5	25.1	26.6	28.2	29.7	31.3
400.00	19.4	21.0	22.6	24.2	25.8	27.4	29.0	30.6	32.3
450.00	19.8	21.5	23.1	24.8	26.5	28.1	29.8	31.5	33.1
500.00	20.2	21.9	23.7	25.4	27.1	28.8	30.5	32.2	33.9
600.00	21.0	22.8	24.6	26.4	28.1	29.9	31.7	33.5	35.3
700.00	21.6	23.5	25.4	27.2	29.1	30.9	32.8	34.7	36.5
800.00	22.2	24.1	26.1	28.0	29.9	31.8	33.7	35.6	37.6
900.00	22.7	24.7	26.7	28.6	30.6	32.6	34.5	36.5	38.5
999.99	23.2	25.2	27.3	29.3	31.3	33.3	35.3	37.3	39.3

TABLE A.7

$f_h/U:g$ *Relationships when* $z = 16$

	Values of g when j of cooling curve is:								
J_h/U	0.40	0.60	0.80	1.00	1.20	1.40	1.60	1.80	2.00
0.20	$3.67{-}05$	$3.95{-}05$	$4.23{-}05$	$4.50{-}05$	$4.78{-}05$	$5.06{-}05$	$5.34{-}05$	$5.61{-}05$	$5.89{-}05$
0.30	$1.60{-}03$	$1.74{-}03$	$1.88{-}03$	$2.02{-}03$	$2.16{-}03$	$2.31{-}03$	$2.45{-}03$	$2.59{-}03$	$2.73{-}03$
0.40	$1.12{-}02$	$1.21{-}02$	$1.31{-}02$	$1.40{-}02$	$1.49{-}02$	$1.58{-}02$	$1.68{-}02$	$1.77{-}02$	$1.86{-}02$
0.50	$3.69{-}02$	$3.97{-}02$	$4.24{-}02$	$4.51{-}02$	$4.79{-}02$	$5.06{-}02$	$5.33{-}02$	$5.61{-}02$	$5.88{-}02$
0.60	$8.28{-}02$	$8.83{-}02$	$9.38{-}02$	$9.93{-}02$	$1.05{-}01$	$1.10{-}01$	$1.16{-}01$	$1.21{-}01$	$1.27{-}01$
0.70	0.148	0.157	0.166	0.175	0.184	0.193	0.202	0.211	0.221
0.80	0.230	0.243	0.256	0.270	0.283	0.296	0.309	0.322	0.335
0.90	0.324	0.342	0.360	0.378	0.395	0.413	0.431	0.448	0.466
1.00	0.427	0.450	0.473	0.496	0.518	0.541	0.564	0.586	0.609
2.00	1.51	1.60	1.69	1.78	1.87	1.96	2.05	2.15	2.24
3.00	2.39	2.56	2.73	2.90	3.07	3.24	3.42	3.59	3.76
4.00	3.11	3.36	3.60	3.85	4.09	4.33	4.58	4.82	5.06
5.00	3.73	4.04	4.35	4.66	4.96	5.27	5.58	5.88	6.19
6.00	4.28	4.65	5.01	5.37	5.73	6.09	6.45	6.81	7.17
7.00	4.78	5.19	5.60	6.00	6.41	6.82	7.23	7.63	8.04
8.00	5.23	5.68	6.13	6.58	7.03	7.47	7.92	8.37	8.82
9.00	5.65	6.13	6.62	7.10	7.58	8.07	8.55	9.04	9.52
10.00	6.04	6.55	7.07	7.58	8.10	8.61	9.13	9.64	10.16
15.00	7.65	8.28	8.91	9.55	10.18	10.81	11.44	12.08	12.71
20.00	8.90	9.61	10.32	11.03	11.75	12.46	13.17	13.88	14.60
25.00	9.91	10.68	11.46	12.23	13.01	13.78	14.56	15.33	16.11
30.00	10.7	11.6	12.4	13.2	14.1	14.9	15.7	16.5	17.4
35.00	11.5	12.3	13.2	14.1	15.0	15.8	16.7	17.6	18.5
40.00	12.1	13.0	13.9	14.8	15.8	16.7	17.6	18.5	19.4
45.00	12.6	13.5	14.5	15.5	16.5	17.4	18.4	19.3	20.3
50.00	13.1	14.1	15.1	16.1	17.1	18.1	19.1	20.1	21.1
60.00	13.9	15.0	16.1	17.1	18.2	19.2	20.3	21.4	22.4
70.00	14.6	15.7	16.8	18.0	19.1	20.2	21.4	22.5	23.6
80.00	15.2	16.4	17.5	18.7	19.9	21.1	22.3	23.5	24.6
90.00	15.7	16.9	18.1	19.4	20.6	21.8	23.1	24.3	25.6
100.00	16.1	17.4	18.7	20.0	21.2	22.5	23.8	25.1	26.4
150.00	17.8	19.3	20.7	22.2	23.7	25.1	26.6	28.1	29.6
200.00	19.0	20.6	22.2	23.8	25.4	27.0	28.6	30.2	31.8
250.00	19.9	21.6	23.4	25.1	26.8	28.5	30.2	31.9	33.6
300.00	20.7	22.5	24.3	26.1	27.9	29.7	31.5	33.3	35.1
350.00	21.4	23.2	25.1	27.0	28.9	30.8	32.6	34.5	36.4
400.00	21.9	23.9	25.8	27.8	29.7	31.7	33.6	35.5	37.5
450.00	22.5	24.5	26.5	28.5	30.5	32.4	34.4	36.4	38.4
500.00	23.0	25.0	27.0	29.1	31.1	33.2	35.2	37.2	39.3
600.00	23.9	26.0	28.1	30.2	32.3	34.4	36.5	38.6	40.7
700.00	24.7	26.8	29.0	31.1	33.3	35.5	37.6	39.8	41.9
800.00	25.4	27.6	29.8	32.0	34.2	36.4	38.6	40.8	43.0
900.00	26.1	28.3	30.5	32.7	35.0	37.2	39.4	41.6	43.9
999.99	26.7	28.9	31.2	33.4	35.7	37.9	40.2	42.4	44.7

TABLE A.8

$f_h/U:g$ *Relationships when* $z = 18$

	Values of g when j of cooling curve is:								
f_h/U	0.40	0.60	0.80	1.00	1.20	1.40	1.60	1.80	2.00
0.20	4.09-05	4.42-05	4.76-05	5.09-05	5.43-05	5.76-05	6.10-05	6.44-05	6.77-05
0.30	2.01-03	2.14-03	2.27-03	2.40-03	2.53-03	2.66-03	2.79-03	2.93-03	3.06-03
0.40	1.33-02	1.43-02	1.52-02	1.62-02	1.71-02	1.80-02	1.90-02	1.99-02	2.09-02
0.50	4.11-02	4.42-02	4.74-02	5.06-02	5.38-02	5.70-02	6.02-02	6.34-02	6.65-02
0.60	8.70-02	9.43-02	1.02-01	1.09-01	1.16-01	1.23-01	1.31-01	1.38-01	1.45-01
0.70	0.150	0.163	0.176	0.189	0.202	0.215	0.228	0.241	0.255
0.80	0.226	0.246	0.267	0.287	0.308	0.328	0.349	0.369	0.390
0.90	0.313	0.342	0.371	0.400	0.429	0.458	0.487	0.516	0.545
1.00	0.408	0.447	0.485	0.523	0.561	0.600	0.638	0.676	0.715
2.00	1.53	1.66	1.80	1.93	2.07	2.21	2.34	2.48	2.61
3.00	2.63	2.84	3.05	3.26	3.47	3.68	3.89	4.10	4.31
4.00	3.61	3.87	4.14	4.41	4.68	4.94	5.21	5.48	5.75
5.00	4.44	4.76	5.08	5.40	5.71	6.03	6.35	6.67	6.99
6.00	5.15	5.52	5.88	6.25	6.61	6.98	7.34	7.71	8.07
7.00	5.77	6.18	6.59	7.00	7.41	7.82	8.23	8.64	9.05
8.00	6.29	6.75	7.20	7.66	8.11	8.56	9.02	9.47	9.93
9.00	6.76	7.26	7.75	8.25	8.74	9.24	9.74	10.23	10.73
10.00	7.17	7.71	8.24	8.78	9.32	9.86	10.39	10.93	11.47
15.00	8.73	9.44	10.16	10.88	11.59	12.31	13.02	13.74	14.45
20.00	9.83	10.69	11.55	12.40	13.26	14.11	14.97	15.82	16.68
25.00	10.7	11.7	12.7	13.6	14.6	15.6	16.5	17.5	18.4
30.00	11.5	12.5	13.6	14.6	15.7	16.8	17.8	18.9	19.9
35.00	12.1	13.3	14.4	15.5	16.7	17.8	18.9	20.0	21.2
40.00	12.8	13.9	15.1	16.3	17.5	18.7	19.9	21.1	22.3
45.00	13.3	14.6	15.8	17.0	18.3	19.5	20.8	22.0	23.2
50.00	13.8	15.1	16.4	17.7	19.0	20.3	21.6	22.8	24.1
60.00	14.8	16.1	17.5	18.9	20.2	21.6	22.9	24.3	25.7
70.00	15.6	17.0	18.4	19.9	21.3	22.7	24.1	25.6	27.0
80.00	16.3	17.8	19.3	20.8	22.2	23.7	25.2	26.7	28.1
90.00	17.0	18.5	20.1	21.6	23.1	24.6	26.1	27.6	29.2
100.00	17.6	19.2	20.8	22.3	23.9	25.4	27.0	28.5	30.1
150.00	20.1	21.8	23.5	25.2	26.8	28.5	30.2	31.9	33.6
200.00	21.7	23.5	25.3	27.1	28.9	30.7	32.5	34.3	36.2
250.00	22.9	24.8	26.7	28.6	30.5	32.4	34.3	36.2	38.1
300.00	23.8	25.8	27.8	29.8	31.8	33.7	35.7	37.7	39.7
350.00	24.5	26.6	28.6	30.7	32.8	34.9	37.0	39.0	41.1
400.00	25.1	27.2	29.4	31.5	33.7	35.9	38.0	40.2	42.3
450.00	25.6	27.8	30.0	32.3	34.5	36.7	38.9	41.2	43.4
500.00	26.0	28.3	30.6	32.9	35.2	37.5	39.8	42.1	44.4
600.00	26.8	29.2	31.6	34.0	36.4	38.8	41.2	43.6	46.0
700.00	27.5	30.0	32.5	35.0	37.5	39.9	42.4	44.9	47.4
800.00	28.1	30.7	33.3	35.8	38.4	40.9	43.5	46.0	48.6
900.00	28.7	31.3	34.0	36.6	39.2	41.8	44.4	47.0	49.7
999.99	29.3	31.9	34.6	37.3	39.9	42.6	45.3	47.9	50.6

TABLE A.9

$f_h/U : g$ *Relationships when* $z = 20$

Values of g when j of cooling curve is:

f_h/U	0.40	0.60	0.80	1.00	1.20	1.40	1.60	1.80	2.00
0.20	4.52–05	4.90–05	5.28–05	5.67–05	6.05–05	6.43–05	6.82–05	7.20–05	7.59–05
0.30	2.08–03	2.26–03	2.43–03	2.61–03	2.78–03	2.96–03	3.13–03	3.31–03	3.48–03
0.40	1.44–02	1.55–02	1.67–02	1.79–02	1.90–02	2.02–02	2.14–02	2.26–02	2.37–02
0.50	4.60–02	4.96–02	5.33–02	5.70–02	6.07–02	6.43–02	6.80–02	7.17–02	7.54–02
0.60	0.100	0.108	0.116	0.124	0.132	0.140	0.148	0.155	0.163
0.70	0.175	0.189	0.203	0.217	0.230	0.244	0.258	0.271	0.285
0.80	0.268	0.289	0.310	0.330	0.351	0.372	0.393	0.414	0.435
0.90	0.373	0.402	0.431	0.461	0.490	0.519	0.548	0.577	0.607
1.00	0.488	0.526	0.564	0.603	0.641	0.679	0.718	0.756	0.794
2.00	1.76	1.90	2.05	2.19	2.34	2.48	2.63	2.77	2.91
3.00	2.90	3.15	3.39	3.64	3.88	4.12	4.37	4.61	4.86
4.00	3.87	4.20	4.53	4.86	5.19	5.52	5.85	6.18	6.51
5.00	4.71	5.11	5.52	5.92	6.32	6.73	7.13	7.53	7.94
6.00	5.44	5.91	6.38	6.84	7.31	7.78	8.24	8.71	9.18
7.00	6.09	6.61	7.14	7.66	8.18	8.71	9.23	9.75	10.28
8.00	6.67	7.24	7.82	8.39	8.96	9.54	10.11	10.69	11.26
9.00	7.19	7.81	8.43	9.05	9.67	10.29	10.91	11.54	12.16
10.00	7.67	8.33	8.99	9.66	10.32	10.98	11.65	12.31	12.97
15.00	9.55	10.39	11.23	12.07	12.91	13.75	14.59	15.42	16.26
20.00	10.9	11.9	12.9	13.9	14.8	15.8	16.8	17.7	18.7
25.00	12.0	13.1	14.2	15.3	16.4	17.4	18.5	19.6	20.7
30.00	12.9	14.1	15.3	16.5	17.6	18.8	20.0	21.2	22.3
35.00	13.7	15.0	16.2	17.5	18.7	20.0	21.2	22.5	23.7
40.00	14.4	15.7	17.1	18.4	19.7	21.0	22.3	23.7	25.0
45.00	15.0	16.4	17.8	19.2	20.6	21.9	23.3	24.7	26.1
50.00	15.6	17.0	18.5	19.9	21.3	22.8	24.2	25.6	27.1
60.00	16.6	18.1	19.6	21.2	22.7	24.2	25.8	27.3	28.8
70.00	17.4	19.0	20.6	22.2	23.9	25.5	27.1	28.7	30.3
80.00	18.2	19.8	21.5	23.2	24.9	26.6	28.2	29.9	31.6
90.00	18.8	20.5	22.3	24.0	25.8	27.5	29.3	31.0	32.7
100.00	19.4	21.2	23.0	24.8	26.6	28.4	30.2	32.0	33.8
150.00	21.7	23.7	25.7	27.7	29.7	31.7	33.7	35.7	37.7
200.00	23.2	25.4	27.6	29.7	31.9	34.1	36.2	38.4	40.5
250.00	24.5	26.8	29.0	31.3	33.6	35.9	38.2	40.4	42.7
300.00	25.5	27.9	30.3	32.6	35.0	37.4	39.8	42.1	44.5
350.00	26.4	28.8	31.3	33.7	36.2	38.7	41.1	43.6	46.0
400.00	27.1	29.7	32.2	34.7	37.3	39.8	42.3	44.8	47.4
450.00	27.8	30.4	33.0	35.6	38.2	40.8	43.4	46.0	48.5
500.00	28.5	31.1	33.7	36.4	39.0	41.7	44.3	46.9	49.6
600.00	29.6	32.3	35.0	37.7	40.5	43.2	45.9	48.6	51.4
700.00	30.5	33.3	36.1	38.9	41.7	44.5	47.3	50.1	52.9

TABLE A.10

$f_h/U : g$ Relationships when $z = 22$

	Values of g when j of cooling curve is:								
f_h/U	0.40	0.60	0.80	1.00	1.20	1.40	1.60	1.80	2.00
0.20	4.80-05	5.25-05	5.71-05	6.16-05	6.61-05	7.06-05	7.52-05	7.97-05	8.42-05
0.30	2.19-03	2.40-03	2.61-03	2.82-03	3.04-03	3.25-03	3.46-03	3.67-03	3.88-03
0.40	1.59-02	1.73-02	1.86-02	2.00-02	2.13-02	2.26-02	2.40-02	2.53-02	2.67-02
0.50	5.32-02	5.71-02	6.10-02	6.50-02	6.89-02	7.29-02	7.68-02	8.08-02	8.47-02
0.60	0.118	0.126	0.135	0.143	0.151	0.159	0.167	0.175	0.183
0.70	0.208	0.222	0.236	0.250	0.264	0.277	0.291	0.305	0.319
0.80	0.317	0.338	0.359	0.380	0.401	0.422	0.443	0.464	0.485
0.90	0.438	0.468	0.497	0.527	0.556	0.586	0.615	0.645	0.674
1.00	0.568	0.607	0.646	0.685	0.725	0.764	0.803	0.842	0.881
2.00	1.91	2.08	2.24	2.41	2.57	2.74	2.91	3.07	3.24
3.00	3.11	3.40	3.69	3.98	4.27	4.56	4.85	5.13	5.42
4.00	4.15	4.54	4.94	5.33	5.72	6.12	6.51	6.90	7.30
5.00	5.07	5.55	6.03	6.51	6.98	7.46	7.94	8.42	8.90
6.00	5.88	6.43	6.98	7.53	8.08	8.63	9.18	9.73	10.28
7.00	6.61	7.22	7.83	8.44	9.06	9.67	10.28	10.89	11.50
8.00	7.27	7.93	8.60	9.26	9.92	10.59	11.25	11.91	12.58
9.00	7.87	8.58	9.29	10.00	10.71	11.42	12.13	12.84	13.55
10.00	8.41	9.17	9.92	10.67	11.43	12.18	12.93	13.69	14.44
15.00	10.6	11.5	12.4	13.4	14.3	15.2	16.2	17.1	18.0
20.00	12.1	13.2	14.2	15.3	16.4	17.5	18.6	19.7	20.8
25.00	13.2	14.5	15.7	16.9	18.1	19.3	20.5	21.8	23.0
30.00	14.2	15.5	16.9	18.2	19.5	20.9	22.2	23.5	24.9
35.00	15.0	16.4	17.9	19.3	20.7	22.2	23.6	25.0	26.5
40.00	15.7	17.2	18.7	20.3	21.8	23.3	24.8	26.4	27.9
45.00	16.3	17.9	19.5	21.1	22.7	24.3	25.9	27.5	29.1
50.00	16.9	18.5	20.2	21.9	23.6	25.2	26.9	28.6	30.2
55.00	17.4	19.1	20.8	22.6	24.3	26.0	27.8	29.5	31.2
60.00	17.8	19.6	21.4	23.2	25.0	26.8	28.6	30.4	32.2
70.00	18.7	20.6	22.5	24.3	26.2	28.1	30.0	31.9	33.8
80.00	19.4	21.4	23.4	25.3	27.3	29.2	31.2	33.2	35.1
90.00	20.1	22.2	24.2	26.2	28.2	30.3	32.3	34.3	36.3
100.00	20.8	22.9	24.9	27.0	29.1	31.2	33.3	35.3	37.4
150.00	23.4	25.7	28.0	30.3	32.6	34.8	37.1	39.4	41.7
200.00	25.5	27.9	30.3	32.7	35.2	37.6	40.0	42.4	44.8
250.00	27.1	29.6	32.2	34.7	37.2	39.8	42.3	44.8	47.4
300.00	28.3	30.9	33.6	36.2	38.9	41.5	44.1	46.8	49.4
350.00	29.2	31.9	34.7	37.4	40.2	42.9	45.6	48.4	51.1
400.00	29.9	32.7	35.6	38.4	41.2	44.0	46.8	49.7	52.5
450.00	30.5	33.4	36.3	39.2	42.1	45.0	47.9	50.8	53.7
460.00	30.6	33.5	36.4	39.3	42.2	45.1	48.1	51.0	53.9

TABLE A.11

$f_h/U : g$ *Relationships when* $z = 24$

	Values of g when j of cooling curve is:								
f_h/U	0.40	0.60	0.80	1.00	1.20	1.40	1.60	1.80	2.00
0.20	5.39-05	5.88-05	6.38-05	6.87-05	7.36-05	7.86-05	8.35-05	8.84-05	9.33-05
0.30	2.47-03	2.70-03	2.92-03	3.15-03	3.38-03	3.60-03	3.83-03	4.06-03	4.28-03
0.40	1.69-02	1.85-02	2.00-02	2.16-02	2.31-02	2.47-02	2.62-02	2.78-02	2.93-02
0.50	5.42-02	5.90-02	6.39-02	6.88-02	7.37-02	7.85-02	8.34-02	8.83-02	9.32-02
0.60	0.118	0.129	0.139	0.150	0.160	0.171	0.181	0.192	0.202
0.70	0.208	0.226	0.244	0.262	0.280	0.298	0.317	0.335	0.353
0.80	0.318	0.346	0.373	0.401	0.428	0.455	0.483	0.510	0.538
0.90	0.445	0.483	0.521	0.559	0.597	0.635	0.673	0.711	0.749
1.00	0.583	0.633	0.682	0.732	0.782	0.831	0.881	0.931	0.980
2.00	2.10	2.29	2.47	2.66	2.85	3.03	3.22	3.40	3.59
3.00	3.43	3.75	4.07	4.39	4.71	5.03	5.35	5.67	5.99
4.00	4.56	4.99	5.43	5.86	6.30	6.73	7.17	7.60	8.04
5.00	5.53	6.06	6.59	7.13	7.66	8.19	8.72	9.26	9.79
6.00	6.38	7.00	7.62	8.23	8.85	9.46	10.08	10.70	11.31
7.00	7.15	7.84	8.53	9.21	9.90	10.59	11.28	11.97	12.66
8.00	7.84	8.59	9.34	10.10	10.85	11.60	12.36	13.11	13.87
9.00	8.46	9.27	10.09	10.90	11.71	12.52	13.33	14.15	14.96
10.00	9.03	9.90	10.76	11.63	12.50	13.36	14.23	15.09	15.96
15.00	11.3	12.4	13.5	14.6	15.6	16.7	17.8	18.9	20.0
20.00	12.9	14.2	15.5	16.7	18.0	19.2	20.5	21.7	23.0
25.00	14.2	15.6	17.0	18.4	19.8	21.2	22.6	24.0	25.4
30.00	15.2	16.8	18.3	19.8	21.3	22.9	24.4	25.9	27.4
35.00	16.1	17.8	19.4	21.0	22.6	24.3	25.9	27.5	29.2
40.00	16.9	18.6	20.3	22.1	23.8	25.5	27.2	28.9	30.7
45.00	17.6	19.4	21.2	23.0	24.8	26.6	28.4	30.2	32.0
50.00	18.2	20.1	22.0	23.8	25.7	27.6	29.4	31.3	33.2
55.00	18.8	20.7	22.7	24.6	26.5	28.5	30.4	32.3	34.3
60.00	19.3	21.3	23.3	25.3	27.3	29.3	31.3	33.3	35.3
70.00	20.3	22.4	24.5	26.6	28.7	30.8	32.8	34.9	37.0
80.00	21.2	23.3	25.5	27.7	29.9	32.0	34.2	36.4	38.6
90.00	21.9	24.2	26.4	28.7	30.9	33.2	35.4	37.7	40.0
100.00	22.6	24.9	27.3	29.6	31.9	34.2	36.5	38.9	41.2
150.00	25.4	28.0	30.5	33.1	35.7	38.2	40.8	43.3	45.9
200.00	27.4	30.1	32.9	35.6	38.3	41.0	43.8	46.5	49.2
250.00	28.9	31.7	34.6	37.5	40.4	43.2	46.1	49.0	51.9

TABLE A.12

$f_h/U : g$ *Relationships when* $z = 26$

	Values of g when j of cooling curve is:								
f_h/U	0.40	0.60	0.80	1.00	1.20	1.40	1.60	1.80	2.00
0.20	5.78-05	6.34-05	6.89-05	7.45-05	8.01-05	8.56-05	9.12-05	9.68-05	1.02-04
0.30	2.74-03	2.98-03	3.22-03	3.47-03	3.71-03	3.96-03	4.20-03	4.44-03	4.69-03
0.40	1.87-02	2.04-02	2.20-02	2.37-02	2.54-02	2.70-02	2.87-02	3.04-02	3.21-02
0.50	5.89-02	6.43-02	6.97-02	7.51-02	8.05-02	8.59-02	9.13-02	9.67-02	1.02-01
0.60	0.127	0.139	0.151	0.163	0.174	0.186	0.198	0.210	0.222
0.70	0.220	0.241	0.262	0.283	0.304	0.325	0.346	0.367	0.388
0.80	0.334	0.366	0.399	0.431	0.463	0.495	0.528	0.560	0.592
0.90	0.464	0.509	0.554	0.600	0.645	0.690	0.736	0.781	0.826
1.00	0.606	0.665	0.725	0.784	0.844	0.903	0.963	1.022	1.082
2.00	2.21	2.43	2.65	2.87	3.08	3.30	3.52	3.74	3.96
3.00	3.71	4.07	4.42	4.78	5.14	5.50	5.86	6.21	6.57
4.00	4.99	5.47	5.94	6.42	6.89	7.37	7.84	8.32	8.80
5.00	6.08	6.66	7.24	7.82	8.39	8.97	9.55	10.13	10.71
6.00	7.02	7.69	8.36	9.03	9.70	10.37	11.04	11.70	12.37
7.00	7.84	8.59	9.34	10.09	10.85	11.60	12.35	13.10	13.85
8.00	8.56	9.39	10.21	11.04	11.87	12.70	13.52	14.35	15.18
9.00	9.20	10.10	10.99	11.89	12.79	13.69	14.59	15.48	16.38
10.00	9.77	10.74	11.70	12.67	13.63	14.59	15.56	16.52	17.48
15.00	12.0	13.3	14.5	15.7	17.0	18.2	19.4	20.7	21.9
20.00	13.6	15.1	16.5	18.0	19.4	20.9	22.3	23.8	25.2
25.00	15.0	16.6	18.2	19.8	21.4	23.0	24.6	26.3	27.9
30.00	16.1	17.8	19.6	21.3	23.1	24.8	26.6	28.3	30.1
35.00	17.1	18.9	20.8	22.6	24.5	26.4	28.2	30.1	31.9
40.00	17.9	19.9	21.8	23.8	25.7	27.7	29.6	31.6	33.6
45.00	18.7	20.8	22.8	24.8	26.9	28.9	30.9	33.0	35.0
50.00	19.5	21.6	23.7	25.8	27.9	30.0	32.1	34.2	36.3
55.00	20.2	22.3	24.5	26.6	28.8	31.0	33.1	35.3	37.5
60.00	20.8	23.0	25.2	27.4	29.7	31.9	34.1	36.3	38.5
70.00	21.9	24.2	26.5	28.9	31.2	33.5	35.8	38.1	40.4
80.00	22.9	25.3	27.7	30.1	32.5	34.9	37.3	39.7	42.1
90.00	23.8	26.2	28.7	31.2	33.7	36.1	38.6	41.1	43.6
100.00	24.5	27.1	29.6	32.2	34.7	37.3	39.8	42.3	44.9
150.00	27.4	30.3	33.1	35.9	38.7	41.5	44.3	47.2	50.0
200.00	29.4	32.4	35.4	38.5	41.5	44.5	47.5	50.5	53.6

TABLE A.13

$f_h/U : g$ Relationships when $z = 28$

	Values of g when j of cooling curve is:								
f_h/U	0.10	0.60	0.80	1.00	1.20	1.40	1.60	1.80	2.00
0.20	6.17-05	6.79-05	7.42-05	8.04-05	8.66-05	9.28-05	9.90-05	1.05-04	1.11-04
0.30	2.86-03	3.15-03	3.43-03	3.72-03	4.00-03	4.29-03	4.57-03	4.86-03	5.14-03
0.40	1.98-02	2.17-02	2.36-02	2.55-02	2.74-02	2.94-02	3.13-02	3.32-02	3.51-02
0.50	6.32-02	6.92-02	7.52-02	8.13-02	8.73-02	9.33-02	9.93-02	1.05-01	1.11-01
0.60	0.138	0.150	0.163	0.176	0.189	0.202	0.215	0.228	0.241
0.70	0.240	0.263	0.285	0.308	0.331	0.353	0.376	0.398	0.421
0.80	0.366	0.401	0.435	0.470	0.504	0.538	0.573	0.607	0.642
0.90	0.510	0.558	0.606	0.654	0.702	0.750	0.799	0.847	0.895
1.00	0.667	0.730	0.793	0.856	0.919	0.982	1.045	1.109	1.172
2.00	2.40	2.64	2.88	3.12	3.36	3.59	3.83	4.07	4.31
3.00	3.96	4.36	4.76	5.16	5.57	5.97	6.37	6.78	7.18
4.00	5.27	5.81	6.36	6.90	7.44	7.98	8.52	9.07	9.61
5.00	6.40	7.06	7.72	8.38	9.04	9.71	10.37	11.03	11.69
6.00	7.38	8.15	8.91	9.68	10.44	11.21	11.97	12.74	13.51
7.00	8.25	9.11	9.96	10.82	11.68	12.54	13.40	14.25	15.11
8.00	9.03	9.97	10.91	11.85	12.79	13.73	14.67	15.61	16.55
9.00	9.73	10.74	11.76	12.78	13.79	14.81	15.83	16.85	17.86
10.00	10.4	11.5	12.5	13.6	14.7	15.8	16.9	18.0	19.1
15.00	12.9	14.3	15.6	17.0	18.4	19.8	21.1	22.5	23.9
20.00	14.7	16.3	17.9	19.5	21.1	22.7	24.3	25.9	27.4
25.00	16.2	17.9	19.7	21.5	23.2	25.0	26.8	28.5	30.3
30.00	17.3	19.3	21.2	23.1	25.0	26.9	28.8	30.7	32.6
35.00	18.4	20.4	22.4	24.5	26.5	28.5	30.6	32.6	34.7
40.00	19.3	21.4	23.5	25.7	27.8	30.0	32.1	34.3	36.4
45.00	20.1	22.3	24.5	26.8	29.0	31.3	33.5	35.7	38.0
50.00	20.8	23.1	25.4	27.8	30.1	32.4	34.7	37.1	39.4
55.00	21.5	23.9	26.3	28.7	31.1	33.5	35.8	38.2	40.6
60.00	22.1	24.6	27.0	29.5	31.9	34.4	36.9	39.3	41.8
70.00	23.2	25.8	28.4	31.0	33.5	36.1	38.7	41.3	43.9
80.00	24.2	26.9	29.6	32.3	34.9	37.6	40.3	43.0	45.6
90.00	25.1	27.9	30.7	33.4	36.2	38.9	41.7	44.5	47.2
100.00	26.0	28.8	31.6	34.5	37.3	40.1	43.0	45.8	48.6

TABLE A.14

$f_h/U:g$ Relationships when $z = 30$

f_h/U	Values of g when j of cooling curve is:								
	0.40	0.60	0.80	1.00	1.20	1.40	1.60	1.80	2.00
0.20	6.59-05	7.28-05	7.96-05	8.65-05	9.33-05	1.00-04	1.07-04	1.14-04	1.21-04
0.30	3.05-03	3.37-03	3.68-03	4.00-03	4.32-03	4.63-03	4.95-03	5.26-03	5.58-03
0.40	2.10-02	2.31-02	2.53-02	2.74-02	2.96-02	3.17-02	3.38-02	3.60-02	3.81-02
0.50	6.71-02	7.38-02	8.06-02	8.73-02	9.40-02	1.01-01	1.07-01	1.14-01	1.21-01
0.60	0.146	0.161	0.175	0.189	0.204	0.218	0.233	0.247	0.262
0.70	0.255	0.281	0.306	0.331	0.356	0.381	0.406	0.432	0.457
0.80	0.390	0.428	0.466	0.505	0.543	0.581	0.620	0.658	0.697
0.90	0.543	0.596	0.650	0.703	0.757	0.810	0.864	0.918	0.971
1.00	0.710	0.780	0.850	0.920	0.991	1.061	1.131	1.201	1.272
2.00	2.56	2.83	3.09	3.36	3.62	3.88	4.15	4.41	4.67
3.00	4.23	4.67	5.12	5.56	6.00	6.45	6.89	7.34	7.78
4.00	5.62	6.22	6.82	7.42	8.02	8.62	9.22	9.82	10.42
5.00	6.80	7.54	8.27	9.01	9.74	10.48	11.21	11.95	12.68
6.00	7.82	8.68	9.53	10.38	11.24	12.09	12.94	13.80	14.65
7.00	8.73	9.68	10.64	11.60	12.56	13.52	14.48	15.44	16.39
8.00	9.53	10.58	11.64	12.69	13.74	14.80	15.85	16.91	17.96
9.00	10.3	11.4	12.5	13.7	14.8	16.0	17.1	18.2	19.4
10.00	10.9	12.1	13.4	14.6	15.8	17.0	18.2	19.5	20.7
15.00	13.6	15.1	16.7	18.2	19.7	21.3	22.8	24.3	25.8
20.00	15.6	17.4	19.1	20.9	22.7	24.4	26.2	27.9	29.7
25.00	17.2	19.2	21.1	23.0	25.0	26.9	28.9	30.8	32.7
30.00	18.5	20.6	22.7	24.8	26.9	29.0	31.1	33.2	35.3
35.00	19.6	21.9	24.1	26.3	28.5	30.8	33.0	35.2	37.4
40.00	20.6	22.9	25.3	27.6	30.0	32.3	34.6	37.0	39.3
45.00	21.4	23.9	26.3	28.8	31.2	33.7	36.1	38.6	41.0
50.00	22.2	24.7	27.3	29.8	32.3	34.9	37.4	40.0	42.5
55.00	22.9	25.5	28.1	30.7	33.4	36.0	38.6	41.2	43.8
60.00	23.5	26.2	28.9	31.6	34.3	37.0	39.7	42.4	45.1
70.00	24.6	27.5	30.3	33.1	36.0	38.8	41.6	44.4	47.3
80.00	25.6	28.6	31.5	34.5	37.4	40.4	43.3	46.2	49.2
90.00	26.5	29.6	32.6	35.7	38.7	41.7	44.8	47.8	50.9
100.00	27.4	30.5	33.6	36.7	39.9	43.0	46.1	49.2	52.3

TABLE A.15

$f_h/U : g$ Relationships when $z = 32$

f_h/U	\multicolumn{9}{c}{Values of g when j of cooling curve is:}
	0.40	0.60	0.80	1.00	1.20	1.40	1.60	1.80	2.00
0.20	7.10-05	7.85-05	8.59-05	9.33-05	1.01-04	1.08-04	1.16-04	1.23-04	1.30-04
0.30	3.26-03	3.60-03	3.95-03	4.29-03	4.63-03	4.98-03	5.32-03	5.67-03	6.01-03
0.40	2.21-02	2.45-02	2.68-02	2.92-02	3.16-02	3.40-02	3.64-02	3.88-02	4.12-02
0.50	7.01-02	7.77-02	8.53-02	9.29-02	1.00-01	1.08-01	1.16-01	1.23-01	1.31-01
0.60	0.152	0.169	0.185	0.202	0.218	0.235	0.251	0.267	0.284
0.70	0.267	0.295	0.324	0.353	0.381	0.410	0.438	0.467	0.495
0.80	0.408	0.451	0.495	0.538	0.582	0.625	0.668	0.712	0.755
0.90	0.570	0.630	0.690	0.751	0.811	0.871	0.932	0.992	1.053
1.00	0.746	0.825	0.904	0.983	1.062	1.141	1.220	1.299	1.378
2.00	2.72	3.01	3.30	3.59	3.88	4.17	4.46	4.75	5.04
3.00	4.49	4.98	5.46	5.95	6.44	6.93	7.42	7.90	8.39
4.00	5.97	6.63	7.29	7.95	8.61	9.26	9.92	10.58	11.24
5.00	7.23	8.04	8.84	9.65	10.46	11.26	12.07	12.88	13.68
6.00	8.31	9.25	10.19	11.13	12.06	13.00	13.94	14.88	15.82
7.00	9.26	10.32	11.37	12.43	13.48	14.53	15.59	16.64	17.70
8.00	10.1	11.3	12.4	13.6	14.7	15.9	17.1	18.2	19.4
9.00	10.9	12.1	13.4	14.6	15.9	17.1	18.4	19.6	20.9
10.00	11.6	12.9	14.2	15.6	16.9	18.3	19.6	20.9	22.3
15.00	14.0	16.0	17.7	19.4	21.1	22.8	24.5	26.1	27.8
20.00	16.4	18.4	20.3	22.3	24.2	26.1	28.1	30.0	32.0
25.00	18.1	20.3	22.4	24.5	26.7	28.8	31.0	33.1	35.3
30.00	19.5	21.8	24.1	26.4	28.7	31.1	33.4	35.7	38.0
35.00	20.7	23.1	25.6	28.0	30.5	32.9	35.4	37.8	40.3
40.00	21.7	24.3	26.8	29.4	32.0	34.6	37.1	39.7	42.3
45.00	22.6	25.3	28.0	30.6	33.3	36.0	38.7	41.4	44.0
50.00	23.4	26.2	28.9	31.7	34.5	37.3	40.0	42.8	45.6
55.00	24.1	27.0	29.8	32.7	35.6	38.4	41.3	44.2	47.0
60.00	24.8	27.7	30.7	33.6	36.6	39.5	42.4	45.4	48.3
70.00	26.0	29.1	32.2	35.3	38.3	41.4	44.5	47.6	50.6
80.00	27.2	30.4	33.6	36.7	39.9	43.1	46.3	49.4	52.6

TABLE A.16

$f_h/U : g$ *Relationships when* $z = 34$

f_h/U	Values of g when j of cooling curve is:								
	0.40	0.60	0.80	1.00	1.20	1.40	1.60	1.80	2.00
0.20	7.54−05	8.31−05	9.09−05	9.86−05	1.06−04	1.14−04	1.22−04	1.29−04	1.37−04
0.30	3.38−03	3.79−03	4.20−03	4.61−03	5.02−03	5.43−03	5.83−03	6.24−03	6.65−03
0.40	2.33−02	2.60−02	2.87−02	3.14−02	3.41−02	3.67−02	3.94−02	4.21−02	4.48−02
0.50	7.56−02	8.34−02	9.12−02	9.90−02	1.07−01	1.15−01	1.23−01	1.30−01	1.38−01
0.60	0.165	0.182	0.198	0.214	0.230	0.247	0.263	0.279	0.295
0.70	0.289	0.317	0.345	0.374	0.402	0.430	0.458	0.486	0.515
0.80	0.439	0.483	0.527	0.570	0.614	0.658	0.701	0.745	0.789
0.90	0.609	0.671	0.734	0.796	0.859	0.922	0.984	1.047	1.109
1.00	0.791	0.876	0.960	1.044	1.129	1.213	1.298	1.382	1.466
2.00	2.85	3.18	3.51	3.83	4.16	4.49	4.82	5.14	5.47
3.00	4.78	5.30	5.82	6.34	6.86	7.38	7.90	8.42	8.94
4.00	6.34	7.05	7.75	8.46	9.17	9.88	10.59	11.30	12.00
5.00	7.61	8.50	9.39	10.28	11.17	12.06	12.95	13.84	14.73
6.00	8.71	9.76	10.81	11.86	12.91	13.96	15.01	16.06	17.11
7.00	9.69	10.87	12.06	13.24	14.42	15.60	16.79	17.97	19.15
8.00	10.6	11.9	13.2	14.5	15.8	17.0	18.3	19.6	20.9
9.00	11.4	12.8	14.2	15.6	16.9	18.3	19.7	21.1	22.5
10.00	12.2	13.6	15.1	16.6	18.0	19.5	21.0	22.4	23.9
15.00	15.1	17.0	18.8	20.6	22.5	24.3	26.1	27.9	29.8
20.00	17.2	19.4	21.5	23.7	25.8	28.0	30.1	32.3	34.4
25.00	19.0	21.3	23.7	26.1	28.4	30.8	33.2	35.6	37.9
30.00	20.5	23.0	25.5	28.0	30.6	33.1	35.6	38.1	40.6
35.00	21.8	24.5	27.1	29.7	32.4	35.0	37.6	40.3	42.9
40.00	22.9	25.7	28.4	31.2	34.0	36.7	39.5	42.3	45.0
45.00	23.8	26.7	29.6	32.5	35.4	38.3	41.3	44.2	47.1
50.00	24.6	27.6	30.7	33.7	36.8	39.8	42.8	45.9	48.9
55.00	25.4	28.5	31.7	34.8	37.9	41.0	44.2	47.3	50.4
60.00	26.3	29.4	32.6	35.8	38.9	42.1	45.2	48.4	51.6

TABLE A.17

$f_h/U : g$ Relationships when $z = 36$

f_h/U	\multicolumn{9}{c}{Values of g when j of cooling curve is:}
	0.40	0.60	0.80	1.00	1.20	1.40	1.60	1.80	2.00
0.20	8.70-05	9.45-05	1.02-04	1.10-04	1.17-04	1.25-04	1.32-04	1.40-04	1.47-04
0.30	3.65-03	4.06-03	4.47-03	4.88-03	5.30-03	5.71-03	6.12-03	6.53-03	6.94-03
0.40	2.30-02	2.61-02	2.92-02	3.23-02	3.55-02	3.86-02	4.17-02	4.48-02	4.80-02
0.50	7.19-02	8.20-02	9.20-02	1.02-01	1.12-01	1.22-01	1.32-01	1.42-01	1.52-01
0.60	0.157	0.179	0.201	0.222	0.244	0.265	0.287	0.308	0.330
0.70	0.281	0.317	0.354	0.391	0.427	0.464	0.501	0.538	0.574
0.80	0.439	0.493	0.547	0.602	0.656	0.710	0.764	0.819	0.873
0.90	0.628	0.701	0.774	0.847	0.920	0.993	1.067	1.140	1.213
1.00	0.842	0.934	1.027	1.120	1.212	1.305	1.397	1.490	1.583
2.00	3.20	3.52	3.84	4.17	4.49	4.81	5.13	5.45	5.77
3.00	4.97	5.56	6.14	6.73	7.32	7.91	8.50	9.09	9.68
4.00	6.40	7.22	8.05	8.87	9.70	10.52	11.34	12.17	12.99
5.00	7.67	8.69	9.71	10.73	11.74	12.76	13.78	14.80	15.82
6.00	8.85	10.02	11.20	12.37	13.55	14.73	15.90	17.08	18.25
7.00	9.94	11.25	12.56	13.86	15.17	16.48	17.78	19.09	20.39
8.00	11.0	12.4	13.8	15.2	16.6	18.1	19.5	20.9	22.3
9.00	12.0	13.5	15.0	16.5	18.0	19.5	21.0	22.5	24.0
10.00	12.9	14.5	16.0	17.6	19.2	20.8	22.4	24.0	25.6
15.00	16.5	18.4	20.3	22.3	24.2	26.1	28.0	29.9	31.8
20.00	18.8	21.1	23.3	25.5	27.7	29.9	32.1	34.3	36.5
25.00	20.4	22.9	25.3	27.8	30.3	32.8	35.3	37.8	40.2
30.00	21.5	24.3	27.0	29.7	32.4	35.1	37.9	40.6	43.3
35.00	22.5	25.5	28.4	31.3	34.2	37.1	40.1	43.0	45.9
40.00	23.5	26.5	29.6	32.7	35.8	38.9	42.0	45.1	48.1
45.00	24.3	27.5	30.8	34.0	37.2	40.4	43.7	46.9	50.1
50.00	25.2	28.5	31.8	35.2	38.5	41.9	45.2	48.5	51.9
55.00	26.0	29.4	32.9	36.3	39.7	43.1	46.6	50.0	53.4

TABLE A.18

$f_h/U:g$ Relationships when $z = 38$

	Values of g when j of cooling curve is:								
f_h/U	0.40	0.60	0.80	1.00	1.20	1.40	1.60	1.80	2.00
0.20	8.24-05	9.21-05	1.02-04	1.11-04	1.21-04	1.31-04	1.41-04	1.50-04	1.60-04
0.30	3.79-03	4.25-03	4.70-03	5.16-03	5.61-03	6.07-03	6.52-03	6.97-03	7.43-03
0.40	2.59-02	2.90-02	3.21-02	3.52-02	3.83-02	4.14-02	4.45-02	4.76-02	5.07-02
0.50	8.27-02	9.24-02	1.02-01	1.12-01	1.22-01	1.31-01	1.41-01	1.51-01	1.61-01
0.60	0.180	0.201	0.222	0.243	0.264	0.285	0.306	0.327	0.348
0.70	0.314	0.351	0.388	0.424	0.461	0.498	0.535	0.571	0.608
0.80	0.479	0.535	0.591	0.647	0.703	0.759	0.816	0.872	0.928
0.90	0.667	0.745	0.824	0.902	0.981	1.059	1.137	1.216	1.294
1.00	0.872	0.975	1.078	1.181	1.284	1.387	1.489	1.592	1.695
2.00	3.16	3.54	3.92	4.30	4.69	5.07	5.45	5.83	6.21
3.00	5.22	5.86	6.49	7.13	7.77	8.41	9.05	9.69	10.33
4.00	6.94	7.80	8.66	9.52	10.38	11.24	12.10	12.96	13.82
5.00	8.39	9.44	10.50	11.55	12.60	13.65	14.70	15.75	16.80
6.00	9.64	10.86	12.08	13.30	14.52	15.74	16.96	18.18	19.40
7.00	10.7	12.1	13.5	14.8	16.2	17.6	18.9	20.3	21.7
8.00	11.7	13.2	14.7	16.2	17.7	19.2	20.7	22.2	23.7
9.00	12.6	14.2	15.8	17.4	19.1	20.7	22.3	24.0	25.6
10.00	13.4	15.1	16.8	18.6	20.3	22.1	23.8	25.5	27.3
15.00	16.5	18.7	20.9	23.1	25.3	27.5	29.6	31.8	34.0
20.00	19.0	21.5	24.0	26.4	28.9	31.4	33.9	36.4	38.9
25.00	20.9	23.6	26.4	29.1	31.9	34.6	37.3	40.1	42.8
30.00	22.5	25.5	28.4	31.3	34.3	37.2	40.1	43.1	46.0
35.00	23.9	27.0	30.1	33.2	36.3	39.4	42.5	45.6	48.7
40.00	25.2	28.4	31.6	34.9	38.1	41.3	44.6	47.8	51.0
45.00	26.2	29.6	32.9	36.3	39.7	43.0	46.4	49.7	53.1

TABLE A.19

$f_h/U : g$ Relationships when $z = 40$

	Values of g when j of cooling curve is:								
f_h/U	0.40	0.60	0.80	1.00	1.20	1.40	1.60	1.80	2.00
0.20	8.58-05	9.64-05	1.07-04	1.18-04	1.28-04	1.39-04	1.49-04	1.60-04	1.70-04
0.30	3.98-03	4.47-03	4.96-03	5.46-03	5.95-03	6.44-03	6.93-03	7.43-03	7.92-03
0.40	2.72-02	3.06-02	3.39-02	3.73-02	4.07-02	4.40-02	4.74-02	5.08-02	5.42-02
0.50	8.67-02	9.72-02	1.08-01	1.18-01	1.29-01	1.39-01	1.50-01	1.61-01	1.71-01
0.60	0.188	0.211	0.234	0.256	0.279	0.302	0.324	0.347	0.370
0.70	0.329	0.368	0.408	0.447	0.487	0.527	0.566	0.606	0.645
0.80	0.501	0.562	0.622	0.682	0.743	0.803	0.864	0.924	0.984
0.90	0.698	0.782	0.867	0.952	1.036	1.121	1.205	1.290	1.374
1.00	0.912	1.023	1.135	1.246	1.357	1.469	1.580	1.691	1.802
2.00	3.30	3.72	4.13	4.55	4.96	5.38	5.79	6.21	6.62
3.00	5.46	6.15	6.84	7.53	8.22	8.91	9.60	10.29	10.98
4.00	7.25	8.18	9.11	10.04	10.97	11.90	12.83	13.77	14.70
5.00	8.75	9.89	11.04	12.18	13.32	14.46	15.61	16.75	17.89
6.00	10.1	11.4	12.7	14.0	15.4	16.7	18.0	19.3	20.7
7.00	11.2	12.7	14.2	15.7	17.1	18.6	20.1	21.6	23.1
8.00	12.2	13.8	15.5	17.1	18.7	20.4	22.0	23.6	25.2
9.00	13.1	14.9	16.6	18.4	20.2	21.9	23.7	25.4	27.2
10.00	14.0	15.8	17.7	19.6	21.5	23.3	25.2	27.1	29.0
15.00	17.3	19.6	22.0	24.3	26.7	29.0	31.4	33.8	36.1
20.00	19.8	22.5	25.2	27.8	30.5	33.2	35.9	38.6	41.3
25.00	21.8	24.8	27.7	30.6	33.6	36.5	39.4	42.4	45.3
30.00	23.5	26.6	29.8	32.9	36.1	39.2	42.4	45.6	48.7
35.00	24.9	28.3	31.6	34.9	38.2	41.5	44.9	48.2	51.5

TABLE A.20

$f_h/U : g$ *Relationships when* $z = 42$

f_h/U	Values of g when j of cooling curve is:								
	0.40	0.60	0.80	1.00	1.20	1.40	1.60	1.80	2.00
0.20	8.94-05	1.01-04	1.12-04	1.24-04	1.35-04	1.47-04	1.58-04	1.70-04	1.81-04
0.30	4.16-03	4.69-03	5.22-03	5.75-03	6.28-03	6.81-03	7.34-03	7.87-03	8.40-03
0.40	2.85-02	3.21-02	3.57-02	3.93-02	4.29-02	4.65-02	5.01-02	5.37-02	5.73-02
0.50	9.05-02	1.02-01	1.13-01	1.25-01	1.36-01	1.48-01	1.59-01	1.70-01	1.82-01
0.60	0.196	0.221	0.245	0.270	0.295	0.320	0.345	0.370	0.395
0.70	0.341	0.385	0.428	0.472	0.515	0.559	0.602	0.646	0.689
0.80	0.520	0.586	0.652	0.719	0.785	0.852	0.918	0.985	1.051
0.90	0.724	0.816	0.909	1.002	1.095	1.187	1.280	1.373	1.466
1.00	0.947	1.069	1.190	1.311	1.433	1.554	1.676	1.797	1.919
2.00	3.46	3.90	4.35	4.79	5.24	5.69	6.13	6.58	7.03
3.00	5.69	6.44	7.19	7.94	8.69	9.43	10.18	10.93	11.68
4.00	7.55	8.56	9.56	10.57	11.58	12.59	13.59	14.60	15.61
5.00	9.11	10.34	11.58	12.81	14.04	15.27	16.51	17.74	18.97
6.00	10.5	11.9	13.3	14.7	16.2	17.6	19.0	20.5	21.9
7.00	11.6	13.2	14.8	16.4	18.1	19.7	21.3	22.9	24.5
8.00	12.7	14.5	16.2	18.0	19.7	21.5	23.2	25.0	26.8
9.00	13.6	15.5	17.4	19.3	21.2	23.1	25.0	26.9	28.8
10.00	14.5	16.5	18.6	20.6	22.6	24.6	26.7	28.7	30.7
10.25	14.7	16.8	18.8	20.9	22.9	25.0	27.1	29.1	31.2
10.50	14.9	17.0	19.1	21.2	23.3	25.3	27.4	29.5	31.6
10.75	15.1	17.2	19.3	21.5	23.6	25.7	27.8	29.9	32.0
11.00	15.3	17.5	19.6	21.7	23.9	26.0	28.2	30.3	32.4
12.00	16.1	18.3	20.6	22.8	25.0	27.3	29.5	31.8	34.0
13.00	16.8	19.1	21.4	23.8	26.1	28.5	30.8	33.2	35.5
14.00	17.4	19.8	22.3	24.7	27.1	29.6	32.0	34.5	36.9
15.00	18.0	20.5	23.1	25.6	28.1	30.6	33.1	35.7	38.2
16.00	18.6	21.2	23.8	26.4	29.0	31.6	34.2	36.8	39.4
17.00	19.1	21.8	24.5	27.2	29.8	32.5	35.2	37.9	40.5
18.00	19.7	22.4	25.1	27.9	30.6	33.4	36.1	38.9	41.6
19.00	20.2	23.0	25.8	28.6	31.4	34.2	37.0	39.8	42.6
20.00	20.6	23.5	26.4	29.2	32.1	35.0	37.9	40.7	43.6
21.00	21.1	24.0	26.9	29.9	32.8	35.7	38.7	41.6	44.5
22.00	21.5	24.5	27.5	30.5	33.5	36.5	39.5	42.4	45.4
23.00	21.9	25.0	28.0	31.1	34.1	37.2	40.2	43.2	46.3
24.00	22.3	25.4	28.5	31.6	34.7	37.8	40.9	44.0	47.1
25.00	22.7	25.9	29.0	32.2	35.3	38.4	41.6	44.7	47.9
26.00	23.1	26.3	29.5	32.7	35.9	39.1	42.2	45.4	48.6
27.00	23.5	26.7	29.9	33.2	36.4	39.6	42.9	46.1	49.3
28.00	23.8	27.1	30.4	33.6	36.9	40.2	43.5	46.7	50.0
29.00	24.2	27.5	30.8	34.1	37.4	40.7	44.1	47.4	50.7
30.00	24.5	27.9	31.2	34.6	37.9	41.3	44.6	48.0	51.3
31.00	24.8	28.2	31.6	35.0	38.4	41.8	45.2	48.6	52.0
32.00	25.1	28.6	32.0	35.4	38.8	42.3	45.7	49.1	52.6
33.00	25.4	28.9	32.4	35.8	39.3	42.8	46.2	49.7	53.1

TABLE A.21

$f_h/U:g$ Relationships when $z = 44$

	Values of g when j of cooling curve is:								
f_h/U	0.40	0.60	0.80	1.00	1.20	1.40	1.60	1.80	2.00
0.20	9.35−05	1.06−04	1.18−04	1.30−04	1.43−04	1.55−04	1.67−04	1.80−04	1.92−04
0.30	4.33−03	4.90−03	5.48−03	6.05−03	6.62−03	7.19−03	7.76−03	8.33−03	8.91−03
0.40	2.96−02	3.35−02	3.74−02	4.13−02	4.52−02	4.91−02	5.30−02	5.69−02	6.08−02
0.50	9.40−02	1.06−01	1.19−01	1.31−01	1.43−01	1.56−01	1.68−01	1.81−01	1.93−01
0.60	0.204	0.231	0.257	0.284	0.311	0.338	0.365	0.391	0.418
0.70	0.355	0.402	0.449	0.496	0.543	0.590	0.637	0.683	0.730
0.80	0.542	0.613	0.685	0.756	0.828	0.899	0.971	1.042	1.114
0.90	0.755	0.855	0.955	1.054	1.154	1.254	1.354	1.453	1.553
1.00	0.988	1.119	1.250	1.380	1.511	1.641	1.772	1.903	2.033
2.00	3.60	4.08	4.56	5.04	5.52	6.00	6.48	6.96	7.45
3.00	5.92	6.72	7.53	8.34	9.14	9.95	10.76	11.56	12.37
4.00	7.84	8.93	10.02	11.10	12.19	13.27	14.36	15.44	16.53
5.00	9.47	10.79	12.12	13.45	14.77	16.10	17.43	18.75	20.08
6.00	10.9	12.4	13.9	15.5	17.0	18.6	20.1	21.6	23.2
7.00	12.1	13.8	15.5	17.3	19.0	20.7	22.4	24.2	25.9
8.00	13.2	15.1	17.0	18.9	20.7	22.6	24.5	26.4	28.3
9.00	14.2	16.2	18.3	20.3	22.3	24.4	26.4	28.4	30.5
10.00	15.1	17.3	19.4	21.6	23.8	25.9	28.1	30.3	32.5
10.25	15.3	17.5	19.7	21.9	24.1	26.3	28.5	30.7	32.9
10.50	15.5	17.7	20.0	22.2	24.4	26.7	28.9	31.1	33.4
10.75	15.7	18.0	20.2	22.5	24.8	27.0	29.3	31.6	33.8
11.00	15.9	18.2	20.5	22.8	25.1	27.4	29.7	32.0	34.3
12.00	16.7	19.1	21.5	23.9	26.3	28.7	31.1	33.5	35.9
13.00	17.4	19.9	22.4	24.9	27.5	30.0	32.5	35.0	37.5
14.00	18.1	20.7	23.3	25.9	28.5	31.1	33.7	36.3	38.9
15.00	18.7	21.4	24.1	26.8	29.5	32.2	34.9	37.6	40.3
16.00	19.3	22.1	24.9	27.7	30.4	33.2	36.0	38.8	41.6
17.00	19.9	22.7	25.6	28.5	31.3	34.2	37.0	39.9	42.8
18.00	20.4	23.4	26.3	29.2	32.2	35.1	38.0	40.9	43.9
19.00	20.9	23.9	26.9	29.9	32.9	35.9	38.9	41.9	44.9
20.00	21.4	24.5	27.6	30.6	33.7	36.8	39.8	42.9	46.0
21.00	21.9	25.0	28.2	31.3	34.4	37.5	40.7	43.8	46.9
22.00	22.4	25.5	28.7	31.9	35.1	38.3	41.5	44.7	47.8
23.00	22.8	26.0	29.3	32.5	35.8	39.0	42.2	45.5	48.7
24.00	23.2	26.5	29.8	33.1	36.4	39.7	43.0	46.3	49.6
25.00	23.6	27.0	30.3	33.7	37.0	40.3	43.7	47.0	50.4
26.00	24.0	27.4	30.8	34.2	37.6	41.0	44.4	47.8	51.2
27.00	24.4	27.8	31.3	34.7	38.2	41.6	45.0	48.5	51.9
28.00	24.8	28.2	31.7	35.2	38.7	42.2	45.7	49.1	52.6
29.00	25.1	28.6	32.2	35.7	39.2	42.7	46.3	49.8	53.3

TABLE A.22

$f_h/U : g$ *Relationships when* $z = 46$

Values of g when j of cooling curve is:

f_h/U	0.40	0.60	0.80	1.00	1.20	1.40	1.60	1.80	2.00
0.20	9.72-05	1.10-04	1.24-04	1.37-04	1.50-04	1.63-04	1.76-04	1.90-04	2.03-04
0.30	4.50-03	5.11-03	5.73-03	6.34-03	6.96-03	7.58-03	8.19-03	8.81-03	9.42-03
0.40	3.08-02	3.49-02	3.91-02	4.33-02	4.75-02	5.17-02	5.59-02	6.01-02	6.43-02
0.50	9.77-02	1.11-01	1.24-01	1.38-01	1.51-01	1.64-01	1.77-01	1.91-01	2.04-01
0.60	0.212	0.241	0.269	0.298	0.327	0.356	0.385	0.413	0.442
0.70	0.370	0.420	0.470	0.521	0.571	0.621	0.671	0.722	0.772
0.80	0.564	0.641	0.718	0.794	0.871	0.948	1.024	1.101	1.178
0.90	0.787	0.894	1.001	1.107	1.214	1.321	1.428	1.535	1.642
1.00	1.03	1.17	1.31	1.45	1.59	1.73	1.87	2.01	2.15
2.00	3.73	4.25	4.77	5.29	5.80	6.32	6.84	7.35	7.87
3.00	6.13	7.00	7.87	8.74	9.60	10.47	11.34	12.21	13.07
4.00	8.13	9.30	10.46	11.63	12.80	13.96	15.13	16.30	17.46
5.00	9.82	11.24	12.67	14.09	15.51	16.94	18.36	19.78	21.21
6.00	11.3	12.9	14.6	16.2	17.9	19.5	21.2	22.8	24.4
7.00	12.6	14.4	16.2	18.1	19.9	21.8	23.6	25.5	27.3
8.00	13.7	15.7	17.7	19.7	21.8	23.8	25.8	27.8	29.8
9.00	14.7	16.9	19.1	21.2	23.4	25.6	27.8	30.0	32.1
10.00	15.7	18.0	20.3	22.6	24.9	27.2	29.6	31.9	34.2
10.25	15.9	18.2	20.6	22.9	25.3	27.6	30.0	32.3	34.7
10.50	16.1	18.5	20.9	23.3	25.6	28.0	30.4	32.8	35.2
10.75	16.3	18.7	21.1	23.6	26.0	28.4	30.8	33.2	35.6
11.00	16.5	19.0	21.4	23.9	26.3	28.8	31.2	33.7	36.1
12.00	17.3	19.9	22.5	25.0	27.6	30.2	32.7	35.3	37.8
13.00	18.1	20.7	23.4	26.1	28.8	31.4	34.1	36.8	39.5
14.00	18.8	21.5	24.3	27.1	29.9	32.7	35.4	38.2	41.0
15.00	19.4	22.3	25.2	28.0	30.9	33.8	36.7	39.5	42.4
16.00	20.0	23.0	26.0	28.9	31.9	34.8	37.8	40.8	43.7
17.00	20.6	23.7	26.7	29.8	32.8	35.8	38.9	41.9	45.0
18.00	21.2	24.3	27.4	30.5	33.7	36.8	39.9	43.0	46.1
19.00	21.7	24.9	28.1	31.3	34.5	37.7	40.9	44.1	47.2
20.00	22.2	25.5	28.8	32.0	35.3	38.5	41.8	45.0	48.3
21.00	22.7	26.0	29.4	32.7	36.0	39.3	42.7	46.0	49.3
22.00	23.2	26.6	30.0	33.3	36.7	40.1	43.5	46.9	50.3
23.00	23.7	27.1	30.5	34.0	37.4	40.8	44.3	47.7	51.2
24.00	24.1	27.6	31.1	34.6	38.1	41.6	45.1	48.5	52.0
25.00	24.5	28.1	31.6	35.1	38.7	42.2	45.8	49.3	52.9

TABLE A.23

$f_h/U : g$ *Relationships when* $z = 48$

Values of g when j of cooling curve is:

f_h/U	0.40	0.60	0.80	1.00	1.20	1.40	1.60	1.80	2.00
0.20	1.01-04	1.15-04	1.29-04	1.43-04	1.57-04	1.71-04	1.85-04	2.00-04	2.14-04
0.30	4.67-03	5.33-03	6.00-03	6.66-03	7.32-03	7.98-03	8.64-03	9.30-03	9.96-03
0.40	3.16-02	3.61-02	4.07-02	4.52-02	4.97-02	5.42-02	5.88-02	6.33-02	6.78-02
0.50	0.100	0.115	0.129	0.143	0.158	0.172	0.186	0.201	0.215
0.60	0.218	0.249	0.280	0.311	0.342	0.373	0.404	0.435	0.466
0.70	0.383	0.437	0.491	0.545	0.599	0.653	0.707	0.761	0.815
0.80	0.586	0.668	0.750	0.832	0.915	0.997	1.079	1.161	1.243
0.90	0.820	0.934	1.048	1.162	1.276	1.391	1.505	1.619	1.733
1.00	1.07	1.22	1.37	1.52	1.67	1.82	1.97	2.12	2.27
2.00	3.88	4.43	4.98	5.54	6.09	6.64	7.19	7.75	8.30
3.00	6.34	7.27	8.21	9.14	10.07	11.00	11.93	12.86	13.79
4.00	8.39	9.65	10.90	12.15	13.41	14.66	15.92	17.17	18.42
5.00	10.1	11.7	13.2	14.7	16.2	17.8	19.3	20.8	22.4
6.00	11.6	13.4	15.2	16.9	18.7	20.5	22.2	24.0	25.8
7.00	13.0	14.9	16.9	18.9	20.8	22.8	24.8	26.8	28.7
8.00	14.1	16.3	18.5	20.6	22.8	24.9	27.1	29.2	31.4
9.00	15.2	17.6	19.9	22.2	24.5	26.8	29.2	31.5	33.8
10.00	16.2	18.7	21.2	23.6	26.1	28.6	31.0	33.5	36.0
10.25	16.5	19.0	21.5	24.0	26.5	29.0	31.5	34.0	36.5
10.50	16.7	19.2	21.8	24.3	26.8	29.4	31.9	34.4	37.0
10.75	16.9	19.5	22.1	24.6	27.2	29.8	32.3	34.9	37.5
11.00	17.1	19.7	22.3	24.9	27.5	30.1	32.7	35.3	37.9
12.00	18.0	20.7	23.4	26.2	28.9	31.6	34.3	37.1	39.8
13.00	18.8	21.6	24.5	27.3	30.1	33.0	35.8	38.6	41.5
14.00	19.5	22.5	25.4	28.3	31.3	34.2	37.2	40.1	43.0
15.00	20.2	23.2	26.3	29.3	32.4	35.4	38.4	41.5	44.5
16.00	20.9	24.0	27.1	30.2	33.4	36.5	39.6	42.7	45.9
17.00	21.5	24.7	27.9	31.1	34.3	37.5	40.7	44.0	47.2
18.00	22.0	25.3	28.6	31.9	35.2	38.5	41.8	45.1	48.4
19.00	22.6	26.0	29.3	32.7	36.1	39.4	42.8	46.2	49.5
20.00	23.1	26.6	30.0	33.4	36.9	40.3	43.7	47.2	50.6
21.00	23.6	27.1	30.6	34.1	37.6	41.1	44.6	48.2	51.7
22.00	24.1	27.7	31.2	34.8	38.4	41.9	45.5	49.1	52.6

TABLE A.24

$f_h/U : g$ *Relationships when* $z = 50$

	Values of g when j of cooling curve is:								
f_h/U	0.40	0.60	0.80	1.00	1.20	1.40	1.60	1.80	2.00
0.20	1.04−04	1.19−04	1.34−04	1.50−04	1.65−04	1.80−04	1.95−04	2.10−04	2.25−04
0.30	4.81−03	5.52−03	6.23−03	6.94−03	7.65−03	8.36−03	9.07−03	9.78−03	1.05−02
0.40	3.31−02	3.79−02	4.27−02	4.75−02	5.23−02	5.71−02	6.18−02	6.66−02	7.14−02
0.50	0.106	0.121	0.136	0.151	0.166	0.181	0.196	0.211	0.226
0.60	0.230	0.262	0.295	0.328	0.360	0.393	0.426	0.458	0.491
0.70	0.401	0.458	0.515	0.572	0.629	0.686	0.743	0.800	0.857
0.80	0.611	0.698	0.785	0.872	0.960	1.047	1.134	1.221	1.308
0.90	0.850	0.972	1.094	1.216	1.337	1.459	1.581	1.703	1.825
1.00	1.11	1.27	1.43	1.59	1.75	1.91	2.07	2.23	2.39
2.00	3.99	4.58	5.18	5.77	6.37	6.96	7.56	8.15	8.75
3.00	6.56	7.55	8.55	9.54	10.54	11.53	12.53	13.52	14.52
4.00	8.71	10.05	11.38	12.71	14.04	15.37	16.71	18.04	19.37
5.00	10.5	12.2	13.8	15.4	17.0	18.6	20.3	21.9	23.5
6.00	12.1	14.0	15.8	17.7	19.6	21.5	23.3	25.2	27.1
7.00	13.5	15.6	17.7	19.7	21.8	23.9	26.0	28.1	30.2
8.00	14.7	17.0	19.3	21.5	23.8	26.1	28.4	30.7	33.0
9.00	15.8	18.2	20.7	23.2	25.6	28.1	30.5	33.0	35.5
10.00	16.8	19.4	22.0	24.6	27.3	29.9	32.5	35.1	37.7
10.25	17.0	19.7	22.3	25.0	27.6	30.3	33.0	35.6	38.3
10.50	17.2	19.9	22.6	25.3	28.0	30.7	33.4	36.1	38.8
10.75	17.5	20.2	22.9	25.7	28.4	31.1	33.8	36.6	39.3
11.00	17.7	20.4	23.2	26.0	28.7	31.5	34.3	37.0	39.8
12.00	18.5	21.4	24.3	27.2	30.1	33.0	35.9	38.8	41.7
13.00	19.3	22.3	25.4	28.4	31.4	34.4	37.4	40.4	43.5
14.00	20.1	23.2	26.3	29.4	32.6	35.7	38.8	42.0	45.1
15.00	20.8	24.0	27.2	30.5	33.7	36.9	40.1	43.4	46.6
16.00	21.4	24.7	28.1	31.4	34.7	38.1	41.4	44.7	48.0
17.00	22.1	25.5	28.9	32.3	35.7	39.1	42.5	46.0	49.4
18.00	22.7	26.2	29.6	33.1	36.6	40.1	43.6	47.1	50.6
19.00	23.2	26.8	30.4	34.0	37.5	41.1	44.7	48.2	51.8
20.00	23.8	27.4	31.1	34.7	38.4	42.0	45.7	49.3	52.9

TABLE A.25

$f_h/U : g$ Relationships when $z = 52$

f_h/U	\multicolumn{9}{c}{Values of g when j of cooling curve is:}
	0.40	0.60	0.80	1.00	1.20	1.40	1.60	1.80	2.00
0.20	1.08-04	1.24-04	1.40-04	1.56-04	1.73-04	1.89-04	2.05-04	2.21-04	2.37-04
0.30	5.00-03	5.75-03	6.51-03	7.26-03	8.01-03	8.76-03	9.51-03	1.03-02	1.10-02
0.40	3.42-02	3.93-02	4.44-02	4.95-02	5.46-02	5.98-02	6.49-02	7.00-02	7.51-02
0.50	0.108	0.125	0.141	0.157	0.174	0.190	0.206	0.222	0.239
0.60	0.235	0.270	0.306	0.341	0.376	0.412	0.447	0.482	0.518
0.70	0.410	0.472	0.534	0.595	0.657	0.719	0.780	0.842	0.904
0.80	0.626	0.720	0.814	0.908	1.002	1.096	1.190	1.284	1.378
0.90	0.873	1.004	1.135	1.266	1.397	1.527	1.658	1.789	1.920
1.00	1.14	1.31	1.48	1.66	1.83	2.00	2.17	2.34	2.51
2.00	4.14	4.77	5.40	6.03	6.67	7.30	7.93	8.56	9.19
3.00	6.80	7.85	8.91	9.96	11.02	12.08	13.13	14.19	15.25
4.00	8.99	10.41	11.83	13.25	14.67	16.09	17.51	18.93	20.34
5.00	10.8	12.6	14.3	16.0	17.8	19.5	21.2	22.9	24.7
6.00	12.4	14.4	16.4	18.4	20.4	22.4	24.4	26.4	28.4
7.00	13.9	16.1	18.3	20.5	22.8	25.0	27.2	29.4	31.7
8.00	15.1	17.5	20.0	22.4	24.8	27.3	29.7	32.1	34.6
9.00	16.2	18.9	21.5	24.1	26.7	29.3	31.9	34.6	37.2
10.00	17.3	20.1	22.8	25.6	28.4	31.2	34.0	36.7	39.5
10.25	17.5	20.3	23.2	26.0	28.8	31.6	34.4	37.3	40.1
10.50	17.8	20.6	23.5	26.3	29.2	32.1	34.9	37.8	40.6
10.75	18.0	20.9	23.8	26.7	29.6	32.5	35.4	38.3	41.1
11.00	18.2	21.2	24.1	27.0	30.0	32.9	35.8	38.7	41.7
12.00	19.1	22.2	25.3	28.3	31.4	34.4	37.5	40.6	43.6
13.00	20.0	23.1	26.3	29.5	32.7	35.9	39.1	42.3	45.4
14.00	20.7	24.0	27.3	30.6	33.9	37.2	40.5	43.8	47.1
15.00	21.5	24.9	28.3	31.7	35.1	38.5	41.9	45.3	48.7
16.00	22.2	25.7	29.2	32.7	36.2	39.7	43.2	46.7	50.2
17.00	22.8	26.4	30.0	33.6	37.2	40.8	44.4	48.0	51.5
18.00	23.5	27.1	30.8	34.5	38.1	41.8	45.5	49.2	52.8

TABLE A.26

$f_h/U : g$ *Relationships when* $z = 54$

Values of g when j of cooling curve is:

f_h/U	0.40	0.60	0.80	1.00	1.20	1.40	1.60	1.80	2.00
0.20	1.12-04	1.29-04	1.46-04	1.63-04	1.80-04	1.98-04	2.15-04	2.32-04	2.49-04
0.30	5.19-03	5.99-03	6.78-03	7.57-03	8.37-03	9.16-03	9.96-03	1.08-02	1.15-02
0.40	3.52-02	4.06-02	4.61-02	5.16-02	5.70-02	6.25-02	6.80-02	7.35-02	7.89-02
0.50	0.111	0.129	0.146	0.164	0.181	0.199	0.216	0.234	0.251
0.60	0.240	0.278	0.316	0.354	0.392	0.430	0.468	0.506	0.544
0.70	0.420	0.486	0.552	0.619	0.685	0.751	0.817	0.884	0.950
0.80	0.641	0.742	0.843	0.944	1.044	1.145	1.246	1.347	1.448
0.90	0.895	1.035	1.176	1.316	1.456	1.596	1.737	1.877	2.017
1.00	1.17	1.36	1.54	1.72	1.91	2.09	2.27	2.46	2.64
2.00	4.28	4.95	5.62	6.29	6.96	7.63	8.31	8.98	9.65
3.00	7.02	8.14	9.26	10.39	11.51	12.63	13.75	14.87	15.99
4.00	9.28	10.79	12.29	13.80	15.30	16.81	18.31	19.82	21.33
5.00	11.2	13.0	14.8	16.7	18.5	20.3	22.2	24.0	25.8
6.00	12.8	14.9	17.0	19.2	21.3	23.4	25.5	27.6	29.7
7.00	14.2	16.6	19.0	21.3	23.7	26.1	28.4	30.8	33.1
8.00	15.5	18.1	20.7	23.3	25.8	28.4	31.0	33.6	36.2
9.00	16.7	19.5	22.2	25.0	27.8	30.6	33.3	36.1	38.9
10.00	17.8	20.7	23.7	26.6	29.5	32.5	35.4	38.4	41.3
10.25	18.0	21.0	24.0	27.0	30.0	32.9	35.9	38.9	41.9
10.50	18.3	21.3	24.3	27.3	30.4	33.4	36.4	39.4	42.4
10.75	18.5	21.6	24.6	27.7	30.8	33.8	36.9	39.9	43.0
11.00	18.8	21.9	25.0	28.1	31.1	34.2	37.3	40.4	43.5
12.00	19.7	22.9	26.2	29.4	32.6	35.9	39.1	42.3	45.6
13.00	20.6	23.9	27.3	30.6	34.0	37.4	40.7	44.1	47.4
14.00	21.4	24.8	28.3	31.8	35.3	38.7	42.2	45.7	49.2
15.00	22.1	25.7	29.3	32.9	36.5	40.0	43.6	47.2	50.8
16.00	22.9	26.6	30.2	33.9	37.6	41.3	44.9	48.6	52.3

TABLE A.27

$f_h/U:g$ Relationships when $z = 56$

	Values of g when j of cooling curve is:								
f_h/U	0.40	0.60	0.80	1.00	1.20	1.40	1.60	1.80	2.00
0.20	1.14−04	1.32−04	1.51−04	1.69−04	1.88−04	2.06−04	2.24−04	2.43−04	2.61−04
0.30	5.35−03	6.19−03	7.04−03	7.89−03	8.73−03	9.58−03	1.04−02	1.13−02	1.21−02
0.40	3.66−02	4.24−02	4.81−02	5.39−02	5.96−02	6.54−02	7.11−02	7.69−02	8.27−02
0.50	0.116	0.134	0.152	0.171	0.189	0.207	0.226	0.244	0.262
0.60	0.250	0.290	0.330	0.370	0.410	0.450	0.490	0.530	0.570
0.70	0.435	0.505	0.575	0.645	0.715	0.785	0.855	0.925	0.995
0.80	0.662	0.769	0.876	0.983	1.090	1.197	1.304	1.411	1.518
0.90	0.921	1.070	1.220	1.369	1.519	1.668	1.817	1.967	2.116
1.00	1.20	1.40	1.60	1.79	1.99	2.18	2.38	2.57	2.77
2.00	4.40	5.11	5.83	6.54	7.25	7.97	8.68	9.40	10.11
3.00	7.24	8.43	9.62	10.81	11.99	13.18	14.37	15.56	16.74
4.00	9.58	11.17	12.76	14.36	15.95	17.54	19.13	20.72	22.31
5.00	11.5	13.5	15.4	17.3	19.3	21.2	23.2	25.1	27.0
6.00	13.2	15.4	17.7	19.9	22.1	24.4	26.6	28.9	31.1
7.00	14.7	17.2	19.7	22.2	24.7	27.2	29.6	32.1	34.6
8.00	16.0	18.7	21.4	24.2	26.9	29.6	32.3	35.0	37.8
9.00	17.2	20.1	23.0	26.0	28.9	31.8	34.7	37.6	40.6
10.00	18.3	21.4	24.5	27.6	30.7	33.8	36.9	40.0	43.1
10.25	18.5	21.7	24.8	28.0	31.1	34.3	37.4	40.5	43.7
10.50	18.8	22.0	25.2	28.3	31.5	34.7	37.9	41.1	44.3
10.75	.0	22.3	25.5	28.7	31.9	35.2	38.4	41.6	44.8
11.00	19.3	22.6	25.8	29.1	32.3	35.6	38.9	42.1	45.4
12.00	20.2	23.6	27.1	30.5	33.9	37.3	40.7	44.1	47.5
13.00	21.1	24.7	28.2	31.7	35.3	38.8	42.3	45.9	49.4
14.00	22.0	25.6	29.3	32.9	36.6	40.2	43.9	47.5	51.2
15.00	22.8	26.5	30.3	34.1	37.8	41.6	45.3	49.1	52.9

TABLE A.28

$f_h/U : g$ *Relationships when* $z = 58$

	Values of g when j of cooling curve is:								
f_h/U	0.40	0.60	0.80	1.00	1.20	1.40	1.60	1.80	2.00
0.20	1.18-04	1.38-04	1.57-04	1.76-04	1.96-04	2.15-04	2.35-04	2.54-04	2.73-04
0.30	5.48-03	6.38-03	7.29-03	8.19-03	9.09-03	9.99-03	1.09-02	1.18-02	1.27-02
0.40	3.74-02	4.36-02	4.97-02	5.59-02	6.20-02	6.82-02	7.43-02	8.05-02	8.66-02
0.50	0.119	0.138	0.158	0.177	0.197	0.216	0.236	0.255	0.275
0.60	0.257	0.300	0.342	0.384	0.427	0.469	0.512	0.554	0.596
0.70	0.449	0.523	0.597	0.671	0.745	0.819	0.893	0.967	1.041
0.80	0.685	0.798	0.911	1.024	1.137	1.249	1.362	1.475	1.588
0.90	0.955	1.113	1.270	1.427	1.584	1.742	1.899	2.056	2.214
1.00	1.25	1.46	1.66	1.87	2.07	2.28	2.49	2.69	2.90
2.00	4.52	5.28	6.04	6.80	7.55	8.31	9.07	9.82	10.58
3.00	7.41	8.68	9.94	11.20	12.47	13.73	14.99	16.26	17.52
4.00	9.81	11.50	13.19	14.88	16.57	18.26	19.95	21.64	23.33
5.00	11.8	13.9	15.9	18.0	20.0	22.1	24.1	26.2	28.2
6.00	13.6	15.9	18.3	20.7	23.0	25.4	27.7	30.1	32.5
7.00	15.1	17.7	20.4	23.0	25.6	28.2	30.9	33.5	36.1
8.00	16.5	19.3	22.2	25.1	27.9	30.8	33.6	36.5	39.4
9.00	17.7	20.8	23.9	26.9	30.0	33.1	36.1	39.2	42.3
10.00	18.9	22.1	25.4	28.6	31.9	35.1	38.4	41.6	44.9
10.25	19.1	22.4	25.7	29.0	32.3	35.6	38.9	42.2	45.5
10.50	19.4	22.7	26.1	29.4	32.7	36.1	39.4	42.7	46.1
10.75	19.7	23.0	26.4	29.8	33.2	36.5	39.9	43.3	46.7
11.00	19.9	23.3	26.7	30.2	33.6	37.0	40.4	43.8	47.2
12.00	20.9	24.5	28.0	31.6	35.1	38.7	42.3	45.8	49.4
13.00	21.8	25.5	29.2	32.9	36.6	40.3	44.0	47.7	51.4
14.00	22.7	26.5	30.3	34.1	37.9	41.8	45.6	49.4	53.2

TABLE A.29

$f_h/U : g$ *Relationships when* $z = 60$

f_h/U	Values of g when j of cooling curve is:								
	0.40	0.60	0.80	1.00	1.20	1.40	1.60	1.80	2.00
0.20	1.20-04	1.41-04	1.61-04	1.82-04	2.03-04	2.24-04	2.45-04	2.66-04	2.86-04
0.30	5.64-03	6.59-03	7.55-03	8.50-03	9.46-03	1.04-02	1.14-02	1.23-02	1.33-02
0.40	3.90-02	4.54-02	5.18-02	5.83-02	6.47-02	7.11-02	7.76-02	8.40-02	9.04-02
0.50	0.124	0.144	0.165	0.185	0.205	0.226	0.246	0.267	0.287
0.60	0.268	0.312	0.357	0.401	0.445	0.490	0.534	0.578	0.623
0.70	0.466	0.543	0.621	0.699	0.777	0.854	0.932	1.010	1.088
0.80	0.707	0.826	0.945	1.064	1.183	1.302	1.422	1.541	1.660
0.90	0.982	1.149	1.315	1.482	1.648	1.815	1.981	2.148	2.314
1.00	1.28	1.50	1.72	1.94	2.16	2.37	2.59	2.81	3.03
2.00	4.63	5.43	6.24	7.04	7.85	8.65	9.45	10.26	11.06
3.00	7.62	8.96	10.29	11.63	12.96	14.29	15.63	16.96	18.29
4.00	10.1	11.9	13.7	15.4	17.2	19.0	20.8	22.6	24.3
5.00	12.2	14.3	16.5	18.7	20.8	23.0	25.1	27.3	29.4
6.00	14.0	16.5	18.9	21.4	23.9	26.4	28.9	31.3	33.8
7.00	15.5	18.3	21.1	23.8	26.6	29.3	32.1	34.9	37.6
8.00	16.9	19.9	22.9	26.0	29.0	32.0	35.0	38.0	41.0
9.00	18.2	21.4	24.6	27.9	31.1	34.3	37.5	40.7	44.0
10.00	19.4	22.8	26.2	29.6	33.0	36.4	39.8	43.2	46.6
10.25	19.6	23.1	26.6	30.0	33.5	36.9	40.4	43.8	47.3
10.50	19.9	23.4	26.9	30.4	33.9	37.4	40.9	44.4	47.9
10.75	20.2	23.7	27.3	30.8	34.3	37.9	41.4	44.9	48.5
11.00	20.4	24.0	27.6	31.2	34.8	38.3	41.9	45.5	49.1
12.00	21.5	25.2	28.9	32.6	36.4	40.1	43.8	47.6	51.3
13.00	22.4	26.3	30.1	34.0	37.9	41.7	45.6	49.5	53.3

TABLE A.30

$f_h/U : g$ *Relationships when* $z = 62$

f_h/U	Values of g when j of cooling curve is:								
	0.40	0.60	0.80	1.00	1.20	1.40	1.60	1.80	2.00
0.20	1.24-04	1.46-04	1.68-04	1.90-04	2.12-04	2.33-04	2.55-04	2.77-04	2.99-04
0.30	5.80-03	6.80-03	7.81-03	8.82-03	9.83-03	1.08-02	1.18-02	1.29-02	1.39-02
0.40	3.95-02	4.64-02	5.33-02	6.02-02	6.71-02	7.39-02	8.08-02	8.77-02	9.46-02
0.50	0.125	0.147	0.169	0.191	0.213	0.235	0.257	0.279	0.300
0.60	0.271	0.319	0.366	0.414	0.461	0.509	0.556	0.604	0.651
0.70	0.474	0.557	0.640	0.723	0.806	0.888	0.971	1.054	1.137
0.80	0.723	0.850	0.976	1.102	1.229	1.355	1.481	1.607	1.734
0.90	1.01	1.18	1.36	1.54	1.71	1.89	2.06	2.24	2.42
1.00	1.32	1.55	1.78	2.01	2.24	2.47	2.70	2.93	3.16
2.00	4.77	5.62	6.46	7.31	8.16	9.00	9.85	10.69	11.54
2.50	6.36	7.50	8.64	9.78	10.92	12.06	13.20	14.34	15.48
3.00	7.81	9.22	10.63	12.04	13.45	14.86	16.27	17.67	19.08
3.50	9.13	10.78	12.44	14.09	15.75	17.40	19.06	20.72	22.37
4.00	10.3	12.2	14.1	16.0	17.9	19.7	21.6	23.5	25.4
4.50	11.4	13.5	15.6	17.7	19.8	21.9	24.0	26.0	28.1
5.00	12.5	14.7	17.0	19.3	21.6	23.8	26.1	28.4	30.7
5.20	12.8	15.2	17.5	19.9	22.2	24.6	26.9	29.3	31.6
5.40	13.2	15.6	18.1	20.5	22.9	25.3	27.7	30.1	32.6
5.60	13.6	16.1	18.6	21.0	23.5	26.0	28.5	31.0	33.5
5.80	13.9	16.5	19.0	21.6	24.1	26.7	29.2	31.8	34.3
6.00	14.3	16.9	19.5	22.1	24.7	27.4	30.0	32.6	35.2
6.20	14.6	17.3	20.0	22.7	25.3	28.0	30.7	33.4	36.0
6.40	15.0	17.7	20.4	23.2	25.9	28.6	31.4	34.1	36.8
6.60	15.3	18.1	20.9	23.7	26.5	29.2	32.0	34.8	37.6
6.80	15.6	18.5	21.3	24.1	27.0	29.8	32.7	35.5	38.4
7.00	15.9	18.8	21.7	24.6	27.5	30.4	33.3	36.2	39.1
7.20	16.2	19.2	22.1	25.1	28.0	31.0	33.9	36.9	39.9
7.40	16.5	19.5	22.5	25.5	28.5	31.5	34.5	37.6	40.6
7.60	16.8	19.9	22.9	26.0	29.0	32.1	35.1	38.2	41.2
7.80	17.1	20.2	23.3	26.4	29.5	32.6	35.7	38.8	41.9
8.00	17.4	20.5	23.7	26.8	30.0	33.1	36.3	39.4	42.6
8.20	17.6	20.8	24.0	27.2	30.4	33.6	36.8	40.0	43.2
8.40	17.9	21.2	24.4	27.6	30.9	34.1	37.4	40.6	43.9
8.60	18.2	21.5	24.7	28.0	31.3	34.6	37.9	41.2	44.5
8.80	18.4	21.8	25.1	28.4	31.7	35.1	38.4	41.7	45.1
9.00	18.7	22.1	25.4	28.8	32.2	35.5	38.9	42.3	45.7
9.20	18.9	22.3	25.8	29.2	32.6	36.0	39.4	42.8	46.2
9.40	19.2	22.6	26.1	29.5	33.0	36.4	39.9	43.3	46.8
9.60	19.4	22.9	26.4	29.9	33.4	36.9	40.4	43.8	47.3
9.80	19.7	23.2	26.7	30.2	33.8	37.3	40.8	44.4	47.9
10.00	19.9	23.5	27.0	30.6	34.2	37.7	41.3	44.8	48.4
11.00	21.0	24.8	28.5	32.2	36.0	39.7	43.4	47.2	50.9
12.00	22.1	26.0	29.8	33.7	37.6	41.5	45.4	49.3	53.2

TABLE A.31

$f_h/U:g$ *Relationships when* $z = 64$

Values of g when j of cooling curve is:

f_h/U	0.40	0.60	0.80	1.00	1.20	1.40	1.60	1.80	2.00
0.20	1.28-04	1.51-04	1.74-04	1.97-04	2.20-04	2.43-04	2.66-04	2.88-04	3.11-04
0.30	5.94-03	7.00-03	8.07-03	9.13-03	1.02-02	1.13-02	1.23-02	1.34-02	1.45-02
0.40	4.05-02	4.78-02	5.51-02	6.23-02	6.96-02	7.69-02	8.42-02	9.14-02	9.87-02
0.50	0.129	0.152	0.175	0.198	0.221	0.244	0.267	0.290	0.313
0.60	0.279	0.329	0.379	0.429	0.479	0.529	0.579	0.630	0.680
0.70	0.487	0.574	0.661	0.749	0.836	0.924	1.011	1.099	1.186
0.80	0.742	0.876	1.009	1.142	1.275	1.409	1.542	1.675	1.809
0.90	1.03	1.22	1.41	1.59	1.78	1.96	2.15	2.33	2.52
1.00	1.35	1.60	1.84	2.08	2.33	2.57	2.81	3.06	3.30
2.00	4.89	5.78	6.68	7.57	8.46	9.35	10.24	11.14	12.03
2.50	6.52	7.73	8.93	10.13	11.33	12.53	13.73	14.93	16.13
3.00	8.01	9.49	10.97	12.46	13.94	15.43	16.91	18.39	19.88
3.50	9.35	11.10	12.84	14.58	16.32	18.07	19.81	21.55	23.29
4.00	10.6	12.6	14.5	16.5	18.5	20.5	22.5	24.4	26.4
4.50	11.7	13.9	16.1	18.3	20.5	22.7	24.9	27.1	29.3
5.00	12.0	15.2	17.5	19.9	22.3	24.7	27.1	29.5	31.9
5.20	13.2	15.6	18.1	20.6	23.0	25.5	28.0	30.4	32.9
5.40	13.5	16.1	18.7	21.2	23.7	26.2	28.8	31.3	33.9
5.60	13.9	16.5	19.1	21.7	24.4	27.0	29.6	32.2	34.8
5.80	14.3	17.0	19.6	22.3	25.0	27.7	30.3	33.0	35.7
6.00	14.6	17.4	20.1	22.9	25.6	28.4	31.1	33.8	36.6
6.20	15.0	17.8	20.6	23.4	26.2	29.0	31.8	34.6	37.4
6.40	15.3	18.2	21.1	23.9	26.8	29.7	32.5	35.4	38.3
6.60	15.7	18.6	21.5	24.4	27.4	30.3	33.2	36.1	39.1
6.80	16.0	19.0	22.0	24.9	27.9	30.9	33.9	36.9	39.9
7.00	16.3	19.4	22.4	25.4	28.5	31.5	34.5	37.6	40.6
7.20	16.6	19.7	22.8	25.9	29.0	32.1	35.2	38.3	41.4
7.40	16.9	20.1	23.2	26.4	29.5	32.7	35.8	39.0	42.1
7.60	17.2	20.4	23.6	26.8	30.0	33.2	36.4	39.6	42.8
7.80	17.5	20.8	24.0	27.3	30.5	33.8	37.0	40.3	43.5
8.00	17.8	21.1	24.4	27.7	31.0	34.3	37.6	40.9	44.2
8.20	18.1	21.4	24.8	28.1	31.5	34.8	38.2	41.5	44.8
8.40	18.4	21.8	25.1	28.5	31.9	35.3	38.7	42.1	45.5
8.60	18.6	22.1	25.5	28.9	32.4	35.8	39.2	42.7	46.1
8.80	18.9	22.4	25.9	29.3	32.8	36.3	39.8	43.3	46.7
9.00	19.2	22.7	26.2	29.7	33.3	36.8	40.3	43.8	47.3
9.20	19.4	23.0	26.5	30.1	33.7	37.2	40.8	44.4	47.9
9.40	19.7	23.3	26.9	30.5	34.1	37.7	41.3	44.9	48.5
9.60	19.9	23.6	27.2	30.9	34.6	38.1	41.8	45.4	49.1
9.80	20.2	23.9	27.5	31.2	34.9	38.6	42.3	45.9	49.6
10.00	20.4	24.1	27.9	31.6	35.3	39.0	42.7	46.4	50.2
11.00	21.6	25.5	29.4	33.2	37.1	41.0	44.9	48.8	52.7

TABLE A.32

$f_h/U : g$ *Relationships when* $z = 66$

	Values of g when j of cooling curve is:								
f_h/U	0.40	0.60	0.80	1.00	1.20	1.40	1.60	1.80	2.00
0.20	1.31-04	1.55-04	1.79-04	2.04-04	2.28-04	2.52-04	2.76-04	3.00-04	3.25-04
0.30	6.08-03	7.21-03	8.33-03	9.45-03	1.06-02	1.17-02	1.28-02	1.40-02	1.51-02
0.40	4.16-02	4.92-02	5.69-02	6.45-02	7.22-02	7.99-02	8.75-02	9.52-02	1.03-01
0.50	0.132	0.156	0.181	0.205	0.229	0.254	0.278	0.302	0.326
0.60	0.286	0.339	0.391	0.444	0.497	0.550	0.602	0.655	0.708
0.70	0.499	0.591	0.683	0.775	0.867	0.959	1.052	1.144	1.236
0.80	0.760	0.901	1.041	1.182	1.323	1.463	1.604	1.744	1.885
0.90	1.06	1.26	1.45	1.65	1.84	2.04	2.24	2.43	2.63
1.00	1.39	1.64	1.90	2.16	2.41	2.67	2.93	3.18	3.44
2.00	5.01	5.95	6.89	7.83	8.77	9.71	10.65	11.59	12.53
2.50	6.68	7.95	9.21	10.47	11.73	13.00	14.26	15.52	16.79
3.00	8.20	9.76	11.32	12.88	14.44	16.00	17.56	19.12	20.68
3.50	9.58	11.41	13.24	15.07	16.90	18.73	20.56	22.39	24.22
4.00	10.8	12.9	15.0	17.1	19.1	21.2	23.3	25.4	27.5
4.50	12.0	14.3	16.6	18.9	21.2	23.5	25.8	28.1	30.4
5.00	13.1	15.6	18.1	20.6	23.1	25.6	28.1	30.6	33.1
5.20	13.5	16.1	18.6	21.2	23.8	26.4	29.0	31.6	34.2
5.40	13.9	16.5	19.2	21.9	24.5	27.2	29.8	32.5	35.2
5.60	14.3	17.0	19.7	22.5	25.2	27.9	30.6	33.4	36.1
5.80	14.6	17.4	20.2	23.0	25.8	28.6	31.4	34.2	37.0
6.00	15.0	17.9	20.7	23.6	26.5	29.3	32.2	35.1	38.0
6.20	15.4	18.3	21.2	24.2	27.1	30.0	33.0	35.9	38.8
6.40	15.7	18.7	21.7	24.7	27.7	30.7	33.7	36.7	39.7
6.60	16.1	19.1	22.2	25.2	28.3	31.3	34.4	37.5	40.5
6.80	16.4	19.5	22.6	25.7	28.9	32.0	35.1	38.2	41.3
7.00	16.7	19.9	23.1	26.2	29.4	32.6	35.8	38.9	42.1
7.20	17.0	20.3	23.5	26.7	30.0	33.2	36.4	39.7	42.9
7.40	17.3	20.6	23.9	27.2	30.5	33.8	37.1	40.4	43.6
7.60	17.6	21.0	24.3	27.7	31.0	34.4	37.7	41.0	44.4
7.80	17.9	21.3	24.7	28.1	31.5	34.9	38.3	41.7	45.1
8.00	18.2	21.7	25.1	28.6	32.0	35.5	38.9	42.3	45.8
8.20	18.5	22.0	25.5	29.0	32.5	36.0	39.5	43.0	46.5
8.40	18.8	22.4	25.9	29.4	33.0	36.5	40.0	43.6	47.1
8.60	19.1	22.7	26.3	29.8	33.4	37.0	40.6	44.2	47.8
8.80	19.4	23.0	26.6	30.3	33.9	37.5	41.1	44.8	48.4
9.00	19.6	23.3	27.0	30.7	34.3	38.0	41.7	45.3	49.0
9.20	19.9	23.6	27.3	31.0	34.8	38.5	42.2	45.9	49.6
9.40	20.2	23.9	27.7	31.4	35.2	38.9	42.7	46.4	50.2
9.60	20.4	24.2	28.0	31.8	35.6	39.4	43.2	47.0	50.8
9.80	20.7	24.5	28.3	32.2	36.0	39.8	43.7	47.5	51.3
10.00	20.9	24.8	28.7	32.5	36.4	40.3	44.1	48.0	51.9

TABLE A.33

$f_h/U : g$ *Relationships when* $z = 68$

	Values of g when j of cooling curve is:								
f_h/U	0.40	0.60	0.80	1.00	1.20	1.40	1.60	1.80	2.00
0.20	1.34-04	1.60-04	1.85-04	2.11-04	2.36-04	2.62-04	2.87-04	3.12-04	3.38-04
0.30	6.23-03	7.41-03	8.59-03	9.78-03	1.10-02	1.21-02	1.33-02	1.45-02	1.57-02
0.40	4.25-02	5.06-02	5.86-02	6.67-02	7.48-02	8.29-02	9.09-02	9.90-02	1.07-01
0.50	0.135	0.160	0.186	0.212	0.237	0.263	0.289	0.314	0.340
0.60	0.293	0.348	0.404	0.459	0.515	0.570	0.626	0.681	0.737
0.70	0.511	0.608	0.705	0.802	0.899	0.996	1.093	1.190	1.286
0.80	0.779	0.927	1.075	1.223	1.370	1.518	1.666	1.814	1.962
0.90	1.09	1.29	1.50	1.70	1.91	2.12	2.32	2.53	2.73
1.00	1.42	1.69	1.96	2.23	2.50	2.77	3.04	3.31	3.58
2.00	5.13	6.11	7.10	8.09	9.08	10.07	11.05	12.04	13.03
2.50	6.84	8.16	9.49	10.82	12.14	13.47	14.80	16.13	17.45
3.00	8.39	10.03	11.67	13.30	14.94	16.58	18.22	19.85	21.49
3.50	9.80	11.72	13.64	15.56	17.48	19.40	21.32	23.24	25.16
4.00	11.1	13.3	15.4	17.6	19.8	22.0	24.2	26.3	28.5
4.50	12.3	14.7	17.1	19.5	21.9	24.3	26.7	29.2	31.6
5.00	13.4	16.0	18.6	21.2	23.9	26.5	29.1	31.7	34.4
5.20	13.8	16.5	19.2	21.9	24.6	27.3	30.0	32.7	35.4
5.40	14.2	17.0	19.8	22.5	25.3	28.1	30.9	33.7	36.4
5.60	14.6	17.4	20.3	23.2	26.0	28.9	31.7	34.6	37.4
5.80	15.0	17.9	20.8	23.8	26.7	29.6	32.5	35.5	38.4
6.00	15.3	18.3	21.3	24.3	27.3	30.3	33.3	36.3	39.3
6.20	15.7	18.8	21.8	24.9	28.0	31.0	34.1	37.2	40.2
6.40	16.1	19.2	22.3	25.5	28.6	31.7	34.9	38.0	41.1
6.60	16.4	19.6	22.8	26.0	29.2	32.4	35.6	38.8	42.0
6.80	16.8	20.0	23.3	26.5	29.8	33.0	36.3	39.6	42.8
7.00	17.1	20.4	23.7	27.0	30.4	33.7	37.0	40.3	43.6
7.20	17.4	20.8	24.2	27.5	30.9	34.3	37.7	41.0	44.4
7.40	17.7	21.2	24.6	28.0	31.5	34.9	38.3	41.8	45.2
7.60	18.1	21.5	25.0	28.5	32.0	35.5	39.0	42.4	45.9
7.80	18.4	21.9	25.4	29.0	32.5	36.1	39.6	43.1	46.7
8.00	18.7	22.3	25.8	29.4	33.0	36.6	40.2	43.8	47.4
8.20	19.0	22.6	26.2	29.9	33.5	37.2	40.8	44.4	48.1
8.40	19.3	22.9	26.6	30.3	34.0	37.7	41.4	45.1	48.7
8.60	19.5	23.3	27.0	30.7	34.5	38.2	41.9	45.7	49.4
8.80	19.8	23.6	27.4	31.2	34.9	38.7	42.5	46.3	50.0
9.00	20.1	23.9	27.8	31.6	35.4	39.2	43.0	46.9	50.7
9.20	20.4	24.2	28.1	32.0	35.8	39.7	43.6	47.4	51.3
9.40	20.7	24.6	28.5	32.4	36.3	40.2	44.1	48.0	51.9
9.60	20.9	24.9	28.8	32.8	36.7	40.6	44.6	48.5	52.5
9.80	21.2	25.2	29.1	33.1	37.1	41.1	45.1	49.1	53.0

TABLE A.34

$f_h/U : g$ *Relationships when* $z = 70$

	Values of g when j of cooling curve is:								
f_h/U	0.40	0.60	0.80	1.00	1.20	1.40	1.60	1.80	2.00
0.20	1.37-04	1.64-04	1.91-04	2.18-04	2.44-04	2.71-04	2.98-04	3.25-04	3.52-04
0.30	6.37-03	7.62-03	8.86-03	1.01-02	1.13-02	1.26-02	1.38-02	1.51-02	1.63-02
0.40	4.35-02	5.20-02	6.04-02	6.89-02	7.74-02	8.59-02	9.44-02	1.03-01	1.11-01
0.50	0.138	0.165	0.192	0.219	0.246	0.273	0.299	0.326	0.353
0.60	0.299	0.358	0.416	0.474	0.533	0.591	0.650	0.708	0.766
0.70	0.522	0.624	0.726	0.828	0.930	1.032	1.134	1.236	1.338
0.80	0.797	0.952	1.108	1.263	1.419	1.574	1.730	1.885	2.040
0.90	1.11	1.33	1.54	1.76	1.98	2.19	2.41	2.63	2.84
1.00	1.45	1.74	2.02	2.30	2.59	2.87	3.15	3.44	3.72
2.00	5.24	6.28	7.32	8.35	9.39	10.43	11.46	12.50	13.54
2.50	6.99	8.38	9.77	11.17	12.56	13.95	15.34	16.73	18.12
3.00	8.58	10.29	12.01	13.73	15.44	17.16	18.87	20.59	22.31
3.50	10.0	12.0	14.0	16.1	18.1	20.1	22.1	24.1	26.1
4.00	11.3	13.6	15.9	18.2	20.4	22.7	25.0	27.3	29.6
4.50	12.5	15.1	17.6	20.1	22.6	25.2	27.7	30.2	32.7
5.00	13.7	16.4	19.2	21.9	24.6	27.4	30.1	32.9	35.6
5.20	14.1	16.9	19.7	22.6	25.4	28.2	31.0	33.9	36.7
5.40	14.5	17.4	20.3	23.2	26.1	29.0	31.9	34.8	37.7
5.60	14.9	17.9	20.9	23.9	26.8	29.8	32.8	35.8	38.8
5.80	15.3	18.4	21.4	24.5	27.5	30.6	33.6	36.7	39.8
6.00	15.7	18.8	21.9	25.1	28.2	31.3	34.5	37.6	40.7
6.20	16.1	19.3	22.5	25.7	28.9	32.1	35.3	38.5	41.7
6.40	16.4	19.7	23.0	26.2	29.5	32.8	36.0	39.3	42.6
6.60	16.8	20.1	23.5	26.8	30.1	33.4	36.8	40.1	43.4
6.80	17.1	20.5	23.9	27.3	30.7	34.1	37.5	40.9	44.3
7.00	17.5	20.9	24.4	27.9	31.3	34.8	38.2	41.7	45.1
7.20	17.8	21.3	24.9	28.4	31.9	35.4	38.9	42.4	45.9
7.40	18.2	21.7	25.3	28.9	32.4	36.0	39.6	43.1	46.7
7.60	18.5	22.1	25.7	29.4	33.0	36.6	40.2	43.9	47.5
7.80	18.8	22.5	26.2	29.8	33.5	37.2	40.9	44.5	48.2
8.00	19.1	22.8	26.6	30.3	34.0	37.8	41.5	45.2	48.9
8.20	19.4	23.2	27.0	30.8	34.5	38.3	42.1	45.9	49.7
8.40	19.7	23.5	27.4	31.2	35.0	38.9	42.7	46.5	50.3
8.60	20.0	23.9	27.8	31.6	35.5	39.4	43.3	47.1	51.0
8.80	20.3	24.2	28.1	32.1	36.0	39.9	43.8	47.8	51.7
9.00	20.6	24.5	28.5	32.5	36.5	40.4	44.4	48.4	52.3
9.20	20.9	24.9	28.9	32.9	36.9	40.9	44.9	48.9	52.9

TABLE A.35

$f_h/U : g$ Relationships when $z = 72$

	Values of g when j of cooling curve is:								
f_h/U	0.40	0.60	0.80	1.00	1.20	1.40	1.60	1.80	2.00
0.20	1.40-04	1.68-04	1.96-04	2.25-04	2.53-04	2.81-04	3.09-04	3.37-04	3.65-04
0.30	6.51-03	7.82-03	9.12-03	1.04-02	1.17-02	1.30-02	1.43-02	1.56-02	1.70-02
0.40	4.45-02	5.34-02	6.23-02	7.12-02	8.01-02	8.90-02	9.79-02	1.07-01	1.16-01
0.50	0.141	0.169	0.198	0.226	0.254	0.282	0.311	0.339	0.367
0.60	0.306	0.367	0.428	0.490	0.551	0.612	0.674	0.735	0.796
0.70	0.534	0.641	0.748	0.855	0.962	1.069	1.176	1.283	1.390
0.80	0.814	0.977	1.140	1.303	1.467	1.630	1.793	1.956	2.119
0.90	1.13	1.36	1.59	1.82	2.04	2.27	2.50	2.73	2.95
1.00	1.48	1.78	2.08	2.38	2.67	2.97	3.27	3.57	3.86
2.00	5.36	6.44	7.53	8.62	9.70	10.79	11.88	12.96	14.05
2.50	7.14	8.60	10.06	11.51	12.97	14.43	15.89	17.35	18.80
3.00	8.76	10.56	12.35	14.15	15.95	17.74	19.54	21.34	23.13
3.50	10.2	12.3	14.4	16.5	18.6	20.7	22.9	25.0	27.1
4.00	11.6	14.0	16.3	18.7	21.1	23.5	25.9	28.2	30.6
4.50	12.8	15.4	18.1	20.7	23.3	26.0	28.6	31.2	33.9
5.00	14.0	16.8	19.7	22.5	25.4	28.3	31.1	34.0	36.8
5.20	14.4	17.3	20.3	23.2	26.2	29.1	32.1	35.0	38.0
5.40	14.8	17.8	20.9	23.9	26.9	30.0	33.0	36.0	39.1
5.60	15.2	18.3	21.5	24.6	27.7	30.8	33.9	37.0	40.1
5.80	15.6	18.8	22.0	25.2	28.4	31.6	34.7	37.9	41.1
6.00	16.0	19.3	22.5	25.8	29.1	32.3	35.6	38.8	42.1
6.20	16.4	19.7	23.1	26.4	29.7	33.1	36.4	39.7	43.1
6.40	16.8	20.2	23.6	27.0	30.4	33.8	37.2	40.6	44.0
6.60	17.2	20.6	24.1	27.6	31.0	34.5	38.0	41.4	44.9
6.80	17.3	21.0	24.6	28.1	31.6	35.2	38.7	42.2	45.8
7.00	17.9	21.5	25.1	28.6	32.2	35.8	39.4	43.0	46.6
7.20	18.2	21.9	25.5	29.2	32.8	36.5	40.1	43.8	47.4
7.40	18.6	22.3	26.0	29.7	33.4	37.1	40.8	44.5	48.2
7.60	18.9	22.7	26.4	30.2	34.0	37.7	41.5	45.2	49.0
7.80	19.2	23.0	26.9	30.7	34.5	38.3	42.1	46.0	49.8
8.00	19.5	23.4	27.3	31.2	35.0	38.9	42.8	46.6	50.5
8.20	19.8	23.8	27.7	31.6	35.5	39.5	43.4	47.3	51.2
8.40	20.2	24.1	28.1	32.1	36.0	40.0	44.0	48.0	51.9
8.60	20.5	24.5	28.5	32.5	36.5	40.6	44.6	48.6	52.6

TABLE A.36

$f_h/U : g$ *Relationships when* $z = 74$

	Values of g when j of cooling curve is:								
f_h/U	0.40	0.60	0.80	1.00	1.20	1.40	1.60	1.80	2.00
0.20	1.43-04	1.73-04	2.03-04	2.33-04	2.63-04	2.93-04	3.23-04	3.53-04	3.83-04
0.30	6.65-03	8.03-03	9.40-03	1.08-02	1.22-02	1.35-02	1.49-02	1.63-02	1.77-02
0.40	4.54-02	5.47-02	6.41-02	7.34-02	8.28-02	9.21-02	1.01-01	1.11-01	1.20-01
0.50	0.144	0.173	0.203	0.232	0.261	0.290	0.320	0.349	0.378
0.60	0.312	0.376	0.440	0.503	0.567	0.631	0.694	0.758	0.822
0.70	0.545	0.657	0.769	0.881	0.994	1.106	1.218	1.330	1.442
0.80	0.831	1.003	1.175	1.347	1.518	1.690	1.862	2.034	2.206
0.90	1.16	1.40	1.88	2.12	2.36	2.60	2.84	3.07
1.00	1.52	1.83	2.14	2.45	2.77	3.08	3.39	3.71	4.02
2.00	5.47	6.61	7.75	8.89	10.03	11.17	12.31	13.45	14.60
2.50	7.29	8.81	10.34	11.87	13.39	14.92	16.45	17.97	19.50
3.00	8.94	10.81	12.69	14.56	16.43	18.31	20.18	22.05	23.93
3.50	10.4	12.6	14.8	17.0	19.2	21.4	23.6	25.8	28.0
4.00	11.8	14.3	16.8	19.3	21.8	24.3	26.7	29.2	31.7
4.50	13.1	15.8	18.6	21.3	24.1	26.8	29.6	32.3	35.1
5.00	14.3	17.2	20.2	23.2	26.2	29.2	32.2	35.1	38.1
5.20	14.7	17.8	20.8	23.9	27.0	30.0	33.1	36.2	39.3
5.40	15.1	18.3	21.4	24.6	27.7	30.9	34.0	37.2	40.3
5.60	15.6	18.8	22.0	25.2	28.5	31.7	34.9	38.2	41.4
5.80	16.0	19.3	22.6	25.9	29.2	32.5	35.8	39.1	42.4
6.00	16.4	19.7	23.1	26.5	29.9	33.3	36.7	40.0	43.4
6.20	16.8	20.2	23.7	27.1	30.6	34.0	37.5	40.9	44.4
6.40	17.1	20.7	24.2	27.7	31.2	34.8	38.3	41.8	45.3
6.60	17.5	21.1	24.7	28.3	31.9	35.5	39.1	42.7	46.3
6.80	17.9	21.6	25.2	28.9	32.5	36.2	39.9	43.5	47.2
7.00	18.3	22.0	25.7	29.4	33.2	36.9	40.6	44.3	48.1
7.20	18.6	22.4	26.2	30.0	33.8	37.6	41.4	45.1	48.9
7.40	19.0	22.8	26.7	30.5	34.4	38.2	42.1	45.9	49.8
7.60	19.3	23.2	27.1	31.0	35.0	38.9	42.8	46.7	50.6
7.80	19.6	23.6	27.6	31.6	35.5	39.5	43.5	47.4	51.4
8.00	20.0	24.0	28.0	32.0	36.1	40.1	44.1	48.1	52.2
8.20	20.3	24.4	28.4	32.5	36.6	40.7	44.7	48.8	52.9

TABLE A.37

$f_h/U:g$ *Relationships when* $z = 76$

	Values of g when j of cooling curve is:								
f_h/U	0.40	0.60	0.80	1.00	1.20	1.40	1.60	1.80	2.00
0.20	1.46-04	1.77-04	2.08-04	2.39-04	2.70-04	3.01-04	3.31-04	3.62-04	3.93-04
0.30	6.78-03	8.22-03	9.65-03	1.11-02	1.25-02	1.40-02	1.54-02	1.68-02	1.83-02
0.40	4.63-02	5.61-02	6.59-02	7.56-02	8.54-02	9.52-02	1.05-01	1.15-01	1.25-01
0.50	0.147	0.178	0.209	0.240	0.271	0.302	0.333	0.364	0.395
0.60	0.319	0.386	0.454	0.521	0.588	0.656	0.723	0.790	0.858
0.70	0.557	0.674	0.792	0.909	1.027	1.144	1.262	1.379	1.497
0.80	0.849	1.028	1.207	1.386	1.565	1.744	1.923	2.102	2.281
0.90	1.18	1.43	1.68	1.93	2.18	2.43	2.68	2.93	3.18
1.00	1.55	1.87	2.20	2.53	2.85	3.18	3.51	3.83	4.16
2.00	5.58	6.77	7.96	9.15	10.34	11.53	12.72	13.91	15.09
2.50	7.43	9.03	10.62	12.21	13.81	15.40	17.00	18.59	20.18
3.00	9.12	11.08	13.04	15.00	16.96	18.92	20.88	22.84	24.80
3.50	10.6	12.9	15.2	17.5	19.8	22.1	24.4	26.7	29.0
4.00	12.0	14.6	17.2	19.8	22.4	25.0	27.6	30.2	32.8
4.50	13.3	16.2	19.1	21.9	24.8	27.6	30.5	33.3	36.2
5.00	14.5	17.6	20.7	23.8	26.9	30.0	33.1	36.2	39.3
5.20	15.0	18.2	21.4	24.6	27.8	30.9	34.1	37.3	40.5
5.40	15.4	18.7	22.0	25.3	28.5	31.8	35.1	38.4	41.7
5.60	15.9	19.2	22.6	26.0	29.3	32.7	36.0	39.4	42.8
5.80	16.3	19.7	23.2	26.6	30.1	33.5	36.9	40.4	43.8
6.00	16.7	20.2	23.7	27.3	30.8	34.3	37.8	41.3	44.8
6.20	17.1	20.7	24.3	27.9	31.5	35.1	38.7	42.3	45.8
6.40	17.5	21.2	24.8	28.5	32.2	35.8	39.5	43.1	46.8
6.60	17.9	21.6	25.4	29.1	32.8	36.6	40.3	44.0	47.7
6.80	18.3	22.1	25.9	29.7	33.5	37.3	41.1	44.9	48.7
7.00	18.6	22.5	26.4	30.2	34.1	38.0	41.8	45.7	49.5
7.20	19.0	22.9	26.9	30.8	34.7	38.6	42.6	46.5	50.4
7.40	19.4	23.3	27.3	31.3	35.3	39.3	43.3	47.3	51.2
7.60	19.7	23.8	27.8	31.8	35.9	39.9	44.0	48.0	52.0
7.80	20.1	24.2	28.2	32.3	36.4	40.5	44.6	48.7	52.8

TABLE A.38

$f_h/U : g$ Relationships when $z = 78$

	Values of g when j of cooling curve is:								
f_h/U	0.40	0.60	0.80	1.00	1.20	1.40	1.60	1.80	2.00
0.20	1.49-04	1.81-04	2.14-04	2.46-04	2.78-04	3.11-04	3.43-04	3.75-04	4.07-04
0.30	6.92-03	8.42-03	9.92-03	1.14-02	1.29-02	1.44-02	1.59-02	1.74-02	1.89-02
0.40	4.72-02	5.74-02	6.77-02	7.79-02	8.82-02	9.84-02	1.09-01	1.19-01	1.29-01
0.50	0.150	0.182	0.215	0.247	0.280	0.312	0.345	0.377	0.410
0.60	0.325	0.395	0.466	0.536	0.607	0.677	0.748	0.818	0.889
0.70	0.567	0.690	0.813	0.936	1.059	1.182	1.305	1.428	1.551
0.80	0.865	1.052	1.240	1.427	1.615	1.802	1.989	2.177	2.364
0.90	1.21	1.47	1.73	1.99	2.25	2.51	2.77	3.03	3.29
1.00	1.58	1.92	2.26	2.60	2.94	3.28	3.63	3.97	4.31
2.00	5.68	6.93	8.17	9.41	10.66	11.90	13.14	14.38	15.63
2.50	7.57	9.24	10.90	12.56	14.23	15.89	17.55	19.22	20.88
3.00	9.29	11.33	13.38	15.42	17.47	19.51	21.56	23.60	25.65
3.50	10.8	13.2	15.6	18.0	20.4	22.8	25.2	27.6	29.9
4.00	12.3	15.0	17.7	20.4	23.1	25.8	28.4	31.1	33.8
4.50	13.6	16.6	19.5	22.5	25.5	28.5	31.4	34.4	37.4
5.00	14.8	18.0	21.3	24.5	27.7	30.9	34.1	37.4	40.6
5.20	15.3	18.6	21.9	25.2	28.5	31.8	35.2	38.5	41.8
5.40	15.7	19.1	22.5	25.9	29.3	32.7	36.1	39.5	42.9
5.60	16.2	19.7	23.2	26.6	30.1	33.6	37.1	40.6	44.1
5.80	16.6	20.2	23.8	27.3	30.9	34.5	38.0	41.6	45.2
6.00	17.0	20.7	24.3	28.0	31.6	35.3	38.9	42.6	46.2
6.20	17.5	21.2	24.9	28.6	32.3	36.1	39.8	43.5	47.2
6.40	17.9	21.7	25.4	29.2	33.0	36.8	40.6	44.4	48.2
6.60	18.3	22.1	26.0	29.8	33.7	37.6	41.4	45.3	49.2
6.80	18.6	22.6	26.5	30.4	34.4	38.3	42.2	46.2	50.1
7.00	19.0	23.0	27.0	31.0	35.0	39.0	43.0	47.0	51.0
7.20	19.4	23.5	27.5	31.6	35.6	39.7	43.7	47.8	51.9
7.40	19.8	23.9	28.0	32.1	36.2	40.4	44.5	48.6	52.7

TABLE A.39

$f_h/U : g$ *Relationships when* $z = 80$

	Values of g when j of cooling curve is:								
f_h/U	0.40	0.60	0.80	1.00	1.20	1.40	1.60	1.80	2.00
0.20	1.52-04	1.86-04	2.19-04	2.53-04	2.87-04	3.21-04	3.54-04	3.88-04	4.22-04
0.30	7.05-03	8.62-03	1.02-02	1.18-02	1.33-02	1.49-02	1.65-02	1.80-02	1.96-02
0.40	4.81-02	5.88-02	6.95-02	8.02-02	9.09-02	1.02-01	1.12-01	1.23-01	1.34-01
0.50	0.153	0.187	0.221	0.255	0.289	0.323	0.356	0.390	0.424
0.60	0.331	0.405	0.478	0.552	0.626	0.699	0.773	0.847	0.920
0.70	0.578	0.706	0.835	0.963	1.092	1.220	1.349	1.477	1.606
0.80	0.881	1.077	1.273	1.469	1.665	1.860	2.056	2.252	2.448
0.90	1.23	1.50	1.77	2.05	2.32	2.59	2.86	3.14	3.41
1.00	1.61	1.96	2.32	2.68	3.03	3.39	3.75	4.10	4.46
2.00	5.79	7.09	8.38	9.68	10.98	12.27	13.57	14.87	16.16
2.50	7.71	9.45	11.18	12.92	14.65	16.38	18.12	19.85	21.58
3.00	9.46	11.59	13.72	15.85	17.98	20.11	22.24	24.36	26.49
3.50	11.1	13.5	16.0	18.5	21.0	23.5	26.0	28.4	30.9
4.00	12.5	15.3	18.1	20.9	23.7	26.5	29.3	32.1	34.9
4.50	13.9	16.9	20.0	23.1	26.2	29.3	32.4	35.4	38.5
5.00	15.1	18.4	21.8	25.1	28.5	31.8	35.1	38.5	41.8
5.20	15.6	19.0	22.4	25.9	29.3	32.7	36.2	39.6	43.0
5.40	16.0	19.6	23.1	26.6	30.1	33.7	37.2	40.7	44.2
5.60	16.5	20.1	23.7	27.3	30.9	34.6	38.2	41.8	45.4
5.80	16.9	20.6	24.3	28.0	31.7	35.4	39.1	42.8	46.5
6.00	17.4	21.1	24.9	28.7	32.5	36.2	40.0	43.8	47.6
6.20	17.8	21.6	25.5	29.3	33.2	37.0	40.9	44.8	48.6
6.40	18.2	22.1	26.1	30.0	33.9	37.8	41.8	45.7	49.6
6.60	18.6	22.6	26.6	30.6	34.6	38.6	42.6	46.6	50.6
6.80	19.0	23.1	27.1	31.2	35.3	39.3	43.4	47.5	51.5
7.00	19.4	23.5	27.7	31.8	35.9	40.0	44.2	48.3	52.4

TABLE A.40

$f_h/U : g$ *Relationships when* $z = 82$

f_h/U	Values of g when j of cooling curve is:								
	0.40	0.60	0.80	1.00	1.20	1.40	1.60	1.80	2.00
0.20	1.55-04	1.90-04	2.25-04	2.60-04	2.96-04	3.31-04	3.66-04	4.01-04	4.37-04
0.30	7.18-03	8.82-03	1.05-02	1.21-02	1.37-02	1.54-02	1.70-02	1.86-02	2.03-02
0.40	4.89-02	6.01-02	7.13-02	8.25-02	9.36-02	1.05-01	1.16-01	1.27-01	1.38-01
0.50	0.155	0.191	0.226	0.262	0.297	0.333	0.368	0.404	0.439
0.60	0.337	0.414	0.491	0.568	0.645	0.721	0.798	0.875	0.952
0.70	0.569	0.723	0.857	0.991	1.125	1.259	1.393	1.527	1.661
0.80	0.898	1.102	1.307	1.511	1.715	1.920	2.124	2.328	2.532
0.82	0.966	1.185	1.405	1.625	1.844	2.064	2.284	2.503	2.723
0.84	1.04	1.27	1.51	1.74	1.98	2.21	2.45	2.68	2.92
0.86	1.11	1.36	1.61	1.86	2.11	2.36	2.61	2.87	3.12
0.88	1.18	1.45	1.71	1.98	2.25	2.52	2.79	3.05	3.32
0.90	1.25	1.54	1.82	2.11	2.39	2.67	2.96	3.24	3.53
0.92	1.33	1.63	1.93	2.23	2.53	2.83	3.14	3.44	3.74
0.94	1.40	1.72	2.04	2.36	2.68	3.00	3.32	3.63	3.95
0.96	1.48	1.82	2.15	2.49	2.83	3.16	3.50	3.83	4.17
0.98	1.56	1.91	2.27	2.62	2.98	3.33	3.68	4.04	4.39
1.00	1.64	2.01	2.38	2.75	3.13	3.50	3.87	4.24	4.62
1.20	2.47	3.03	3.59	4.15	4.71	5.28	5.84	6.40	6.96
1.40	3.33	4.09	4.85	5.61	6.37	7.13	7.89	8.65	9.41
1.60	4.20	5.16	6.12	7.08	8.05	9.01	9.97	10.93	11.89
1.80	5.06	6.22	7.38	8.53	9.69	10.85	12.01	13.17	14.33
2.00	5.89	7.24	8.59	9.95	11.30	12.65	14.00	15.35	16.70
2.20	6.70	8.23	9.77	11.31	12.85	14.39	15.93	17.46	19.00
2.40	7.47	9.19	10.91	12.63	14.35	16.06	17.78	19.50	21.22
2.60	8.22	10.11	12.00	13.89	15.78	17.67	19.56	21.46	23.35
2.80	8.94	10.99	13.05	15.11	17.16	19.22	21.28	23.33	25.39
3.00	9.63	11.84	14.06	16.27	18.49	20.70	22.92	25.13	27.35
3.20	10.3	12.7	15.0.	17.4	19.8	22.1	24.5	26.9	29.2
3.40	10.9	13.4	16.0	18.5	21.0	23.5	26.0	28.5	31.0
3.60	11.6	14.2	16.9	19.5	22.2	24.8	27.4	30.1	32.7
3.80	12.2	14.9	17.7	20.5	23.3	26.1	28.8	31.6	34.4
4.00	12.7	15.6	18.5	21.5	24.4	27.3	30.2	33.1	36.0
4.20	13.3	16.3	19.3	22.4	25.4	28.4	31.5	34.5	37.5
4.40	13.8	17.0	20.1	23.3	26.4	29.6	32.7	35.8	39.0
4.60	14.4	17.6	20.9	24.1	27.4	30.6	33.9	37.1	40.4
4.80	14.9	18.2	21.6	25.0	28.3	31.7	35.0	38.4	41.7
5.00	15.4	18.8	22.3	25.8	29.2	32.7	36.1	39.6	43.0
5.20	15.9	19.4	23.0	26.5	30.1	33.6	37.2	40.8	44.3
5.40	16.3	20.0	23.6	27.3	30.9	34.6	38.2	41.9	45.5
5.60	16.8	20.5	24.3	28.0	31.7	35.5	39.2	43.0	46.7
5.80	17.3	21.1	24.9	28.7	32.5	36.4	40.2	44.0	47.8
6.00	17.7	21.6	25.5	29.4	33.3	37.2	41.1	45.0	48.9
6.20	18.1	22.1	26.1	30.1	34.1	38.0	42.0	46.0	50.0
6.40	18.6	22.6	26.7	30.7	34.8	38.8	42.9	46.9	51.0
6.60	19.0	23.1	27.2	31.3	35.5	39.6	43.7	47.8	52.0

TABLE A.41

$f_h/U : g$ *Relationships when* $z = 84$

f_h/U	Values of g when j of cooling curve is:								
	0.40	0.60	0.80	1.00	1.20	1.40	1.60	1.80	2.00
0.20	1.57-04	1.94-04	2.31-04	2.67-04	3.04-04	3.40-04	3.77-04	4.13-04	4.50-04
0.30	7.32-03	9.02-03	1.07-02	1.24-02	1.41-02	1.58-02	1.75-02	1.92-02	2.09-02
0.40	4.98-02	6.15-02	7.32-02	8.48-02	9.65-02	1.08-01	1.20-01	1.31-01	1.43-01
0.50	0.158	0.195	0.232	0.269	0.306	0.343	0.380	0.417	0.454
0.60	0.342	0.423	0.503	0.583	0.663	0.744	0.824	0.904	0.984
0.70	0.598	0.738	0.878	1.018	1.158	1.298	1.438	1.578	1.718
0.80	0.911	1.125	1.338	1.552	1.766	1.979	2.193	2.406	2.620
0.82	0.980	1.209	1.439	1.669	1.898	2.128	2.357	2.587	2.817
0.84	1.05	1.30	1.54	1.79	2.03	2.28	2.53	2.77	3.02
0.86	1.12	1.38	1.65	1.91	2.17	2.44	2.70	2.96	3.22
0.88	1.20	1.48	1.76	2.04	2.32	2.60	2.88	3.15	3.43
0.90	1.27	1.57	1.86	2.16	2.46	2.76	3.05	3.35	3.65
0.92	1.35	1.66	1.98	2.29	2.61	2.92	3.24	3.55	3.87
0.94	1.42	1.76	2.09	2.42	2.76	3.09	3.42	3.76	4.09
0.96	1.50	1.85	2.20	2.56	2.91	3.26	3.61	3.96	4.31
0.98	1.58	1.95	2.32	2.69	3.06	3.43	3.80	4.17	4.54
1.00	1.66	2.05	2.44	2.83	3.22	3.61	3.99	4.38	4.77
1.20	2.51	3.09	3.68	4.27	4.85	5.44	6.02	6.61	7.19
1.40	3.39	4.18	4.98	5.77	6.56	7.35	8.14	8.94	9.73
1.60	4.28	5.28	6.28	7.28	8.28	9.28	10.28	11.28	12.28
1.80	5.15	6.36	7.56	8.77	9.98	11.18	12.39	13.59	14.80
2.00	6.00	7.41	8.81	10.22	11.63	13.03	14.44	15.84	17.25
2.20	6.82	8.42	10.02	11.62	13.22	14.82	16.42	18.02	19.62
2.40	7.61	9.40	11.18	12.97	14.75	16.54	18.33	20.11	21.90
2.60	8.37	10.33	12.30	14.26	16.23	18.19	20.16	22.12	24.09
2.80	9.10	11.23	13.37	15.51	17.64	19.78	21.91	24.05	26.19
3.00	9.80	12.10	14.40	16.70	19.00	21.30	23.60	25.90	28.20
3.20	10.5	12.9	15.4	17.8	20.3	22.8	25.2	27.7	30.1
3.40	11.1	13.7	16.3	18.9	21.6	24.2	26.8	29.4	32.0
3.60	11.8	14.5	17.3	20.0	22.7	25.5	28.2	31.0	33.7
3.80	12.4	15.2	18.1	21.0	23.9	26.8	29.7	32.6	35.4
4.00	13.0	16.0	19.0	22.0	25.0	28.0	31.0	34.1	37.1
4.20	13.5	16.7	19.8	22.9	26.1	29.2	32.4	35.5	38.6
4.40	14.1	17.3	20.6	23.8	27.1	30.4	33.6	36.9	40.1
4.60	14.6	18.0	21.4	24.7	28.1	31.5	34.8	38.2	41.6
4.80	15.1	18.6	22.1	25.6	29.0	32.5	36.0	39.5	42.9
5.00	15.6	19.2	22.8	26.4	30.0	33.5	37.1	40.7	44.3
5.20	16.1	19.8	23.5	27.2	30.9	34.5	38.2	41.9	45.6
5.40	16.6	20.4	24.2	27.9	31.7	35.5	39.3	43.0	46.8
5.60	17.1	21.0	24.8	28.7	32.5	36.4	40.3	44.1	48.0
5.80	17.6	21.5	25.5	29.4	33.4	37.3	41.2	45.2	49.1
6.00	18.0	22.1	26.1	30.1	34.1	38.2	42.2	46.2	50.2
6.20	18.5	22.6	26.7	30.8	34.9	39.0	43.1	47.2	51.3
6.40	18.9	23.1	27.3	31.4	35.6	39.8	44.0	48.2	52.3

TABLE A.42

$f_h/U:g$ *Relationships when* $z = 86$

	Values of g when j of cooling curve is:								
f_h/U	0.40	0.60	0.80	1.00	1.20	1.40	1.60	1.80	2.00
0.20	1.60-04	1.98-04	2.36-04	2.75-04	3.13-04	3.51-04	3.89-04	4.28-04	4.66-04
0.30	7.43-03	9.21-03	1.10-02	1.28-02	1.45-02	1.63-02	1.81-02	1.99-02	2.17-02
0.40	5.07-02	6.28-02	7.50-02	8.71-02	9.93-02	1.11-01	1.24-01	1.36-01	1.48-01
0.50	0.161	0.199	0.238	0.277	0.315	0.354	0.392	0.431	0.469
0.60	0.349	0.432	0.516	0.600	0.683	0.767	0.850	0.934	1.017
0.70	0.609	0.755	0.900	1.046	1.192	1.338	1.484	1.629	1.775
0.80	0.928	1.150	1.373	1.595	1.817	2.039	2.261	2.483	2.706
0.82	0.998	1.237	1.476	1.715	1.953	2.192	2.431	2.670	2.909
0.84	1.07	1.33	1.58	1.84	2.09	2.35	2.61	2.86	3.12
0.86	1.14	1.42	1.69	1.96	2.24	2.51	2.78	3.06	3.33
0.88	1.22	1.51	1.80	2.09	2.38	2.67	2.96	3.26	3.55
0.90	1.29	1.60	1.91	2.22	2.53	2.84	3.15	3.46	3.77
0.92	1.37	1.70	2.03	2.35	2.68	3.01	3.34	3.67	3.99
0.94	1.45	1.80	2.14	2.49	2.84	3.18	3.53	3.88	4.22
0.96	1.53	1.90	2.26	2.63	2.99	3.36	3.72	4.09	4.46
0.98	1.61	2.00	2.38	2.77	3.15	3.54	3.92	4.31	4.69
1.00	1.69	2.10	2.50	2.91	3.31	3.72	4.12	4.52	4.93
1.20	2.55	3.16	3.77	4.38	4.99	5.60	6.21	6.82	7.43
1.40	3.45	4.27	5.10	5.92	6.75	7.57	8.40	9.22	10.05
1.60	4.35	5.39	6.43	7.47	8.51	9.56	10.60	11.64	12.68
1.80	5.23	6.49	7.74	9.00	10.25	11.51	12.76	14.02	15.27
2.00	6.10	7.56	9.02	10.48	11.95	13.41	14.87	16.34	17.80
2.20	6.93	8.59	10.26	11.92	13.58	15.25	16.91	18.57	20.24
2.40	7.73	9.59	11.44	13.30	15.16	17.01	18.87	20.73	22.58
2.60	8.50	10.54	12.59	14.63	16.67	18.71	20.75	22.79	24.84
2.80	9.25	11.46	13.68	15.90	18.12	20.34	22.56	24.78	26.99
3.00	9.96	12.35	14.74	17.12	19.51	21.90	24.29	26.67	29.06
3.20	10.7	13.2	15.7	18.3	20.8	23.4	25.9	28.5	31.0
3.40	11.3	14.0	16.7	19.4	22.1	24.8	27.5	30.2	32.9
3.60	12.0	14.8	17.7	20.5	23.3	26.2	29.0	31.9	34.7
3.80	12.0	15.6	18.6	21.5	24.5	27.5	30.5	33.5	36.5
4.00	13.2	16.3	19.4	22.5	25.7	28.8	31.9	35.0	38.1
4.20	13.8	17.0	20.3	23.5	26.7	30.0	33.2	36.5	39.7
4.40	14.3	17.7	21.1	24.4	27.8	31.2	34.5	37.9	41.3
4.60	14.9	18.4	21.8	25.3	28.8	32.3	35.8	39.2	42.7
4.80	15.4	19.0	22.6	26.2	29.8	33.4	37.0	40.6	44.1
5.00	15.9	19.6	23.3	27.0	30.7	34.4	38.1	41.8	45.5
5.20	16.4	20.2	24.0	27.8	31.6	35.4	39.2	43.0	46.8
5.40	16.9	20.8	24.7	28.6	32.5	36.4	40.3	44.2	48.1
5.60	17.4	21.4	25.4	29.4	33.3	37.3	41.3	45.3	49.3
5.80	17.9	22.0	26.0	30.1	34.2	38.2	42.3	46.4	50.4
6.00	18.4	22.5	26.7	30.8	35.0	39.1	43.3	47.4	51.6
6.20	18.8	23.0	27.3	31.5	35.7	40.0	44.2	48.4	52.6

TABLE A.43

$f_h/U : g$ Relationships when $z = 88$

	Values of g when j of cooling curve is:								
f_h/U	0.40	0.60	0.80	1.00	1.20	1.40	1.60	1.80	2.00
0.20	1.62-04	2.02-04	2.42-04	2.82-04	3.22-04	3.62-04	4.02-04	4.42-04	4.82-04
0.30	7.54-03	9.40-03	1.13-02	1.31-02	1.50-02	1.68-02	1.87-02	2.05-02	2.24-02
0.40	5.15-02	6.41-02	7.68-02	8.94-02	1.02-01	1.15-01	1.27-01	1.40-01	1.53-01
0.50	0..63	0.204	0.244	0.284	0.324	0.364	0.404	0.444	0.485
0.60	0.355	0.442	0.529	0.616	0.703	0.790	0.877	0.964	1.050
0.70	0.619	0.771	0.923	1.074	1.226	1.378	1.529	1.681	1.833
0.80	0.944	1.175	1.406	1.638	1.869	2.100	2.331	2.562	2.793
0.82	1.02	1.26	1.51	1.76	2.01	2.26	2.51	2.75	3.00
0.84	1.09	1.35	1.62	1.89	2.15	2.42	2.69	2.95	3.22
0.86	1.16	1.45	1.73	2.02	2.30	2.58	2.87	3.15	3.44
0.88	1.24	1.54	1.84	2.15	2.45	2.75	3.06	3.36	3.66
0.90	1.32	1.64	1.96	2.28	2.60	2.92	3.25	3.57	3.89
0.92	1.39	1.74	2.08	2.42	2.76	3.10	3.44	3.78	4.12
0.94	1.47	1.84	2.20	2.56	2.92	3.28	3.64	4.00	4.36
0.96	1.56	1.94	2.32	2.70	3.08	3.46	3.84	4.22	4.60
0.98	1.64	2.04	2.44	2.84	3.24	3.64	4.04	4.44	4.84
1.00	1.72	2.14	2.56	2.98	3.40	3.83	4.25	4.67	5.09
1.20	2.59	3.23	3.86	4.50	5.13	5.77	6.40	7.03	7.67
1.40	3.51	4.36	5.22	6.08	6.94	7.79	8.65	9.51	10.37
1.60	4.42	5.50	6.58	7.67	8.75	9.83	10.92	12.00	13.08
1.80	5.32	6.62	7.93	9.23	10.54	11.84	13.14	14.45	15.75
2.00	6.19	7.71	9.23	10.75	12.27	13.79	15.31	16.83	18.35
2.20	7.04	8.77	10.49	12.22	13.95	15.68	17.41	19.13	20.86
2.40	7.85	9.78	11.71	13.64	15.56	17.49	19.42	21.35	23.27
2.60	8.64	10.76	12.88	14.99	17.11	19.23	21.35	23.47	25.59
2.80	9.39	11.69	14.00	16.30	18.60	20.90	23.20	25.50	27.80
3.00	10.1	12.6	15.1	17.5	20.0	22.5	25.0	27.4	29.9
3.20	10.8	13.5	16.1	18.7	21.4	24.0	26.7	29.3	31.9
3.40	11.5	14.3	17.1	19.9	22.7	25.5	28.3	31.1	33.9
3.60	12.2	15.1	18.0	21.0	23.9	26.9	29.8	32.8	35.7
3.80	12.8	15.9	19.0	22.1	25.1	28.2	31.3	34.4	37.5
4.00	13.4	16.6	19.9	23.1	26.3	29.5	32.8	36.0	39.2
4.20	14.0	17.3	20.7	24.1	27.4	30.8	34.1	37.5	40.8
4.40	14.6	18.0	21.5	25.0	28.5	32.0	35.4	38.9	42.4
4.60	15.1	18.7	22.3	25.9	29.5	33.1	36.7	40.3	43.9
4.80	15.7	19.4	23.1	26.8	30.5	34.2	37.9	41.6	45.3
5.00	16.2	20.0	23.8	27.6	31.5	35.3	39.1	42.9	46.7
5.20	16.7	20.6	24.6	28.5	32.4	36.3	40.2	44.1	48.0
5.40	17.2	21.2	25.3	29.3	33.3	37.3	41.3	45.3	49.3
5.60	17.7	21.8	25.9	30.0	34.1	38.2	42.3	46.4	50.5
5.80	18.2	22.4	26.6	30.8	35.0	39.2	43.3	47.5	51.7
6.00	18.7	23.0	27.2	31.5	35.8	40.0	44.3	48.6	52.8

TABLE A.44

$f_h/U : g$ *Relationships when* $z = 90$

	Values of g when j of cooling curve is:								
f_h/U	0.40	0.60	0.80	1.00	1.20	1.40	1.60	1.80	2.00
0.20	1.65-04	2.06-04	2.48-04	2.89-04	3.31-04	3.73-04	4.14-04	4.56-04	4.97-04
0.30	7.67-03	9.59-03	1.15-02	1.34-02	1.54-02	1.73-02	1.92-02	2.12-02	2.31-02
0.40	5.25-02	6.56-02	7.87-02	9.19-02	1.05-01	1.18-01	1.31-01	1.44-01	1.58-01
0.50	0.166	0.208	0.250	0.292	0.333	0.375	0.417	0.458	0.500
0.60	0.360	0.451	0.541	0.632	0.722	0.813	0.903	0.994	1.084
0.70	0.627	0.785	0.943	1.101	1.259	1.418	1.576	1.734	1.892
0.80	0.956	1.196	1.437	1.678	1.919	2.160	2.401	2.642	2.883
0.82	1.03	1.29	1.55	1.80	2.06	2.32	2.58	2.84	3.10
0.84	1.10	1.38	1.66	1.93	2.21	2.49	2.77	3.04	3.32
0.86	1.18	1.47	1.77	2.07	2.36	2.66	2.95	3.25	3.55
0.88	1.25	1.57	1.88	2.20	2.52	2.83	3.15	3.46	3.78
0.90	1.33	1.67	2.00	2.34	2.67	3.01	3.34	3.68	4.02
0.92	1.41	1.77	2.12	2.48	2.83	3.19	3.54	3.90	4.26
0.94	1.49	1.87	2.24	2.62	3.00	3.37	3.75	4.12	4.50
0.96	1.57	1.97	2.37	2.76	3.16	3.56	3.95	4.35	4.75
0.98	1.66	2.07	2.49	2.91	3.33	3.74	4.16	4.58	5.00
1.00	1.74	2.18	2.62	3.06	3.50	3.93	4.37	4.81	5.25
1.20	2.63	3.29	3.95	4.61	5.27	5.93	6.59	7.25	7.91
1.40	3.56	4.45	5.34	6.23	7.13	8.02	8.91	9.80	10.69
1.60	4.50	5.62	6.74	7.87	8.99	10.11	11.24	12.36	13.48
1.80	5.41	6.77	8.12	9.47	10.82	12.18	13.53	14.88	16.23
2.00	6.31	7.88	9.46	11.03	12.61	14.18	15.76	17.33	18.91
2.20	7.17	8.96	10.75	12.54	14.33	16.12	17.91	19.70	21.49
2.40	7.99	9.99	11.99	13.98	15.98	17.98	19.97	21.97	23.96
2.60	8.79	10.98	13.18	15.37	17.56	19.76	21.95	24.15	26.34
2.80	9.55	11.93	14.32	16.70	19.08	21.46	23.85	26.23	28.61
3.00	10.3	12.8	15.4	18.0	20.5	23.1	25.7	28.2	30.8
3.20	11.0	13.7	16.5	19.2	21.9	24.7	27.4	30.1	32.9
3.40	11.7	14.6	17.5	20.4	23.3	26.2	29.1	31.9	34.8
3.60	12.3	15.4	18.4	21.5	24.5	27.6	30.6	33.7	36.7
3.80	13.0	16.2	19.4	22.6	25.8	29.0	32.2	35.4	38.6
4.00	13.6	16.9	20.3	23.6	26.9	30.3	33.6	36.9	40.3
4.20	14.2	17.7	21.1	24.6	28.1	31.5	35.0	38.5	41.9
4.40	14.8	18.4	22.0	25.6	29.2	32.7	36.3	39.9	43.5
4.60	15.4	19.1	22.8	26.5	30.2	33.9	37.6	41.3	45.1
4.80	15.9	19.7	23.6	27.4	31.2	35.0	38.9	42.7	46.5
5.00	16.5	20.4	24.3	28.2	32.2	36.1	40.0	44.0	47.9
5.20	17.0	21.0	25.0	29.1	33.1	37.2	41.2	45.2	49.3
5.40	17.5	21.6	25.8	29.9	34.0	38.2	42.3	46.4	50.5
5.60	18.0	22.2	26.5	30.7	34.9	39.1	43.3	47.6	51.8

TABLE A.45

$f_h/U:g$ *Relationships when* $z = 92$

	Values of g when j of cooling curve is:								
f_h/U	0.40	0.60	0.80	1.00	1.20	1.40	1.60	1.80	2.00
0.20	1.68-04	2.11-04	2.54-04	2.97-04	3.40-04	3.83-04	4.26-04	4.69-04	5.12-04
0.30	7.79-03	9.79-03	1.18-02	1.38-02	1.58-02	1.78-02	1.98-02	2.18-02	2.38-02
0.40	5.31-02	6.67-02	8.04-02	9.41-02	1.08-01	1.21-01	1.35-01	1.49-01	1.63-01
0.50	0.168	0.212	0.255	0.299	0.342	0.386	0.429	0.472	0.516
0.60	0.365	0.459	0.553	0.648	0.742	0.836	0.930	1.024	1.118
0.70	0.638	0.802	0.966	1.130	1.294	1.458	1.622	1.786	1.950
0.80	0.973	1.223	1.473	1.723	1.972	2.222	2.472	2.722	2.972
0.82	1.05	1.31	1.58	1.85	2.12	2.39	2.66	2.93	3.20
0.84	1.12	1.41	1.70	1.98	2.27	2.56	2.85	3.14	3.42
0.86	1.20	1.51	1.81	2.12	2.43	2.74	3.04	3.35	3.66
0.88	1.28	1.60	1.93	2.26	2.59	2.91	3.24	3.57	3.90
0.90	1.36	1.70	2.05	2.40	2.75	3.10	3.44	3.79	4.14
0.92	1.44	1.81	2.17	2.54	2.91	3.28	3.65	4.02	4.39
0.94	1.52	1.91	2.30	2.69	3.08	3.47	3.86	4.25	4.64
0.96	1.60	2.01	2.43	2.84	3.25	3.66	4.07	4.48	4.89
0.98	1.69	2.12	2.55	2.99	3.42	3.85	4.28	4.72	5.15
1.00	1.77	2.23	2.68	3.14	3.59	4.05	4.50	4.96	5.41
1.20	2.67	3.36	4.04	4.73	5.41	6.10	6.78	7.47	8.15
1.40	3.61	4.54	5.46	6.39	7.32	8.24	9.17	10.09	11.02
1.60	4.56	5.72	6.89	8.06	9.22	10.39	11.56	12.73	13.89
1.80	5.48	6.89	8.29	9.70	11.10	12.51	13.91	15.32	16.72
2.00	6.39	8.02	9.66	11.29	12.93	14.56	16.20	17.83	19.47
2.20	7.26	9.12	10.97	12.83	14.69	16.55	18.40	20.26	22.12
2.40	8.10	10.17	12.24	14.31	16.38	18.45	20.52	22.59	24.66
2.60	8.91	11.18	13.45	15.73	18.00	20.28	22.55	24.83	27.10
2.80	9.69	12.15	14.62	17.09	19.56	22.02	24.49	26.96	29.43
3.00	10.4	13.1	15.7	18.4	21.0	23.7	26.3	29.0	31.6
3.20	11.2	14.0	16.8	19.6	22.5	25.3	28.1	30.9	33.8
3.40	11.9	14.9	17.8	20.8	23.8	26.8	29.8	32.8	35.8
3.60	12.5	15.7	18.8	22.0	25.1	28.3	31.4	34.6	37.7
3.80	13.2	16.5	19.8	23.1	26.4	29.7	33.0	36.3	39.6
4.00	13.8	17.3	20.7	24.1	27.6	31.0	34.5	37.9	41.3
4.20	14.4	18.0	21.6	25.2	28.7	32.3	35.9	39.5	43.0
4.40	15.1	18.7	22.4	26.1	29.8	33.5	37.2	40.9	44.6
4.60	15.6	19.5	23.3	27.1	30.9	34.7	38.6	42.4	46.2
4.80	16.2	20.1	24.1	28.0	31.9	35.9	39.8	43.7	47.7
5.00	16.8	20.8	24.8	28.9	32.9	37.0	41.0	45.1	49.1
5.20	17.3	21.4	25.6	29.7	33.9	38.0	42.2	46.3	50.5
5.40	17.8	22.1	26.3	30.6	34.8	39.0	43.3	47.5	51.0

TABLE A.46

$f_h/U : g$ Relationships when $z = 94$

f_h/U	Values of g when j of cooling curve is:								
	0.40	0.60	0.80	1.00	1.20	1.40	1.60	1.80	2.00
0.20	1.69-04	2.14-04	2.60-04	3.05-04	3.50-04	3.95-04	4.40-04	4.85-04	5.30-04
0.30	7.88-03	9.97-03	1.21-02	1.41-02	1.62-02	1.83-02	2.04-02	2.25-02	2.46-02
0.40	5.39-02	6.81-02	8.23-02	9.65-02	1.11-01	1.25-01	1.39-01	1.53-01	1.67-01
0.50	0.171	0.216	0.261	0.306	0.351	0.396	0.441	0.486	0.531
0.60	0.371	0.469	0.566	0.664	0.762	0.859	0.957	1.055	1.152
0.70	0.648	0.818	0.988	1.158	1.329	1.499	1.669	1.840	2.010
0.80	0.988	1.247	1.506	1.766	2.025	2.284	2.544	2.803	3.063
0.82	1.06	1.34	1.62	1.90	2.18	2.46	2.73	3.01	3.29
0.84	1.14	1.44	1.74	2.03	2.33	2.63	2.93	3.23	3.53
0.86	1.22	1.53	1.85	2.17	2.49	2.81	3.13	3.45	3.77
0.88	1.30	1.64	1.98	2.31	2.65	2.99	3.33	3.67	4.01
0.90	1.38	1.74	2.10	2.46	2.82	3.18	3.54	3.90	4.26
0.92	1.46	1.84	2.22	2.61	2.99	3.37	3.75	4.14	4.52
0.94	1.54	1.95	2.35	2.76	3.16	3.56	3.97	4.37	4.78
0.96	1.63	2.05	2.48	2.91	3.33	3.76	4.19	4.61	5.04
0.98	1.71	2.16	2.61	3.06	3.51	3.96	4.41	4.86	5.31
1.00	1.80	2.27	2.74	3.22	3.69	4.16	4.63	5.10	5.58
1.20	2.71	3.42	4.14	4.85	5.56	6.27	6.98	7.69	8.40
1.40	3.67	4.63	5.59	6.55	7.51	8.47	9.43	10.39	11.35
1.60	4.62	5.83	7.04	8.25	9.46	10.67	11.88	13.09	14.30
1.80	5.56	7.02	8.48	9.93	11.39	12.84	14.30	15.76	17.21
2.00	6.48	8.17	9.87	11.56	13.26	14.95	16.65	18.34	20.04
2.20	7.36	9.29	11.21	13.14	15.06	16.98	18.91	20.83	22.75
2.40	8.22	10.36	12.50	14.65	16.79	18.93	21.08	23.22	25.36
2.60	9.04	11.39	13.74	16.10	18.45	20.80	23.15	25.51	27.86
2.80	9.83	12.38	14.93	17.48	20.04	22.59	25.14	27.69	30.24
3.00	10.6	13.3	16.1	18.8	21.6	24.3	27.0	29.8	32.5
3.20	11.3	14.2	17.2	20.1	23.0	25.9	28.8	31.8	34.7
3.40	12.0	15.1	18.2	21.3	24.4	27.5	30.6	33.7	36.8
3.60	12.7	16.0	19.2	22.5	25.7	29.0	32.2	35.5	38.7
3.80	13.4	16.8	20.2	23.6	27.0	30.4	33.8	37.2	40.6
4.00	14.0	17.6	21.1	24.7	28.2	31.8	35.3	38.9	42.4
4.20	14.7	18.4	22.0	25.7	29.4	33.1	36.8	40.4	44.1
4.40	15.3	19.1	22.9	26.7	30.5	34.3	38.1	42.0	45.8
4.60	15.9	19.8	23.7	27.7	31.6	35.5	39.5	43.4	47.3
4.80	16.5	20.5	24.6	28.6	32.7	36.7	40.7	44.8	48.8
5.00	17.0	21.2	25.3	29.5	33.7	37.8	42.0	46.1	50.3
5.20	17.6	21.8	26.1	30.4	34.6	38.9	43.1	47.4	51.7

TABLE A.47

$f_h/U : g$ *Relationships when* $z = 96$

f_h/U	Values of g when j of cooling curve is:								
	0.40	0.60	0.80	1.00	1.20	1.40	1.60	1.80	2.00
0.20	1.72-04	2.19-04	2.65-04	3.12-04	3.58-04	4.05-04	4.52-04	4.98-04	5.45-04
0.30	8.01-03	1.02-02	1.23-02	1.45-02	1.66-02	1.88-02	2.10-02	2.31-02	2.53-02
0.40	5.46-02	6.93-02	8.41-02	9.88-02	1.14-01	1.28-01	1.43-01	1.58-01	1.73-01
0.50	0.173	0.220	0.267	0.314	0.360	0.407	0.454	0.501	0.548
0.60	0.376	0.477	0.579	0.680	0.781	0.883	0.984	1.086	1.187
0.70	0.656	0.833	1.010	1.187	1.363	1.540	1.717	1.894	2.070
0.80	1.00	1.27	1.54	1.81	2.08	2.35	2.62	2.89	3.15
0.82	1.08	1.37	1.66	1.94	2.23	2.52	2.81	3.10	3.39
0.84	1.15	1.46	1.77	2.08	2.39	2.70	3.01	3.32	3.63
0.86	1.23	1.56	1.90	2.23	2.56	2.89	3.22	3.55	3.88
0.88	1.31	1.67	2.02	2.37	2.72	3.08	3.43	3.78	4.13
0.90	1.40	1.77	2.14	2.52	2.89	3.27	3.64	4.02	4.39
0.92	1.48	1.88	2.27	2.67	3.07	3.46	3.86	4.26	4.65
0.94	1.56	1.98	2.40	2.82	3.24	3.66	4.08	4.50	4.92
0.96	1.65	2.09	2.54	2.98	3.42	3.86	4.31	4.75	5.19
0.98	1.74	2.20	2.67	3.14	3.60	4.07	4.53	5.00	5.46
1.00	1.83	2.32	2.81	3.29	3.78	4.27	4.76	5.25	5.74
1.20	2.75	3.49	4.23	4.96	5.70	6.44	7.17	7.91	8.65
1.40	3.72	4.71	5.71	6.70	7.70	8.69	9.69	10.68	11.68
1.60	4.69	5.94	7.20	8.45	9.70	10.96	12.21	13.46	14.72
1.80	5.64	7.15	8.66	10.17	11.67	13.18	14.69	16.20	17.71
2.00	6.57	8.32	10.08	11.83	13.59	15.34	17.09	18.85	20.60
2.20	7.47	9.46	11.45	13.44	15.43	17.42	19.41	21.40	23.39
2.40	8.33	10.55	12.77	14.98	17.20	19.42	21.63	23.85	26.06
2.60	9.17	11.60	14.03	16.46	18.89	21.32	23.76	26.19	28.62
2.80	9.97	12.61	15.24	17.88	20.51	23.15	25.78	28.42	31.06
3.00	10.7	13.6	16.4	19.2	22.1	24.9	27.7	30.6	33.4
3.20	11.5	14.5	17.5	20.5	23.5	26.6	29.6	32.6	35.6
3.40	12.2	15.4	18.6	21.8	25.0	28.1	31.3	34.5	37.7
3.60	12.9	16.3	19.6	23.0	26.3	29.7	33.0	36.4	39.7
3.80	13.6	17.1	20.6	24.1	27.6	31.1	34.6	38.1	41.6
4.00	14.3	17.9	21.6	25.2	28.9	32.5	36.2	39.8	43.5
4.20	14.9	18.7	22.5	26.3	30.1	33.8	37.6	41.4	45.2
4.40	15.5	19.4	23.4	27.3	31.2	35.1	39.0	43.0	46.9
4.60	16.1	20.2	24.2	28.3	32.3	36.3	40.4	44.4	48.5
4.80	16.7	20.9	25.0	29.2	33.4	37.5	41.7	45.8	50.0
5.00	17.3	21.6	25.8	30.1	34.4	38.6	42.9	47.2	51.4
5.20	17.9	22.2	26.6	31.0	35.4	39.7	44.1	48.5	52.8

TABLE A.48

$f_h/U : g$ Relationships when $z = 98$

f_h/U	Values of g when j of cooling curve is:								
	0.40	0.60	0.80	1.00	1.20	1.40	1.60	1.80	2.00
0.20	1.75-04	2.23-04	2.71-04	3.19-04	3.67-04	4.15-04	4.63-04	5.11-04	5.59-04
0.30	8.12-03	1.04-02	1.26-02	1.48-02	1.71-02	1.93-02	2.15-02	2.38-02	2.60-02
0.40	5.54-02	7.07-02	8.59-02	1.01-01	1.17-01	1.32-01	1.47-01	1.62-01	1.78-01
0.50	0.176	0.224	0.273	0.321	0.370	0.418	0.467	0.515	0.564
0.60	0.381	0.486	0.591	0.697	0.802	0.907	1.012	1.117	1.222
0.70	0.666	0.849	1.032	1.215	1.398	1.582	1.765	1.948	2.131
0.80	1.01	1.29	1.57	1.85	2.13	2.41	2.69	2.97	3.25
0.82	1.09	1.39	1.69	1.99	2.29	2.59	2.89	3.19	3.49
0.84	1.17	1.49	1.81	2.13	2.45	2.78	3.10	3.42	3.74
0.86	1.25	1.59	1.94	2.28	2.62	2.97	3.31	3.65	4.00
0.88	1.33	1.70	2.06	2.43	2.79	3.16	3.52	3.89	4.26
0.90	1.41	1.80	2.19	2.58	2.97	3.36	3.74	4.13	4.52
0.92	1.50	1.91	2.32	2.73	3.14	3.56	3.97	4.38	4.79
0.94	1.58	2.02	2.45	2.89	3.32	3.76	4.19	4.63	5.06
0.96	1.67	2.13	2.59	3.05	3.51	3.97	4.42	4.88	5.34
0.98	1.76	2.24	2.73	3.21	3.69	4.18	4.66	5.14	5.62
1.00	1.85	2.36	2.86	3.37	3.88	4.39	4.89	5.40	5.91
1.20	2.79	3.55	4.32	5.08	5.84	6.61	7.37	8.13	8.90
1.40	3.77	4.80	5.83	6.86	7.89	8.92	9.95	10.98	12.01
1.60	4.75	6.05	7.35	8.65	9.94	11.24	12.54	13.84	15.14
1.80	5.72	7.28	8.84	10.40	11.96	13.52	15.08	16.64	18.20
2.00	6.66	8.47	10.29	12.10	13.92	15.73	17.54	19.36	21.17
2.20	7.57	9.63	11.69	13.74	15.80	17.86	19.92	21.97	24.03
2.40	8.45	10.74	13.03	15.32	17.61	19.90	22.19	24.48	26.77
2.60	9.30	11.81	14.32	16.83	19.34	21.85	24.36	26.87	29.38
2.80	10.1	12.8	15.6	18.3	21.0	23.7	26.4	29.2	31.9
3.00	10.9	13.8	16.7	19.7	22.6	25.5	28.4	31.3	34.2
3.20	11.7	14.8	17.9	21.0	24.1	27.2	30.3	33.4	36.5
3.40	12.4	15.7	19.0	22.2	25.5	28.8	32.1	35.4	38.7
3.60	13.1	16.6	20.0	23.5	26.9	30.4	33.8	37.3	40.7
3.80	13.8	17.4	21.0	24.6	28.2	31.8	35.4	39.0	42.6
4.00	14.5	18.2	22.0	25.7	29.5	33.2	37.0	40.8	44.5
4.20	15.1	19.0	22.9	26.8	30.7	34.6	38.5	42.4	46.3
4.40	15.8	19.8	23.8	27.8	31.9	35.9	39.9	43.9	48.0
4.60	16.4	20.5	24.7	28.8	33.0	37.1	41.3	45.4	49.6
4.80	17.0	21.3	25.5	29.8	34.1	38.3	42.6	46.9	51.1
5.00	17.6	22.0	26.3	30.7	35.1	39.5	43.8	48.2	52.6

TABLE A.49

$f_h/U:g$ *Relationships when* $z = 100$

	Values of g when j of cooling curve is:								
f_h/U	0.40	0.60	0.80	1.00	1.20	1.40	1.60	1.80	2.00
0.20	1.79-04	2.29-04	2.78-04	3.28-04	3.78-04	4.27-04	4.77-04	5.26-04	5.76-04
0.30	8.15-03	1.05-02	1.28-02	1.51-02	1.75-02	1.98-02	2.21-02	2.45-02	2.68-02
0.40	5.55-02	7.14-02	8.74-02	1.03-01	1.19-01	1.35-01	1.51-01	1.67-01	1.83-01
0.50	0.177	0.228	0.278	0.329	0.379	0.429	0.480	0.530	0.580
0.60	0.388	0.497	0.605	0.714	0.822	0.931	1.040	1.148	1.257
0.70	0.681	0.870	1.059	1.247	1.436	1.625	1.813	2.002	2.191
0.80	1.04	1.33	1.62	1.90	2.19	2.48	2.76	3.05	3.34
0.82	1.12	1.43	1.74	2.05	2.35	2.66	2.97	3.28	3.59
0.84	1.20	1.53	1.86	2.19	2.52	2.85	3.18	3.51	3.84
0.86	1.29	1.64	1.99	2.34	2.69	3.05	3.40	3.75	4.10
0.88	1.37	1.74	2.12	2.50	2.87	3.25	3.62	4.00	4.37
0.90	1.46	1.85	2.25	2.65	3.05	3.45	3.85	4.25	4.64
0.92	1.54	1.96	2.39	2.81	3.23	3.65	4.08	4.50	4.92
0.94	1.63	2.08	2.52	2.97	3.42	3.86	4.31	4.76	5.20
0.96	1.72	2.19	2.66	3.13	3.60	4.08	4.55	5.02	5.49
0.98	1.81	2.31	2.80	3.30	3.79	4.29	4.79	5.28	5.78
1.00	1.90	2.42	2.94	3.47	3.99	4.51	5.03	5.55	6.07
1.20	2.85	3.64	4.43	5.21	6.00	6.78	7.57	8.36	9.14
1.40	3.84	4.90	5.96	7.03	8.09	9.15	10.22	11.28	12.34
1.60	4.82	6.16	7.50	6.84	10.19	11.53	12.87	14.21	15.56
1.80	5.78	7.39	9.01	10.63	12.24	13.86	15.47	17.09	18.70
2.00	6.71	8.59	10.47	12.35	14.23	16.11	17.99	19.87	21.75
2.20	7.62	9.76	11.89	14.02	16.15	18.29	20.42	22.55	24.68
2.40	8.50	10.88	13.25	15.62	17.99	20.37	22.74	25.11	27.49
2.60	9.35	11.95	14.55	17.16	19.76	22.36	24.96	27.56	30.16
2.80	10.2	13.0	15.8	18.6	21.4	24.3	27.1	29.9	32.7
3.00	11.0	14.0	17.0	20.0	23.0	26.1	29.1	32.1	35.1
3.20	11.8	15.0	18.2	21.4	24.6	27.8	31.0	34.2	37.4
3.40	12.5	15.9	19.3	22.7	26.1	29.4	32.8	36.2	39.6
3.60	13.2	16.8	20.4	23.9	27.5	31.0	34.6	38.1	41.7
3.80	14.0	17.7	21.4	25.1	28.8	32.5	36.2	40.0	43.7
4.00	14.7	18.5	22.4	26.2	30.1	34.0	37.8	41.7	45.5
4.20	15.3	19.3	23.3	27.3	31.3	35.3	39.3	43.3	47.3
4.40	16.0	20.1	24.3	28.4	32.5	36.7	40.8	44.9	49.1
4.60	16.6	20.9	25.2	29.4	33.7	37.9	42.2	46.4	50.7
4.80	17.3	21.6	26.0	30.4	34.8	39.1	43.5	47.9	52.2

TABLE A.50

$f_h/U : g$ *Relationships when* $z = 110$

	Values of g when j of cooling curve is:								
f_h/U	0.40	0.60	0.80	1.00	1.20	1.40	1.60	1.80	2.00
0.20	1.88-04	2.47-04	3.06-04	3.65-04	4.24-04	4.83-04	5.42-04	6.01-04	6.60-04
0.30	8.72-03	1.15-02	1.42-02	1.69-02	1.97-02	2.24-02	2.52-02	2.79-02	3.07-02
0.40	5.93-02	7.81-02	9.69-02	1.16-01	1.34-01	1.53-01	1.72-01	1.91-01	2.09-01
0.50	0.189	0.248	0.308	0.367	0.427	0.486	0.545	0.605	0.664
0.60	0.410	0.539	0.667	0.796	0.924	1.053	1.182	1.310	1.439
0.70	0.716	0.940	1.164	1.388	1.613	1.837	2.061	2.285	2.509
0.80	1.09	1.43	1.77	2.12	2.46	2.80	3.14	3.48	3.82
0.82	1.18	1.54	1.91	2.27	2.64	3.01	3.37	3.74	4.11
0.84	1.26	1.65	2.04	2.44	2.83	3.22	3.61	4.01	4.40
0.86	1.35	1.76	2.18	2.60	3.02	3.44	3.86	4.28	4.70
0.88	1.43	1.88	2.33	2.77	3.22	3.67	4.11	4.56	5.00
0.90	1.52	2.00	2.47	2.95	3.42	3.89	4.37	4.84	5.32
0.92	1.61	2.12	2.62	3.12	3.62	4.13	4.63	5.13	5.63
0.94	1.71	2.24	2.77	3.30	3.83	4.36	4.89	5.42	5.95
0.96	1.80	2.36	2.92	3.48	4.04	4.60	5.16	5.72	6.28
0.98	1.90	2.48	3.07	3.66	4.25	4.84	5.43	6.02	6.61
1.00	1.99	2.61	3.23	3.85	4.47	5.09	5.71	6.32	6.94
1.20	3.00	3.93	4.86	5.79	6.72	7.65	8.58	9.51	10.44
1.40	4.05	5.30	6.56	7.81	9.06	10.31	11.56	12.82	14.07
1.60	5.11	6.68	8.25	9.83	11.40	12.97	14.55	16.12	17.69
1.80	6.15	8.03	9.92	11.80	13.69	15.57	17.46	19.34	21.23
2.00	7.17	9.35	11.53	13.72	15.90	18.09	20.27	22.46	24.64
2.20	8.15	10.62	13.09	15.56	18.03	20.50	22.97	25.44	27.90
2.40	9.10	11.84	14.58	17.32	20.06	22.80	25.53	28.27	31.01
2.60	10.0	13.0	16.0	19.0	22.0	25.0	28.0	31.0	34.0
2.80	10.9	14.1	17.4	20.6	23.8	27.1	30.3	33.5	36.8
3.00	11.8	15.2	18.7	22.1	25.6	29.0	32.5	35.9	39.4
3.10	12.2	15.8	19.3	22.9	26.4	30.0	33.5	37.1	40.7
3.20	12.6	16.3	19.9	23.6	27.3	30.9	34.6	38.2	41.9
3.30	13.0	16.8	20.5	24.3	28.1	31.8	35.6	39.3	43.1
3.40	13.4	17.3	21.1	25.0	28.8	32.7	36.5	40.4	44.3
3.50	13.8	17.8	21.7	25.7	29.6	33.5	37.5	41.4	45.4
3.60	14.2	18.2	22.3	26.3	30.3	34.4	38.4	42.5	46.5
3.70	14.6	18.7	22.8	27.0	31.1	35.2	39.3	43.4	47.6
3.80	15.0	19.2	23.4	27.6	31.8	36.0	40.2	44.4	48.6
3.90	15.4	19.6	23.9	28.2	32.5	36.8	41.0	45.3	49.6
4.00	15.7	20.1	24.4	28.8	33.2	37.5	41.9	46.2	50.6
4.10	16.1	20.5	25.0	29.4	33.8	38.3	42.7	47.1	51.6
4.20	16.5	21.0	25.5	30.0	34.5	39.0	43.5	48.0	52.5

TABLE A.51

$f_h/U:g$ *Relationships when* $z = 120$

f_h/U	Values of g when j of cooling curve is:								
	0.40	0.60	0.80	1.00	1.20	1.40	1.60	1.80	2.00
0.20	1.97-04	2.66-04	3.35-04	4.04-04	4.73-04	5.41-04	6.10-04	6.79-04	7.48-04
0.30	9.16-03	1.24-02	1.55-02	1.87-02	2.19-02	2.51-02	2.83-02	3.15-02	3.47-02
0.40	6.24-02	8.42-02	1.06-01	1.28-01	1.50-01	1.72-01	1.93-01	2.15-01	2.37-01
0.50	0.198	0.267	0.337	0.406	0.475	0.544	0.614	0.683	0.752
0.60	0.431	0.580	0.730	0.880	1.029	1.179	1.329	1.479	1.628
0.70	0.752	1.013	1.274	1.534	1.795	2.056	2.317	2.577	2.838
0.80	1.15	1.54	1.94	2.34	2.73	3.13	3.53	3.92	4.32
0.82	1.23	1.66	2.09	2.51	2.94	3.36	3.79	4.22	4.64
0.84	1.32	1.78	2.23	2.69	3.15	3.60	4.06	4.52	4.97
0.86	1.41	1.90	2.39	2.87	3.36	3.85	4.34	4.82	5.31
0.88	1.50	2.02	2.54	3.06	3.58	4.10	4.62	5.14	5.66
0.90	1.60	2.15	2.70	3.25	3.80	4.35	4.90	5.46	6.01
0.92	1.69	2.28	2.86	3.45	4.03	4.61	5.20	5.78	6.36
0.94	1.79	2.41	3.03	3.64	4.26	4.88	5.49	6.11	6.73
0.96	1.89	2.54	3.19	3.84	4.49	5.14	5.79	6.44	7.09
0.98	1.99	2.68	3.36	4.04	4.73	5.41	6.10	6.78	7.46
1.00	2.09	2.81	3.53	4.25	4.97	5.69	6.40	7.12	7.84
1.20	3.16	4.23	5.31	6.39	7.46	8.54	9.62	10.69	11.77
1.40	4.27	5.71	7.16	8.60	10.05	11.49	12.94	14.39	15.83
1.60	5.38	7.19	9.00	10.81	12.63	14.44	16.25	18.06	19.87
1.80	6.48	8.64	10.81	12.97	15.14	17.30	19.47	21.63	23.80
2.00	7.55	10.06	12.56	15.06	17.56	20.06	22.56	25.06	27.56
2.20	8.60	11.42	14.23	17.05	19.87	22.69	25.51	28.33	31.14
2.40	9.61	12.73	15.84	18.96	22.07	25.19	28.30	31.42	34.53
2.60	10.6	14.0	17.4	20.8	24.2	27.6	30.9	34.3	37.7
2.80	11.5	15.2	18.8	22.5	26.1	29.8	33.4	37.1	40.7
3.00	12.5	16.4	20.2	24.1	28.0	31.9	35.8	39.7	43.6
3.10	12.9	16.9	20.9	24.9	28.9	32.9	36.9	40.9	44.9
3.20	13.4	17.5	21.6	25.7	29.8	33.9	38.0	42.1	46.2
3.30	13.8	18.0	22.2	26.4	30.7	34.9	39.1	43.3	47.5
3.40	14.2	18.6	22.9	27.2	31.5	35.8	40.1	44.4	48.7
3.50	14.7	19.1	23.5	27.9	32.3	36.7	41.1	45.5	49.9
3.60	15.1	19.6	24.1	28.6	33.1	37.6	42.1	46.6	51.1
3.70	15.5	20.1	24.7	29.3	33.9	38.4	43.0	47.6	52.2

TABLE A.52

$f_h/U : g$ *Relationships when* $z = 130$

	Values of g when j of cooling curve is:								
f_h/U	0.40	0.60	0.80	1.00	1.20	1.40	1.60	1.80	2.00
0.20	2.05-04	2.84-04	3.63-04	4.43-04	5.22-04	6.01-04	6.80-04	7.59-04	8.38-04
0.30	9.54-03	1.32-02	1.69-02	2.06-02	2.42-02	2.79-02	3.16-02	3.53-02	3.89-02
0.40	6.50-02	9.01-02	1.15-01	1.40-01	1.65-01	1.90-01	2.16-01	2.41-01	2.66-01
0.50	0.207	0.286	0.366	0.445	0.525	0.604	0.684	0.763	0.843
0.60	0.448	0.620	0.792	0.964	1.136	1.309	1.481	1.653	1.825
0.70	0.783	1.083	1.382	1.681	1.981	2.280	2.580	2.879	3.179
0.80	1.20	1.65	2.11	2.56	3.02	3.47	3.93	4.38	4.83
0.82	1.29	1.77	2.26	2.75	3.24	3.73	4.22	4.71	5.20
0.84	1.38	1.90	2.42	2.95	3.47	3.99	4.52	5.04	5.57
0.86	1.47	2.03	2.59	3.15	3.71	4.27	4.82	5.38	5.94
0.88	1.57	2.16	2.76	3.35	3.95	4.54	5.14	5.73	6.33
0.90	1.67	2.30	2.93	3.56	4.19	4.82	5.46	6.09	6.72
0.92	1.77	2.44	3.10	3.77	4.44	5.11	5.78	6.45	7.12
0.94	1.87	2.58	3.28	3.99	4.69	5.40	6.11	6.81	7.52
0.96	1.97	2.72	3.46	4.21	4.95	5.70	6.44	7.19	7.93
0.98	2.08	2.86	3.64	4.43	5.21	5.99	6.78	7.56	8.34
1.00	2.18	3.01	3.83	4.65	5.47	6.30	7.12	7.94	8.76
1.10	2.73	3.75	4.78	5.80	6.82	7.85	8.87	9.89	10.92
1.20	3.29	4.52	5.75	6.98	8.21	9.44	10.67	11.90	13.13
1.30	3.87	5.31	6.75	8.19	9.62	11.06	12.50	13.94	15.38
1.40	4.46	6.10	7.75	9.40	11.04	12.69	14.33	15.98	17.63
1.50	5.04	6.90	8.75	10.60	12.45	14.31	16.16	18.01	19.87
1.60	5.63	7.68	9.74	11.79	13.85	15.91	17.96	20.02	22.08
1.70	6.21	8.46	10.72	12.97	15.23	17.48	19.74	21.99	24.25
1.80	6.78	9.23	11.68	14.13	16.58	19.03	21.48	23.93	26.38
1.90	7.35	9.99	12.63	15.26	17.90	20.54	23.18	25.82	28.45
2.00	7.91	10.73	13.55	16.37	19.20	22.02	24.84	27.66	30.48
2.10	8.47	11.46	14.46	17.46	20.46	23.45	26.45	29.45	32.45
2.20	9.02	12.18	15.35	18.52	21.68	24.85	28.02	31.19	34.35
2.30	9.56	12.89	16.22	19.55	22.88	26.21	29.54	32.87	36.21
2.40	10.1	13.6	17.1	20.6	24.0	27.5	31.0	34.5	38.0
2.50	10.6	14.3	17.9	21.5	25.2	28.8	32.5	36.1	39.7
2.60	11.1	14.9	18.7	22.5	26.3	30.1	33.8	37.6	41.4
2.70	11.6	15.6	19.5	23.4	27.3	31.3	35.2	39.1	43.0
2.80	12.2	16.2	20.3	24.3	28.4	32.4	36.5	40.5	44.6
2.90	12.7	16.8	21.0	25.2	29.4	33.6	37.8	41.9	46.1
3.00	13.1	17.5	21.8	26.1	30.4	34.7	39.0	43.3	47.6
3.10	13.6	18.1	22.5	26.9	31.3	35.7	40.2	44.6	49.0
3.20	14.1	18.7	23.2	27.7	32.2	36.8	41.3	45.8	50.4
3.30	14.6	19.2	23.9	28.5	33.1	37.8	42.4	47.0	51.7

TABLE A.53

f_h/U : g Relationships when $z = 140$

	Values of g when j of cooling curve is:								
f_h/U	0.40	0.60	0.80	1.00	1.20	1.40	1.60	1.80	2.00
0.20	2.12-04	3.02-04	3.92-04	4.82-04	5.72-04	6.61-04	7.51-04	8.41-04	9.31-04
0.30	7.86-03	1.40-02	1.82-02	2.24-02	2.65-02	3.07-02	3.49-02	3.91-02	4.32-02
0.40	6.73-02	9.58-02	1.24-01	1.53-01	1.81-01	2.10-01	2.38-01	2.67-01	2.95-01
0.50	0.214	0.304	0.394	0.485	0.575	0.666	0.756	0.846	0.937
0.60	0.464	0.659	0.855	1.050	1.245	1.441	1.636	1.831	2.027
0.70	0.811	1.150	1.490	1.830	2.170	2.510	2.849	3.189	3.529
0.80	1.24	1.75	2.27	2.79	3.30	3.82	4.33	4.85	5.36
0.82	1.33	1.89	2.44	2.99	3.55	4.10	4.66	5.21	5.76
0.84	1.43	2.02	2.61	3.21	3.80	4.39	4.99	5.58	6.17
0.86	1.52	2.16	2.79	3.42	4.06	4.69	5.32	5.96	6.59
0.88	1.62	2.30	2.97	3.65	4.32	4.99	5.67	6.34	7.02
0.90	1.73	2.44	3.16	3.87	4.59	5.30	6.02	6.73	7.45
0.92	1.83	2.59	3.35	4.10	4.86	5.62	6.37	7.13	7.89
0.94	1.94	2.74	3.54	4.34	5.14	5.94	6.74	7.53	8.33
0.96	2.04	2.89	3.73	4.57	5.42	6.26	7.10	7.94	8.79
0.98	2.15	3.04	3.93	4.81	5.70	6.58	7.47	8.36	9.24
1.00	2.26	3.19	4.12	5.05	5.98	6.91	7.85	8.78	9.71
1.10	2.83	3.99	5.14	6.30	7.45	8.61	9.77	10.92	12.08
1.20	3.42	4.80	6.19	7.58	8.97	10.35	11.74	13.13	14.51
1.30	4.02	5.64	7.26	8.88	10.50	12.12	13.74	15.36	16.98
1.40	4.63	6.48	8.33	10.18	12.04	13.89	15.74	17.59	19.44
1.50	5.24	7.32	9.40	11.48	13.56	15.64	17.72	19.80	21.89
1.60	5.85	8.15	10.46	12.76	15.07	17.37	19.68	21.99	24.29
1.70	6.46	8.98	11.50	14.03	16.55	19.08	21.60	24.12	26.65
1.80	7.06	9.80	12.53	15.27	18.00	20.74	23.48	26.21	28.95
1.90	7.66	10.60	13.54	16.48	19.42	22.36	25.30	28.25	31.19
2.00	8.25	11.39	14.53	17.67	20.80	23.94	27.08	30.22	33.36
2.10	8.83	12.16	15.49	18.82	22.15	25.48	28.81	32.14	35.47
2.20	9.41	12.92	16.44	19.95	23.46	26.97	30.48	33.99	37.50
2.30	9.98	13.67	17.36	21.04	24.73	28.41	32.10	35.79	39.47
2.40	10.6	14.4	18.3	22.1	26.0	29.8	33.7	37.5	41.4
2.50	11.1	15.1	19.1	23.1	27.2	31.2	35.2	39.2	43.2
2.60	11.7	15.8	20.0	24.2	28.3	32.5	36.6	40.8	45.0
2.70	12.2	16.5	20.8	25.1	29.4	33.7	38.0	42.4	46.7
2.80	12.8	17.2	21.6	26.1	30.5	35.0	39.4	43.9	48.3
2.90	13.3	17.9	22.4	27.0	31.6	36.2	40.7	45.3	49.9
3.00	13.8	18.5	23.2	27.9	32.6	37.3	42.0	46.7	51.4

TABLE A.54

$f_h/U : g$ *Relationships when* $z = 150$

Values of g when j of cooling curve is:

f_h/U	0.40	0.60	0.80	1.00	1.20	1.40	1.60	1.80	2.00
0.20	2.18-04	3.19-04	4.20-04	5.22-04	6.23-04	7.24-04	8.25-04	9.27-04	1.03-03
0.30	1.01-02	1.48-02	1.95-02	2.42-02	2.89-02	3.37-02	3.84-02	4.31-02	4.78-02
0.40	6.93-02	1.01-01	1.33-01	1.65-01	1.97-01	2.29-01	2.61-01	2.93-01	3.25-01
0.50	0.220	0.322	0.423	0.525	0.626	0.728	0.829	0.931	1.032
0.60	0.477	0.697	0.916	1.136	1.356	1.575	1.795	2.014	2.234
0.70	0.834	1.216	1.597	1.979	2.361	2.742	3.124	3.505	3.887
0.80	1.27	1.85	2.43	3.01	3.59	4.17	4.75	5.33	5.90
0.82	1.37	1.99	2.61	3.24	3.86	4.48	5.10	5.72	6.34
0.84	1.47	2.13	2.80	3.47	4.13	4.80	5.46	6.13	6.79
0.86	1.57	2.28	2.99	3.70	4.41	5.12	5.83	6.54	7.25
0.88	1.67	2.43	3.18	3.94	4.70	5.45	6.21	6.96	7.72
0.90	1.78	2.58	3.38	4.18	4.99	5.79	6.59	7.39	8.19
0.92	1.89	2.73	3.58	4.43	5.28	6.13	6.98	7.83	8.67
0.94	1.99	2.89	3.79	4.68	5.58	6.48	7.37	8.27	9.16
0.96	2.11	3.05	3.99	4.94	5.88	6.83	7.77	8.71	9.66
0.98	2.22	3.21	4.20	5.20	6.19	7.18	8.17	9.17	10.16
1.00	2.33	3.37	4.41	5.46	6.50	7.54	8.58	9.62	10.67
1.10	2.92	4.21	5.50	6.79	8.09	9.38	10.67	11.97	13.26
1.20	3.53	5.07	6.62	8.17	9.72	11.27	12.82	14.37	15.91
1.30	4.15	5.96	7.76	9.57	11.37	13.18	14.98	16.79	18.59
1.40	4.78	6.84	8.90	10.96	13.02	15.08	17.14	19.20	21.26
1.50	5.42	7.73	10.04	12.35	14.66	16.97	19.28	21.59	23.90
1.60	6.05	8.61	11.16	13.72	16.27	18.83	21.39	23.94	26.50
1.70	6.69	9.48	12.27	15.06	17.86	20.65	23.44	26.24	29.03
1.80	7.32	10.34	13.36	16.38	19.40	22.43	25.45	28.47	31.49
1.90	7.94	11.18	14.43	17.67	20.91	24.15	27.39	30.64	33.88
2.00	8.56	12.02	15.47	18.92	22.38	25.83	29.28	32.73	36.19
2.10	9.18	12.83	16.49	20.14	23.80	27.45	31.11	34.76	38.41
2.20	9.79	13.64	17.48	21.33	25.18	29.02	32.87	36.72	40.56
2.30	10.4	14.4	18.5	22.5	26.5	30.5	34.6	38.6	42.6
2.40	11.0	15.2	19.4	23.6	27.8	32.0	36.2	40.4	44.6
2.50	11.6	16.0	20.3	24.7	29.1	33.4	37.8	42.2	46.5
2.60	12.2	16.7	21.2	25.7	30.3	34.8	39.3	43.8	48.3
2.70	12.8	17.4	22.1	26.8	31.4	36.1	40.8	45.4	50.1
2.80	13.4	18.2	23.0	27.8	32.6	37.4	42.2	47.0	51.8

TABLE A.55

$f_h/U:g$ Relationships when $z = 160$

	Values of g when j of cooling curve is:								
f_h/U	0.40	0.60	0.80	1.00	1.20	1.40	1.60	1.80	2.00
0.20	2.23-04	3.36-04	4.49-04	5.62-04	6.75-04	7.87-04	9.00-04	1.01-03	1.13-03
0.30	1.04-02	1.56-02	2.08-02	2.61-02	3.13-02	3.66-02	4.18-02	4.70-02	5.23-02
0.40	7.06-02	1.06-01	1.42-01	1.78-01	2.14-01	2.49-01	2.85-01	3.21-01	3.57-01
0.50	0.225	0.338	0.451	0.564	0.678	0.791	0.904	1.017	1.131
0.60	0.488	0.733	0.977	1.222	1.466	1.711	1.955	2.200	2.445
0.70	0.853	1.278	1.703	2.128	2.553	2.977	3.402	3.827	4.252
0.80	1.30	1.95	2.59	3.24	3.88	4.52	5.17	5.81	6.45
0.82	1.40	2.09	2.79	3.48	4.17	4.86	5.55	6.24	6.93
0.84	1.50	2.24	2.98	3.72	4.46	5.20	5.94	6.68	7.42
0.86	1.61	2.40	3.19	3.98	4.76	5.55	6.34	7.13	7.92
0.88	1.71	2.55	3.39	4.23	5.07	5.91	6.75	7.59	8.43
0.90	1.82	2.71	3.60	4.49	5.38	6.27	7.17	8.06	8.95
0.92	1.93	2.88	3.82	4.76	5.70	6.64	7.59	8.53	9.47
0.94	2.05	3.04	4.03	5.03	6.02	7.02	8.01	9.01	10.00
0.96	2.16	3.21	4.25	5.30	6.35	7.40	8.45	9.49	10.54
0.98	2.27	3.38	4.48	5.58	6.68	7.78	8.88	9.98	11.09
1.00	2.39	3.55	4.70	5.86	7.01	8.17	9.32	10.48	11.64
1.10	3.00	4.43	5.86	7.29	8.72	10.15	11.58	13.01	14.45
1.20	3.62	5.34	7.05	8.76	10.47	12.18	13.90	15.61	17.32
1.30	4.27	6.26	8.25	10.25	12.24	14.23	16.22	18.22	20.21
1.40	4.92	7.19	9.46	11.73	14.00	16.27	18.54	20.81	23.08
1.50	5.58	8.12	10.66	13.20	15.75	18.29	20.83	23.37	25.91
1.60	6.24	9.04	11.85	14.65	17.46	20.26	23.07	25.88	28.68
1.70	6.90	9.96	13.02	16.08	19.14	22.20	25.26	28.31	31.37
1.80	7.56	10.86	14.16	17.47	20.77	24.07	27.38	30.68	33.98
1.90	8.21	11.75	15.29	18.82	22.36	25.90	29.43	32.97	36.51
2.00	8.87	12.62	16.38	20.14	23.90	27.66	31.42	35.18	38.93
2.10	9.51	13.46	17.45	21.42	25.39	29.36	33.33	37.30	41.27
2.20	10.2	14.3	18.5	22.7	26.8	31.0	35.2	39.3	43.5
2.30	10.8	15.2	19.5	23.9	28.2	32.6	36.9	41.3	45.7
2.40	11.4	16.0	20.5	25.0	29.6	34.1	38.6	43.2	47.7
2.50	12.1	16.8	21.5	26.2	30.9	35.6	40.3	45.0	49.7
2.60	12.7	17.6	22.4	27.3	32.1	37.0	41.8	46.7	51.5

TABLE A.56

$f_h/U : g$ *Relationships when* $z = 170$

f_h/U	Values of g when j of cooling curve is:								
	0.40	0.60	0.80	1.00	1.20	1.40	1.60	1.80	2.00
0.20	2.27-04	3.52-04	4.76-04	6.01-04	7.26-04	8.51-04	9.76-04	1.10-03	1.23-03
0.30	1.05-02	1.63-02	2.21-02	2.79-02	3.37-02	3.95-02	4.53-02	5.11-02	5.69-02
0.40	7.20-02	1.11-01	1.51-01	1.91-01	2.30-01	2.70-01	3.09-01	3.49-01	3.88-01
0.50	0.229	0.354	0.479	0.604	0.729	0.855	0.980	1.105	1.230
0.60	0.497	0.767	1.038	1.308	1.578	1.848	2.118	2.389	2.659
0.70	0.870	1.339	1.808	2.277	2.746	3.215	3.683	4.152	4.621
0.80	1.33	2.04	2.75	3.46	4.17	4.88	5.59	6.30	7.01
0.81	1.38	2.12	2.85	3.59	4.32	5.06	5.80	6.53	7.27
0.82	1.43	2.19	2.96	3.72	4.48	5.24	6.00	6.77	7.53
0.83	1.48	2.27	3.06	3.85	4.64	5.43	6.21	7.00	7.79
0.84	1.54	2.35	3.17	3.98	4.80	5.61	6.43	7.24	8.06
0.85	1.59	2.43	3.27	4.11	4.96	5.80	6.64	7.48	8.33
0.86	1.64	2.51	3.38	4.25	5.12	5.99	6.86	7.73	8.60
0.87	1.70	2.59	3.49	4.39	5.28	6.18	7.08	7.97	8.87
0.88	1.75	2.67	3.60	4.52	5.45	6.37	7.30	8.22	9.15
0.89	1.81	2.76	3.71	4.66	5.62	6.57	7.52	8.47	9.43
0.90	1.86	2.84	3.82	4.80	5.78	6.76	7.74	8.73	9.71
0.91	1.92	2.93	3.93	4.94	5.95	6.96	7.97	8.98	9.99
0.92	1.97	3.01	4.05	5.09	6.12	7.16	8.20	9.24	10.27
0.93	2.03	3.10	4.16	5.23	6.30	7.36	8.43	9.49	10.56
0.94	2.09	3.18	4.28	5.37	6.47	7.56	8.66	9.75	10.85
0.95	2.15	3.27	4.39	5.52	6.64	7.77	8.89	10.01	11.14
0.96	2.20	3.36	4.51	5.66	6.82	7.97	9.12	10.28	11.43
0.97	2.26	3.45	4.63	5.81	6.99	8.18	9.36	10.54	11.72
0.98	2.32	3.53	4.75	5.96	7.17	8.38	9.59	10.81	12.02
0.99	2.38	3.62	4.86	6.11	7.35	8.59	9.83	11.07	12.31
1.00	2.44	3.71	4.98	6.26	7.53	8.80	10.07	11.34	12.61
1.10	3.06	4.63	6.21	7.78	9.35	10.92	12.49	14.07	15.64
1.20	3.71	5.58	7.46	9.34	11.22	13.09	14.97	16.85	18.73
1.30	4.37	6.55	8.73	10.91	13.10	15.28	17.46	19.64	21.82
1.40	5.05	7.53	10.01	12.49	14.97	17.45	19.93	22.40	24.88
1.50	5.73	8.50	11.27	14.04	16.81	19.58	22.35	25.12	27.89
1.60	6.41	9.46	12.52	15.57	18.62	21.67	24.72	27.78	30.83
1.70	7.10	10.42	13.74	17.06	20.39	23.71	27.03	30.35	33.67
1.80	7.79	11.37	14.94	18.52	22.10	25.68	29.26	32.84	36.42
1.90	8.47	12.30	16.12	19.94	23.77	27.59	31.41	35.23	39.06
2.00	9.16	13.21	17.27	21.32	25.37	29.43	33.48	37.53	41.59
2.10	9.84	14.11	18.39	22.66	26.93	31.20	35.47	39.74	44.01
2.20	10.5	15.0	19.5	24.0	28.4	32.9	37.4	41.9	46.3
2.30	11.2	15.9	20.5	25.2	29.9	34.5	39.2	43.9	48.5
2.40	11.9	16.7	21.6	26.4	31.3	36.1	40.9	45.8	50.6

TABLE A.57

$f_h/U:g$ Relationships when $z = 180$

	Values of g when j of cooling curve is:								
f_h/U	0.40	0.60	0.80	1.00	1.20	1.40	1.60	1.80	2.00
0.20	2.30-04	3.67-04	5.04-04	6.41-04	7.78-04	9.15-04	1.05-03	1.19-03	1.33-03
0.30	1.07-02	1.71-02	2.34-02	2.98-02	3.62-02	4.25-02	4.89-02	5.52-02	6.16-02
0.40	7.28-02	1.16-01	1.60-01	2.03-01	2.47-01	2.90-01	3.33-01	3.77-01	4.20-01
0.50	0.231	0.369	0.506	0.644	0.781	0.919	1.056	1.194	1.331
0.60	0.503	0.800	1.096	1.393	1.690	1.986	2.283	2.579	2.876
0.70	0.882	1.396	1.910	2.424	2.938	3.453	3.967	4.481	4.995
0.80	1.35	2.13	2.91	3.68	4.46	5.24	6.01	6.79	7.57
0.81	1.40	2.21	3.01	3.82	4.62	5.43	6.24	7.04	7.85
0.82	1.45	2.29	3.12	3.96	4.79	5.62	6.46	7.29	8.13
0.83	1.51	2.37	3.23	4.10	4.96	5.82	6.68	7.55	8.41
0.84	1.56	2.45	3.34	4.24	5.13	6.02	6.91	7.81	8.70
0.85	1.61	2.54	3.46	4.38	5.30	6.22	7.14	8.06	8.99
0.86	1.67	2.62	3.57	4.52	5.47	6.42	7.38	8.33	9.28
0.87	1.72	2.70	3.69	4.67	5.65	6.63	7.61	8.59	9.57
0.88	1.78	2.79	3.80	4.81	5.82	6.84	7.85	8.86	9.87
0.89	1.83	2.88	3.92	4.96	6.00	7.04	8.09	9.13	10.17
0.90	1.89	2.96	4.04	5.11	6.18	7.25	8.33	9.40	10.47
0.91	1.95	3.05	4.16	5.26	6.36	7.47	8.57	9.67	10.78
0.92	2.01	3.14	4.27	5.41	6.54	7.68	8.81	9.95	11.08
0.93	2.06	3.23	4.40	5.56	6.73	7.89	9.06	10.22	11.39
0.94	2.12	3.32	4.52	5.71	6.91	8.11	9.30	10.50	11.70
0.95	2.18	3.41	4.64	5.87	7.10	8.32	9.55	10.78	12.01
0.96	2.24	3.50	4.76	6.02	7.28	8.54	9.80	11.06	12.32
0.97	2.30	3.59	4.89	6.18	7.47	8.76	10.05	11.35	12.64
0.98	2.36	3.69	5.01	6.33	7.66	8.98	10.31	11.63	12.95
0.99	2.42	3.78	5.14	6.49	7.85	9.20	10.56	11.91	13.27
1.00	2.49	3.87	5.26	6.65	8.04	9.43	10.81	12.20	13.59
1.10	3.12	4.83	6.55	8.26	9.97	11.69	13.40	15.12	16.83
1.20	3.78	5.82	7.87	9.91	11.95	14.00	16.04	18.08	20.12
1.30	4.46	6.83	9.20	11.57	13.94	16.31	18.68	21.05	23.42
1.40	5.16	7.85	10.54	13.22	15.91	18.60	21.29	23.98	26.67
1.50	5.86	8.86	11.86	14.86	17.85	20.85	23.85	26.85	29.84
1.60	6.58	9.87	13.16	16.46	19.75	23.05	26.34	29.64	32.93
1.70	7.29	10.87	14.45	18.02	21.60	25.18	28.76	32.33	35.91
1.80	8.01	11.86	15.70	19.55	23.39	27.24	31.08	34.93	38.77
1.90	8.73	12.83	16.93	21.02	25.12	29.22	33.32	37.41	41.51
2.00	9.45	13.79	18.12	22.46	26.79	31.13	35.46	39.80	44.13
2.10	10.2	14.7	19.3	23.8	28.4	33.0	37.5	42.1	46.6
2.20	10.9	15.7	20.4	25.2	29.9	34.7	39.5	44.2	49.0
2.30	11.6	16.6	21.5	26.5	31.4	36.4	41.3	46.3	51.2

TABLE A.58

$f_h/U : g$ *Relationships when* $z = 190$

f_h/U	Values of g when j of cooling curve is:								
	0.40	0.60	0.80	1.00	1.20	1.40	1.60	1.80	2.00
0.20	2.32-04	3.81-04	5.31-04	6.81-04	8.30-04	9.80-04	1.13-03	1.28-03	1.43-03
0.30	1.08-02	1.77-02	2.47-02	3.16-02	3.86-02	4.55-02	5.25-02	5.94-02	6.64-02
0.40	7.35-02	1.21-01	1.68-01	2.16-01	2.63-01	3.10-01	3.58-01	4.05-01	4.53-01
0.50	0.234	0.384	0.534	0.684	0.833	0.983	1.133	1.283	1.433
0.60	0.509	0.832	1.155	1.479	1.802	2.125	2.448	2.771	3.094
0.70	0.893	1.453	2.012	2.572	3.132	3.691	4.251	4.811	5.371
0.80	1.37	2.21	3.06	3.90	4.75	5.60	6.44	7.29	8.13
0.81	1.42	2.30	3.17	4.05	4.92	5.80	6.68	7.55	8.43
0.82	1.47	2.38	3.29	4.19	5.10	6.01	6.91	7.82	8.73
0.83	1.53	2.46	3.40	4.34	5.28	6.22	7.16	8.09	9.03
0.84	1.58	2.55	3.52	4.49	5.46	6.43	7.40	8.37	9.34
0.85	1.64	2.64	3.64	4.64	5.64	6.64	7.65	8.65	9.65
0.86	1.69	2.72	3.76	4.79	5.83	6.86	7.89	8.93	9.96
0.87	1.75	2.81	3.88	4.95	6.01	7.08	8.14	9.21	10.28
0.88	1.80	2.90	4.00	5.10	6.20	7.30	8.40	9.50	10.59
0.89	1.86	2.99	4.12	5.26	6.39	7.52	8.65	9.78	10.91
0.90	1.92	3.08	4.25	5.41	6.58	7.74	8.91	10.07	11.24
0.91	1.98	3.17	4.37	5.57	6.77	7.97	9.17	10.36	11.56
0.92	2.03	3.27	4.50	5.73	6.96	8.19	9.43	10.66	11.89
0.93	2.09	3.36	4.62	5.89	7.16	8.42	9.69	10.95	12.22
0.94	2.15	3.45	4.75	6.05	7.35	8.65	9.95	11.25	12.55
0.95	2.21	3.55	4.88	6.21	7.55	8.88	10.21	11.55	12.88
0.96	2.27	3.64	5.01	6.38	7.74	9.11	10.48	11.85	13.21
0.97	2.34	3.74	5.14	6.54	7.94	9.34	10.75	12.15	13.55
0.98	2.40	3.83	5.27	6.71	8.14	9.58	11.01	12.45	13.89
0.99	2.46	3.93	5.40	6.87	8.34	9.81	11.28	12.75	14.23
1.00	2.52	4.03	5.53	7.04	8.54	10.05	11.55	13.06	14.57
1.10	3.17	5.02	6.88	8.74	10.59	12.45	14.30	16.16	18.02
1.20	3.85	6.05	8.26	10.47	12.68	14.89	17.09	19.30	21.51
1.30	4.55	7.10	9.66	12.21	14.77	17.32	19.88	22.44	24.99
1.40	5.26	8.16	11.05	13.95	16.84	19.73	22.63	25.52	28.42
1.50	5.99	9.21	12.43	15.65	18.87	22.09	25.31	28.53	31.75
1.60	6.73	10.26	13.79	17.32	20.85	24.38	27.92	31.45	34.98
1.70	7.47	11.30	15.12	18.95	22.78	26.60	30.43	34.25	38.08
1.80	8.22	12.33	16.43	20.53	24.63	28.74	32.84	36.94	41.04
1.90	8.98	13.34	17.70	22.06	26.42	30.79	35.15	39.51	43.87
2.00	9.73	14.34	18.94	23.54	28.15	32.75	37.35	41.96	46.56
2.10	10.5	15.3	20.1	25.0	29.8	34.6	39.5	44.3	49.1
2.20	11.3	16.3	21.3	26.4	31.4	36.4	41.5	46.5	51.5

TABLE A.59

$f_h/U:g$ *Relationships when* $z = 200$

f_h/U	Values of g when j of cooling curve is:								
	0.40	0.60	0.80	1.00	1.20	1.40	1.60	1.80	2.00
0.20	2.32-04	3.95-04	5.57-04	7.20-04	8.82-04	1.04-03	1.21-03	1.37-03	1.53-03
0.30	1.08-02	1.84-02	2.59-02	3.34-02	4.10-02	4.85-02	5.61-02	6.36-02	7.11-02
0.40	7.39-02	1.25-01	1.77-01	2.28-01	2.79-01	3.31-01	3.82-01	4.34-01	4.85-01
0.50	0.235	0.398	0.560	0.723	0.885	1.048	1.211	1.373	1.536
0.60	0.512	0.862	1.213	1.563	1.913	2.264	2.614	2.964	3.314
0.70	0.899	1.505	2.111	2.718	3.324	3.930	4.536	5.142	5.748
0.80	1.38	2.29	3.21	4.12	5.04	5.95	6.87	7.78	8.69
0.81	1.43	2.38	3.33	4.27	5.22	6.17	7.12	8.06	9.01
0.82	1.49	2.47	3.45	4.43	5.41	6.39	7.37	8.35	9.33
0.83	1.54	2.55	3.57	4.58	5.60	6.61	7.63	8.64	9.66
0.84	1.59	2.64	3.69	4.74	5.79	6.84	7.89	8.93	9.98
0.85	1.65	2.73	3.82	4.90	5.98	7.06	8.15	9.23	10.31
0.86	1.71	2.82	3.94	5.06	6.18	7.29	8.41	9.53	10.65
0.87	1.76	2.91	4.07	5.22	6.37	7.52	8.68	9.83	10.98
0.88	1.82	3.01	4.19	5.38	6.57	7.76	8.94	10.13	11.32
0.89	1.88	3.10	4.32	5.55	6.77	7.99	9.21	10.44	11.66
0.90	1.94	3.19	4.45	5.71	6.97	8.23	9.49	10.74	12.00
0.91	1.99	3.29	4.58	5.88	7.17	8.47	9.76	11.05	12.35
0.92	2.05	3.38	4.71	6.04	7.37	8.71	10.04	11.37	12.70
0.93	2.11	3.48	4.85	6.21	7.58	8.95	10.31	11.68	13.04
0.94	2.18	3.58	4.98	6.38	7.79	9.19	10.59	11.99	13.40
0.95	2.24	3.68	5.11	6.55	7.99	9.43	10.87	12.31	13.75
0.96	2.30	3.77	5.25	6.73	8.20	9.68	11.15	12.63	14.10
0.97	2.36	3.87	5.39	6.90	8.41	9.92	11.44	12.95	14.46
0.98	2.42	3.97	5.52	7.07	8.62	10.17	11.72	13.27	14.82
0.99	2.49	4.07	5.66	7.25	8.83	10.42	12.00	13.59	15.18
1.00	2.55	4.17	5.80	7.42	9.04	10.67	12.29	13.91	15.54
1.10	3.21	5.21	7.21	9.20	11.20	13.20	15.20	17.19	19.19
1.20	3.90	6.28	8.65	11.02	13.39	15.76	18.14	20.51	22.88
1.30	4.62	7.36	10.10	12.84	15.58	18.32	21.06	23.80	26.54
1.40	5.36	8.46	11.55	14.65	17.75	20.84	23.94	27.03	30.13
1.50	6.12	9.55	12.99	16.43	19.86	23.30	26.74	30.17	33.61
1.60	6.88	10.64	14.40	18.16	21.92	25.68	29.44	33.20	36.96
1.70	7.66	11.72	15.78	19.85	23.91	27.98	32.04	36.10	40.17
1.80	8.44	12.79	17.13	21.48	25.83	30.18	34.53	38.87	43.22
1.90	9.23	13.84	18.45	23.06	27.68	32.29	36.90	41.51	46.12
2.00	10.0	14.9	19.7	24.6	29.4	34.3	39.2	44.0	48.9
2.10	10.8	15.9	21.0	26.1	31.1	36.2	41.3	46.4	51.4

TABLE A.60
Temperature Conversion

C. ← F.	To convert degrees or	C. → F.	C. ← F.	To convert degrees or	C. → F.
−17.78	0	32	5.56	42	107.6
−17.22	1	33.8	6.11	43	109.4
−16.67	2	35.6	6.67	44	111.2
−16.11	3	37.4	7.22	45	113
−15.56	4	39.2	7.78	46	114.8
−15	5	41	8.33	47	116.6
−14.44	6	42.8	8.89	48	118.4
−13.89	7	44.6	9.44	49	120.2
−13.33	8	46.4	10	50	122
−12.78	9	48.2	10.56	51	123.8
−12.22	10	50	11.11	52	125.6
−11.67	11	51.8	11.67	53	127.4
−11.11	12	53.6	12.22	54	129.2
−10.56	13	55.4	12.78	55	131
−10	14	57.2	13.33	56	132.8
−9.44	15	59	13.89	57	134.6
−8.89	16	60.8	14.44	58	136.4
−8.33	17	62.6	15	59	138.2
−7.78	18	64.4	15.56	60	140
−7.22	19	66.2	16.11	61	141.8
−6.67	20	68	16.67	62	143.6
−6.11	21	69.8	17.22	63	145.4
−5.56	22	71.6	17.78	64	147.2
−5	23	73.4	18.33	65	149
−4.44	24	75.2	18.89	66	150.8
−3.89	25	77	19.44	67	152.6
−3.33	26	78.8	20	68	154.4
−2.78	27	80.6	20.56	69	156.2
−2.22	28	82.4	21.11	70	158
−1.67	29	84.2	21.67	71	159.8
−1.11	30	86	22.22	72	161.6
−0.56	31	87.8	22.78	73	163.4
0	32	89.6	23.33	74	165.2
0.56	33	91.4	23.89	75	167
1.11	34	93.2	24.44	76	168.8
1.67	35	95	25	77	170.6
2.22	36	96.8	25.56	78	172.4
2.78	37	98.6	26.11	79	174.2
3.33	38	100.4	26.67	80	176
3.89	39	102.2	27.22	81	177.8
4.44	40	104	27.78	82	179.6
5	41	105.8	28.33	83	181.4

TABLE A.60 (Continued)

C. ← F.	To convert degrees or	C. → F.	C. ← F.	To convert degrees or	C. → F.
28.89	84	183.2	52.22	126	258.8
29.44	85	185	52.78	127	260.6
30	86	186.8	53.33	128	262.4
30.56	87	188.6	53.89	129	264.2
31.11	88	190.4	54.44	130	266
31.67	89	192.2	55	131	267.8
32.22	90	194	55.56	132	269.6
32.78	91	195.8	56.11	133	271.4
33.33	92	197.6	56.67	134	273.2
33.89	93	199.4	57.22	135	275
34.44	94	201.2	57.78	136	276.8
35	95	203	58.33	137	278.6
35.56	96	204.8	58.89	138	280.4
36.11	97	206.6	59.44	139	282.2
36.67	98	208.4	60	140	284
37.22	99	210.2	60.56	141	285.8
37.78	100	212	61.11	142	287.6
38.33	101	213.8	61.67	143	289.4
38.89	102	215.6	62.22	144	291.2
39.44	103	217.4	62.78	145	293
40	104	219.2	63.33	146	294.8
40.56	105	221	63.89	147	296.6
41.11	106	222.8	64.44	148	298.4
41.67	107	224.6	65	149	300.2
42.22	108	226.4	65.56	150	302
42.78	109	228.2	66.11	151	303.8
43.33	110	230	66.67	152	305.6
43.89	111	231.8	67.22	153	307.4
44.44	112	233.6	67.78	154	309.2
45	113	235.4	68.33	155	311
45.56	114	237.2	68.89	156	312.8
46.11	115	239	69.44	157	314.6
46.67	116	240.8	70	158	316.4
47.22	117	242.6	70.56	159	318.2
47.78	118	244.4	71.11	160	320
48.33	119	246.2	71.67	161	321.8
48.89	120	248	72.22	162	323.6
49.44	121	249.8	72.78	163	325.4
50	122	251.6	73.33	164	327.2
50.56	123	253.4	73.89	165	329
51.11	124	255.2	74.44	166	330.8
51.67	125	257	75	167	332.6

TABLE A.60 (Continued)

C. ← F.	To convert degrees or	C. → F.	C. ← F.	To convert degrees or	C. → F.
75.56	168	334.4	98.89	210	410
76.11	169	336.2	99.44	211	411.8
76.67	170	338	100	212	413.6
77.22	171	339.8	100.56	213	415.4
77.78	172	341.6	101.11	214	417.2
78.33	173	343.4	101.67	215	419
78.89	174	345.2	102.22	216	420.8
79.44	175	347	102.78	217	422.6
80	176	348.8	103.33	218	424.4
80.56	177	350.6	103.89	219	426.2
81.11	178	352.4	104.44	220	428
81.67	179	354.2	105	221	429.8
82.22	180	356	105.56	222	431.6
82.78	181	357.8	106.11	223	433.4
83.33	182	359.6	106.67	224	435.2
83.89	183	361.4	107.22	225	437
84.44	184	363.2	107.78	226	438.8
85	185	365	108.33	227	440.6
85.56	186	366.8	108.89	228	442.4
86.11	187	368.6	109.44	229	444.2
86.67	188	370.4	110	230	446
87.22	189	372.2	110.56	231	447.8
87.78	190	374	111.11	232	449.6
88.33	191	375.8	111.67	233	451.4
88.89	192	377.6	112.22	234	453.2
89.44	193	379.4	112.78	235	455
90	194	381.2	113.33	236	456.8
90.56	195	383	113.89	237	458.6
91.11	196	384.8	114.44	238	460.4
91.67	197	386.6	115	239	462.2
92.22	198	388.4	115.56	240	464
92.78	199	390.2	116.11	241	465.8
93.33	200	392	116.67	242	467.6
93.89	201	393.8	117.22	243	469.4
94.44	202	395.6	117.78	244	471.2
95	203	397.4	118.33	245	473
95.56	204	399.2	118.89	246	474.8
96.11	205	401	119.44	247	476.6
96.67	206	402.8	120	248	478.4
97.22	207	404.6	120.56	249	480.2
97.78	208	406.4	121.11	250	482
98.33	209	408.2	121.67	251	483.8

TABLE A.60 (Continued)

C. ← F.	To convert degrees or	C. → F.	C. ← F.	To convert degrees or	C. → F.
122.22	252	485.6	145.56	294	561.2
122.78	253	487.4	146.11	295	563
123.33	254	489.2	146.67	296	564.8
123.89	255	491	147.22	297	566.6
124.44	256	492.8	147.78	298	568.4
125	257	494.6	148.33	299	570.2
125.56	258	496.4	148.89	300	572
126.11	259	498.2	149.44	301	573.8
126.67	260	500	150	302	575.6
127.22	261	501.8	150.56	303	577.4
127.78	262	503.6	151.11	304	579.2
128.33	263	505.4	151.67	305	581
128.89	264	507.2	152.22	306	582.8
129.44	265	509	152.78	307	584.6
130	266	510.8	153.33	308	586.4
130.56	267	512.6	153.89	309	588.2
131.11	268	514.4	154.44	310	590
131.67	269	516.2	155	311	591.8
132.22	270	518	155.56	312	593.6
132.78	271	519.8	156.11	313	595.4
133.33	272	521.6	156.67	314	597.2
133.89	273	523.4	157.22	315	599
134.44	274	525.2	157.78	316	600.8
135	275	527	158.33	317	602.6
135.56	276	528.8	158.89	318	604.4
136.11	277	530.6	159.44	319	606.2
136.67	278	532.4	160	320	608
137.22	279	534.2	160.56	321	609.8
137.78	280	536	161.11	322	611.6
138.33	281	537.8	161.67	323	613.4
138.89	282	539.6	162.22	324	615.2
139.44	283	541.4	162.78	325	617
140	284	543.2	163.33	326	618.8
140.56	285	545	163.89	327	620.6
141.11	286	546.8	164.44	328	622.4
141.67	287	548.6	165	329	624.2
142.22	288	550.4	165.56	330	626
142.78	289	552.2	166.11	331	627.8
143.33	290	554	166.67	332	629.6
143.89	291	555.8	167.22	333	631.4
144.44	292	557.6	167.78	334	633.2
145	293	559.4	168.33	335	635

TABLE A.60 (Continued)

C. ← F.	To convert degrees or	C. → F.	C. ← F.	To convert degrees or	C. → F.
168.89	336	636.8	187.22	369	696.2
169.44	337	638.6	187.78	370	698
170	338	640.4	188.33	371	699.8
170.56	339	642.2	188.89	372	701.6
171.11	340	644	189.44	373	703.4
171.67	341	645.8	190	374	705.2
172.22	342	647.6	190.56	375	707
172.78	343	649.4	191.11	376	708.8
173.33	344	651.2	191.67	377	710.6
173.89	345	653	192.22	378	712.4
174.44	346	654.8	192.78	379	714.2
175	347	656.6	193.33	380	716
175.56	348	658.4	193.89	381	717.8
176.11	349	660.2	194.44	382	719.6
176.67	350	662	195	383	721.4
177.22	351	663.8	195.56	384	723.2
177.78	352	665.6	196.11	385	725
178.33	353	667.4	196.67	386	726.8
178.89	354	669.2	197.22	387	728.6
179.44	355	671	197.78	388	730.4
180	356	672.8	198.33	389	732.2
180.56	357	674.6	198.89	390	734
181.11	358	676.4	199.44	391	735.8
181.67	359	678.2	200	392	737.6
182.22	360	680	200.56	393	739.4
182.78	361	681.8	201.11	394	741.2
183.33	362	683.6	201.67	395	743
183.89	363	685.4	202.22	396	744.8
184.44	364	687.2	202.78	397	746.6
185	365	689	203.33	398	748.4
185.56	366	690.8	203.89	399	750.2
186.11	367	692.6	204.44	400	752
186.67	368	694.4			

Derivation of F_s Equation (Lethality)
(See Chapter 12)

$$10^{-F_s/D_r} \cdot 1 = \int_0^1 10^{-F_\lambda/D_r} \, dv \tag{A.1}$$

in which

$$F_\lambda = F_c + mv \tag{A.2}$$

Substitute in equation A.1:

$$10^{-F_s/D_r} = \int_0^1 10^{-(F_c+mv)/D_r} \, dv \tag{A.3}$$

in which

$$10^{-(F_c+mv)/D_r} = 10^{-F_c/D_r} \cdot 10^{-mv/D_r} \tag{A.4}$$

Then

$$10^{-F_s/D_r} = 10^{-F_c/D_r} \int_0^1 10^{-mv/D_r} \, dv \tag{A.5}$$

Differentiate $10^{-mv/D_r}$ according to

$$y = a^v$$
$$dy = a^v \cdot \ln a \cdot dv$$

Then

$$y = 10^{-mv/D_r} \tag{A.6}$$

in which $a = 10$ and $v = -mv/Dr$, and

$$dy = 10^{-mv/D_r} \cdot \ln 10 \cdot \frac{-m}{D_r} \, dv \tag{A.7}$$

Then

$$dv = \frac{dy}{10^{-mv/D_r} \cdot \ln 10 \cdot (-m/D_r)} \tag{A.8}$$

Substitute this value of dv in equation A.5:

$$10^{-F_s/D_r} = 10^{-F_c/D_r} \int_0^1 10^{-mv/D_r} \cdot \frac{dy}{10^{-mv/D_r} \cdot \ln 10 (-m/D_r)} \tag{A.9}$$

Then

$$10^{-F_s/D_r} = 10^{-F_c/D_r} \int_0^1 \frac{dy}{-(m/D_r) \ln 10} \tag{A.10}$$

and

$$10^{-F_s/D_r} = 10^{-F_c/D_r} \cdot - \frac{1}{(m/D_r) \cdot \ln 10} \int_0^1 dy \qquad (A.11)$$

or

$$10^{-F_s/D_r} = -10^{-F_c/D_r} \cdot \frac{D_r}{m \cdot \ln 10} \int_0^1 dy \qquad (A.12)$$

Then

$$10^{-F_s/D_r} = -10^{-F_c/D_r} \cdot \frac{D_r}{m \cdot \ln 10} [y]_{v=0}^{v=1} \qquad (A.13)$$

By equation A.6,

$$y = 10^{-mv/D_r}$$

Therefore

$$10^{-F_s/D_r} = -10^{-F_c/D_r} \cdot \frac{D_r}{m \cdot \ln 10} [10^{-mv/D_r}]_{v=0}^{v=1} \qquad (A.14)$$

Enter limits to obtain

$$10^{-mv/D_r} = 10^{-m/D_r} - 10^0 \qquad (A.15)$$

and

$$10^{-F_s/D_r} = -10^{-F_c/D_r} \cdot \frac{D_r}{m \cdot \ln 10} (10^{-m/D_r} - 1) \qquad (A.16)$$

or

$$10^{-F_s/D_r} = 10^{-F_c/D_r} \cdot \frac{D_r}{m \cdot \ln 10} (1 - 10^{-m/D_r}) \qquad (A.17)$$

By trial it was found that $10^{-m/D_r}$ is always negligible compared with unity. To determine this, equation A.2 was rearranged to obtain

$$m = \frac{F_\lambda - F_c}{v} \qquad (A.18)$$

After substituting and evaluating for a number of extreme processing conditions, the maximum value of $10^{-m/D_r}$ determined was about 0.00001. Therefore, equation A.17 reduces to

$$10^{-F_s/D_r} = 10^{-F_c/D_r} \cdot \frac{D_r}{m \cdot \ln 10} \qquad (A.19)$$

Since ln $10 = 2.303$,

$$10^{-F_s/D_r} = 10^{-F_c/D_r} \cdot \frac{D_r}{2.303m} \tag{A.20}$$

Next, logarithms were taken of all terms in equation A.20 to obtain

$$-\frac{F_s}{D_r} = -\frac{F_c}{D_r} + [\log D_r - (\log 2.303 + \log m)] \tag{A.21}$$

or

$$-\frac{F_s}{D_r} = -\frac{F_c}{D_r} + \log D_r - \log 2.303 - \log m \tag{A.22}$$

Since $m = (F_\lambda - F_c)/v$ (equation A.18),

$$-\frac{F_s}{D_r} = -\frac{F_c}{D_r} + \log D_r - \log 2.303 - [\log(F_\lambda - F_c) - \log v] \tag{A.23}$$

or

$$-\frac{F_s}{D_r} = -\frac{F_c}{D_r} + \log D_r - \log 2.303 - \log(F_\lambda - F_c) + \log v \tag{A.24}$$

Next, signs were changed throughout, and v was taken as 0.19 to obtain

$$\frac{F_s}{D_r} = \frac{F_c}{D_r} - \log D_r + \log 2.303 + \log (F_\lambda - F_c) - \log 0.19 \tag{A.25}$$

Since

$$\log 2.303 = 0.36229$$
$$\log 0.19 = 9.27875 - 10$$

equation A.25 was reduced to

$$\frac{F_s}{D_r} = \frac{F_c}{D_r} + 1.084 + \log \frac{F_\lambda - F_c}{D_r} \tag{A.26}$$

Equation A.26 was next multiplied by D_r to obtain

$$F_s = F_c + D_r \left(1.084 + \log \frac{F_\lambda - F_c}{D_r}\right) \tag{A.27}$$

Why was the value of v taken as 0.19? Because, when it is equal to this value,

$$\frac{j_\lambda}{j_c} = 0.5$$

That is,

$$j_\lambda = 0.5j_c$$

and

$$g_\lambda = 0.5g_c$$

These interrelationships become readily apparent from a study of the equation of the heating curve. For any specified process,

$$B = f_h \log \frac{j_c I}{g_c} = f_h \log \frac{j_\lambda I}{g_\lambda}$$

Then

$$\frac{j_c}{g_c} = \frac{j_\lambda}{g_\lambda}$$

and

$$g_\lambda = \frac{j_\lambda}{j_c} g_c$$

or

$$j_\lambda = \frac{g_\lambda}{g_c} j_c$$

According to relationships shown in Fig. 32, when $v = 0.19$, $j_\lambda/j_c = 0.5$. This relationship adds greatly to the convenience of using equation A.27 in evaluating processes.

Derivation of T_s Equation

(See Chapter 11)

$$T_s \cdot 1 = \int_0^1 T_\lambda \, dv \tag{A.28}$$

in which T_s = mass average temperature.

T_λ = temperature at any isothermal surface.

v = fraction of volume inclosed by any isothermal surface.

According to relationships depicted in Fig. 35,

$$\log (T_\lambda - T_w) = mv + b \tag{A.29}$$

in which T_w = temperature of cooling medium.

m = slope of $(T_\lambda - T_w):v$ curves.

b = constant.

Then

$$T_\lambda - T_w = 10^{mv} \cdot 10^b$$
$$T_\lambda = T_w + (10^{mv} \cdot 10^b)$$

and

$$T_s = \int_0^1 [T_w + (10^{mv} \cdot 10^b)] \, dv \qquad (A.30)$$

or

$$T_s = \int_0^1 T_w \, dv + \int_0^1 10^{mv} \cdot 10^b \, dv \qquad (A.31)$$

$$T_s = T_w \int_0^1 dv + 10^b \int_0^1 10^{mv} \, dv \qquad (A.32)$$

$$\int_0^1 dv = 1$$

Then

$$T_s = T_w + 10^b \int_0^1 10^{mv} \, dv \qquad (A.33)$$

Now, differentiate 10^{mv} according to

$$y = a^v$$
$$dy = a^v \cdot \ln a \, dv$$

Let

$$y = 10^{mv}$$
$$a = 10$$
$$v = mv$$

Then

$$dy = 10^{mv} \cdot \ln 10 \cdot m \, dv$$

and

$$dv = \frac{dy}{10^{mv} \cdot \ln 10 \cdot m}$$

Next, substitute in equation A.33 and solve as follows:

$$T_s = T_w + 10^b \int_{v=0}^{v=1} 10^{mv} \cdot \frac{dy}{10^{mv} \cdot \ln 10 \cdot m}$$

$$T_s = T_w + \frac{10^b}{\ln 10 \cdot m} \int_0^1 dy$$

$$T_s = T_w + \frac{10^b}{\ln 10 \cdot m} [y]_{v=0}^{v=1}$$

But

$$y = 10^{mv}$$

and

$$T_s = T_w + \frac{10^b}{\ln 10 \cdot m} \left[10^{mv}\right]_{v=0}^{v=1}$$

$$T_s = T_w + \frac{10^b}{\ln 10 \cdot m} (10^m - 1)$$

From equation A.29, when $v = 0$,

$$\log (T_\lambda - T_w) = b \qquad \text{(the intercept, when } T_\lambda = T_c\text{)}$$

and

$$10^b = T_c - T_w \qquad \text{(where } T_c = \text{temperature at geometrical center)}$$

Then

$$T_s = T_w + \frac{T_c - T_w}{\ln 10 \cdot m} (10^m - 1)$$

$$m = -1.565 \qquad \text{(see Chapter 11)}$$

Therefore

$$T_s = T_w + \frac{T_c - T_w}{2.303 \cdot -1.565} (10^{-1.565} - 1)$$

$$T_s = T_w + 0.270(T_c - T_w)$$

Derivation of F_s Equation (Lethality and Quality Factor Degradation)
(See Chapters 12 and 15)

$$10^{-F_s/D_r} \cdot 1 = \int_0^1 10^{-F_\lambda/D_r} \cdot dv \qquad (A.35)$$

From relationships of $F_\lambda - F_c$ to $\ln (1 - v)$

$$\ln(1 - v) = -m(F_\lambda - F_c) \qquad (A.36)$$

and

$$F_\lambda = F_c - \frac{1}{m} \ln(1 - v) \qquad (A.37)$$

Now let

$$R = \frac{-1}{m} \qquad (A.38)$$

then

$$F_\lambda = F_c + R \ln(1 - v) \tag{A.39}$$

By substitution of equation A.39 in equation A.35 obtain

$$10^{-F_s/D_r} = \int_0^1 10^{-\frac{F_c + R \ln(1-v)}{D_r}} \cdot dv \tag{A.40}$$

and

$$10^{-F_s/D_r} = 10^{-F_c/D_r} \int_0^1 10^{-\frac{R \ln(1-v)}{D_r}} \cdot dv \tag{A.41}$$

Now let

$$c = -R/D_r \tag{A.42}$$

then equation A.41 becomes

$$10^{-F_s/D_r} = 10^{-F_c/D_r} \int_0^1 10^{\ln(1-v)c} \cdot dv \tag{A.43}$$

Now, by letting

$$y = 10^{\ln(1-v)c} \tag{A.44}$$

differentiate and obtain

$$dv = \frac{dy}{10^{\ln(1-v)c} \cdot \ln 10 \left(\dfrac{-c}{1 - v}\right)} \tag{A.45}$$

Also from equation A.44

$$y = 10^{c \, \ln(1-v)} \tag{A.46}$$

or

$$y = (1 - v)^{2.303c} \tag{A.47}$$

Then for

$$v = 0, y = 1$$
$$v = 1, y = 0$$

Substituting equation A.45 into equation A.43 obtain

$$10^{-F_s/D_r} = 10^{-F_c/D_r} \int_1^0 10^{\ln(1-v)c} \cdot \frac{dy}{10^{\ln(1-v)c} \cdot \ln 10 \left(\dfrac{-c}{1 - v}\right)} \tag{A.48}$$

or

$$10^{-F_s/D_r} = 10^{-F_c/D_r} \int_1^0 \frac{dy}{\ln 10 \left(\frac{-c}{1-v}\right)} \tag{A.49}$$

Because according to equations A.38 and A.42

$$c = \frac{1}{mD_r} \tag{A.50}$$

$$10^{-F_s/D_r} = \frac{mD_r}{2.303} \cdot 10^{-F_c/D_r} \int_1^0 (1-v)dy \tag{A.51}$$

From equation A.47 and equation A.50 find

$$v = 1 - y\frac{mD_r}{2.303} \tag{A.52}$$

By substitution of equation A.52 into equation A.51 obtain

$$10^{-F_s/D_r} = \frac{-mD_r}{2.302} \cdot 10^{-F_c/D_r} \int_0^1 1y\frac{mD_r}{2.303} \cdot dy \tag{A.53}$$

Then

$$10^{-F_s/D_r} = \frac{-mD_r}{2.303} \cdot 10^{-F_c/D_r} \cdot \frac{1}{\frac{mD_r}{2.302}+1} \cdot y\frac{mD_r}{2.303}+1 \Bigg]_1^0$$

and

$$10^{-F_s/D_r} = \frac{mD_r}{2.303} \cdot 10^{-F_c/D_r} \cdot \frac{1}{\frac{mD_r}{2.303}+1} \tag{A.54}$$

Reduce equation A.54 by rearranging and taking logs of both sides to obtain

$$\frac{-F_s}{D_r} = \frac{-F_c}{D_r} + \log(mD_r + 2.303) - \log mD_r \tag{A.55}$$

Rearrange equation A.36 to obtain

$$m = \frac{-\ln(1-v)}{F_\lambda - F_c} \tag{A.56}$$

Take $v = 0.19$, representing a point on the straight line where $j_\lambda = 0.5j_c$ and $g_\lambda = 0.5g_c$, and substitute to obtain

$$m = \frac{0.21072}{F_\lambda - F_c} \qquad (A.57)$$

Substitute equation A.57 into equation A.55 and reduce to obtain

$$F_s = F_c + D_r \log \frac{D_r + 10.93(F_\lambda - F_c)}{D_r} \qquad (A.58)$$

SUBJECT INDEX

Printed and bound by CPI Group (UK) Ltd, Croydon, CR0 4YY

03/10/2024

01040418-0006